Fuels and Petrochemicals Division

Presentations at the 2009 AIChE Annual Meeting

Nashville, Tennessee, USA
8-13 November 2009

ISBN: 978-1-61567-938-6

Fuels and Petrochemicals Division

Presentations at the 2009 AIChE Annual Meeting

Nashville, Tennessee, USA
8-13 November 2009

ISBN: 978-1-61567-938-6

AIChE

Printed from e-media with permission by:

Curran Associates, Inc.
57 Morehouse Lane
Red Hook, NY 12571

CURRAN ASSOCIATES INC. proceedings.com

Some format issues inherent in the e-media version may also appear in this print version.

Printed by Curran Associates, Inc. (2010)

For permission requests, please contact AIChE
at the address below.

AIChE
3 Park Avenue
New York, NY 10016-5991

Phone: (203) 702-7660
Fax: (203) 775-5177

www.aiche.org

Additional copies of this publication are available from:

Curran Associates, Inc.
57 Morehouse Lane
Red Hook, NY 12571 USA
Phone: 845-758-0400
Fax: 845-758-2634
Email: curran@proceedings.com
Web: www.proceedings.com

TABLE OF CONTENTS

Approaches to Renewable Diesel Production

Patrick L. Hanks, ExxonMobil Process Research Labs, Annandale, NJ

Katie Cole, ExxonMobil Process Research Labs, Annandale, NJ

Ernie Lewis, ExxonMobil Process Research Labs, Annandale, NJ

Ed Ellis, ExxonMobil Process Research Labs, Annandale, NJ

Governmental policies throughout the world are mandating the use of biofuels as a complement to traditional petroleum based fuels. Biofuels processing and associated handling of oxygenated compounds in refineries accustomed to handling primarily hydrocarbon streams requires innovation and novel solutions. In addition to being capable of handling a wide array of new bio-feedstocks, solutions must address challenges including: catalyst selection, product cold flow properties, heat release, CO management, and corrosion concerns. In addition, regulatory provisions are needed for a broader range of hydroprocessing solutions, such as FCC pretreatment hydrotreaters.

See more of this Session: Catalytic Biofuel Refining
See more of this Group/Topical: Fuels and Petrochemicals Division

Simultaneous Reforming and Hydrodeoxygenation of Bio-Oil for Liquid Fuels Production

Suchithra T. Gopakumar, Chemical Engineering, Auburn University, Auburn, AL

Harideepan Ravindran, Mechanical Engineering, Auburn University, Auburn, AL

Sushil Adhikari, Biosystems Engineering, Auburn University, Auburn, AL

Ram B. Gupta, Chemical Engineering, Auburn University, Auburn, AL

Oladiran Fasina, Biosystems Engineering, Auburn University, Auburn, AL

Economical utilization of renewable energy sources is one of the viable solutions to reducing our dependency on fossil-based energy. Biomass is the only renewable energy source that can provide liquid fuels. In general, liquid fuels are obtained from biomass using either thermochemical or biochemical conversion method. Fast pyrolysis, a thermochemical process, is getting a lot of attention because all the three products (char, bio-oil and gas) have high potential as biomass energy feedstocks. Fast pyrolysis is a high temperature process in which biomass is thermally cracked into smaller compounds in an inert atmosphere at high heating rate. The vapor from this process is quickly condensed to liquid, which is called pyrolysis oil or bio-oil. The favorable conditions for maximum yield of oil are in the range of 400-600 oC and residence time of a few seconds and particle size less than 2 mm. Bio-oil is a chemically complex mixture of more than 300 compounds. Heating value of bio-oil is comparable with the heating value of other conventional liquid fuels - evidence that it can be used as a transportation fuel. The possible utilization of bio-oil are however limited because of some negative attributes such as low pH, low heating value, high oxygen content, and high viscosity.

Gravity can be used to separate bio-oil into two phases - top aqueous phase and lower viscous tar phase. The aqueous phase contains mainly water and carbohydrates and the tar phase contains primarily lignin derived compounds. An aqueous phase reforming (APR) is a process in which hydrogen is produced from carbohydrate solutions. The uniqueness of the APR is that the reforming occurs at liquid state without volatilizing water. Hence, the energy requirement for the APR is lesser than the conventional steam reforming. In addition, the APR occurs at a low temperature and high pressure (200-300 oC and 10-90 bar) which favors water-gas shift reaction. There are a number of physical and chemical techniques to upgrade tar portion of bio-oil to a high quality fuel such as emulsification, solvent addition, hydro treatment and catalytic cracking. Hydrodeoxygenation (HDO) is one of the promising techniques for bio-oil to improve its heating content. The HDO is a process in which high pressure hydrogen is added into bio-oil in the presence of catalyst. Hydrogen reacts with oxygenated components to produce hydrocarbons and water. The operating conditions of HDO are around 250-450 oC and 150 bar. This study will explore the possibility of simultaneous reforming and hydrodeoxygenation of bio-oil.

Since the composition of bio-oil will be different for different feedstocks, four types of feedstocks have been selected for bio-oil production. The study of simultaneous APR and HDO of bio-oil will be carried out at different temperature, pressure and different types of catalyst. The challenging part for this study is the selection of the best catalyst which favors both the reactions with high conversion and stability. Effect of pressure, temperature and residence time on the quality of bio-oil over different catalysts will be reported. The best catalyst will be selected and the operating parameters will be optimized. This study will be carried out in a high pressure batch reactor. The chemical and physical analysis of bio-oil will be quantified before and after the reaction. Product gas analysis also will be carried out using GC coupled with thermal conductivity detector.

See more of this Session: Catalytic Biofuel Refining
See more of this Group/Topical: Fuels and Petrochemicals Division

Continuous Supercritical-Phase Hydrogenations of Biomass-Derived Compounds

Jackson W. Ford, Center for Environmentally Beneficial Catalysis and Department of Chemical & Petroleum Engineering, University of Kansas, Lawrence, KS

Bala Subramaniam, Center for Environmentally Beneficial Catalysis and Department of Chemical & Petroleum Engineering, University of Kansas, Lawrence, KS

Raghunath V. Chaudhari, Center for Environmentally Beneficial Catalysis and Department of Chemical & Petroleum Engineering, University of Kansas, Lawrence, KS

Development of continuous processes that take advantage of the current petrochemical infrastructure is essential for widespread biomass utilization. Hydrogenations of biomass-related substrates can convert highly-oxygenated biomass components to feedstocks which more closely resemble petrochemicals. The high hydrogen pressures needed for such reactions adversely affect the process economics, but the use of supercritical solvents can overcome this barrier by enhancing the mutual solubilities of reactants as well as the transport of materials and energy in the reaction system. We have used a fixed bed reactor to evaluate supported metal catalysts and supercritical solvent media for intensified hydrogenations of biomass constituents at mild process conditions. We have examined several model biomass compounds to understand better the reaction mechanisms involved in biomass hydrogenations.

See more of this Session: Catalytic Biofuel Refining
See more of this Group/Topical: Fuels and Petrochemicals Division

Catalytic Deoxygenation of Free Fatty Acids Over Pd/C

Jeremy G. Immer, Chemical and Biomolecular Engineering, North Carolina State University, Raleigh, NC

H. Henry Lamb, NC State University, Raleigh, NC

Introduction Environmental concerns and dwindling world petroleum reserves have generated great interest in producing transportation fuels from bio-renewable resources. In this work, we investigate catalytic deoxygenation of free fatty acids (FFAs) [1] to produce linear hydrocarbons for second-generation biofuels. FFAs can be derived from hydrolysis of triglycerides found in edible and non-edible fats and oils. After deoxygenation, the hydrocarbon product can undergo hydroisomerization and hydrocracking to yield fuels in the diesel, kerosene and gasoline boiling ranges. For example, the C17 linear hydrocarbons produced by catalytic deoxygenation of C18 FFAs (stearic, oleic, and linoleic acids) can be used in green diesel after only mild hydroisomerization. Catalytic deoxygenation of FFAs has been reported to occur via two reaction pathways: decarboxylation and decarbonylation. Recent literature indicates that the decarboxylation pathway is predominant over Pd/C [2]. Catalyst deactivation has been inferred to occur via coking and/or CO poisoning [3]. The research presented here seeks to further elucidate the deoxygenation pathways of saturated and unsaturated FFAs over Pd/C and the factors that influence catalyst lifetime. Deoxygenation of stearic, oleic, linoleic, and palmitic acids in alkane solvents (dodecane, pentadecane, and heptadecane) was investigated under He and 10% H2 in He. Materials and Methods Reagent-grade chemicals were purchased from Sigma Aldrich and Fisher Scientific and used as received. Ultra-high-purity compressed gases were supplied by National Welders. Batch experiments were conducted using a 50-mL stirred autoclave (Autoclave Engineers) and 335 mg of suspended 5% Pd/C catalyst powder (Degussa E117). Pulsed CO chemisorption indicated the Pd dispersion to be 23% assuming a stoichiometry of 2:1 Pd:CO. Typically, the catalyst was reduced prior to reaction at 200¨¬C in solvent under 2 atm H2. Reaction conditions were 300¨¬C and 15 atm with 60 mL/min of flowing He or 10% H2. The reactor effluent was analyzed on-line using a quadrupole mass spectrometer (QMS) (Pfeiffer Prismaplus) with a capillary inlet system. The liquid product was analyzed by gas chromatography (GC) using an HP 5890 FID instrument equipped with a EC-5 capillary column (30 m x 0.32 mm x 1.0 mm). Results and Discussion Saturated FFAs, such as stearic and palmitic acids, are readily deoxygenated under a He purge to the corresponding CN-1 linear alkanes in high yields (typically >98%). On-line QMS [Fig. 1] revealed the concomitant generation of CO2 and H2. Blank experiments with solvent and no FFA confirmed that H2 is a reaction product. No significant difference in reaction rates was observed for stearic acid deoxygenation under He in dodecane and heptadecane. Additional experiments demonstrated that the decarboxylation rate (as measured by CO2 evolution) declines markedly with increasing FFA/catalyst ratio. Unsaturated FFAs, such as oleic and linoleic acids, must be hydrogenated before deoxygenation can proceed at a significant rate. Under flowing 10% H2, FFA hydrogenation occurs at lower temperatures [Fig. 2], and saturated and unsaturated FFAs yield equivalent rates and product yields. The initial rate of decarboxylation of stearic acid is approximately 3-fold greater under He than under 10% H2; however, rapid catalyst deactivation occurs under He, as evidenced by repeated batch experiments using the same catalyst. In 10% H2, catalyst lifetime is improved approximately 10-fold. This is consistent with previous research which concluded that H2 was necessary for continuous decarboxylation of FFAs over supported Pd catalysts although H2 was not consumed in the reaction [1]. In contrast, the decarbonylation rate is not affected by H2 partial pressure. We infer that H2 is involved in the decarboxylation pathway but not the decarbonylation pathway. Significance This work provides further insight into the reaction pathways by which FFAs available from edible and non-edible fats and oils are deoxygenated over supported Pd catalysts to produce hydrocarbons for biofuels. References 1. Maier, W.F., Roth, W., Thies, I., and Rague Schleyer, P.v. Chem. Ber. 115, 808-812 (1982) 2. Maki-Arvela, P., Kubickova, I., Snare, M., Eranen, K., and Murzin, D.Y. Energy and Fuels 21, 30-41 (2007). 3. Maki-Arvela, P., Snare, M., Eranen, K., Myllyoja, J., Murzin, D.Y. Fuel 87, 3543-3549 (2008).

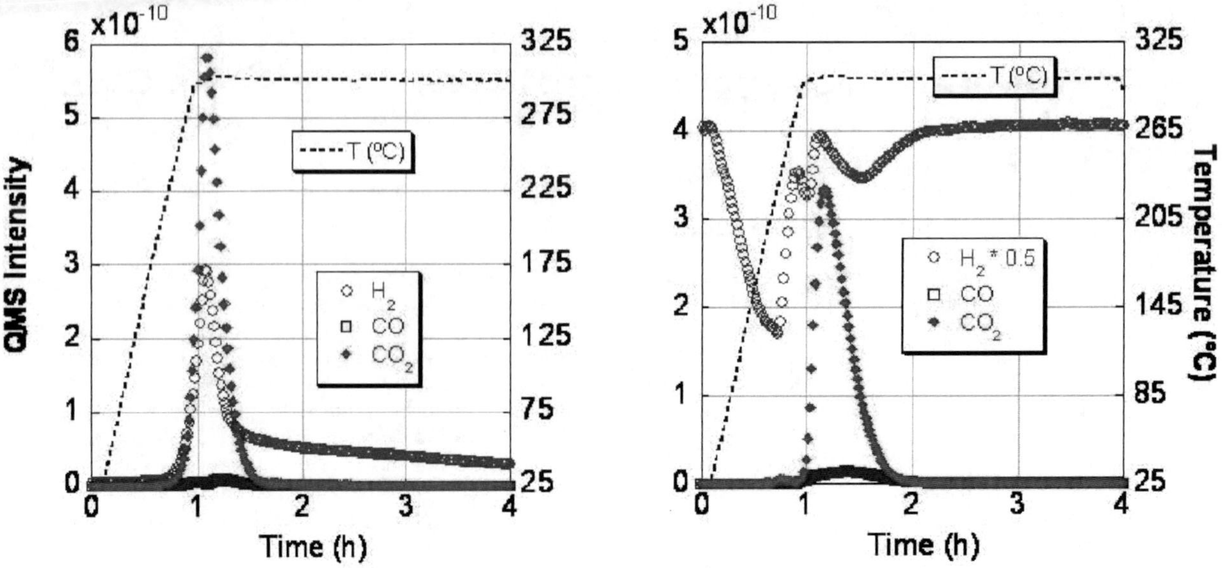

See more of this Session: Catalytic Biofuel Refining
See more of this Group/Topical: Fuels and Petrochemicals Division

Oxidative Stability of Biodiesel From Castor Oil Fatty Acid Methyl Esters

Yguatyara L. Machado, Universidade Federal do Ceará, Fortaleza, CE, Brazil

Monica C. G. Albuquerque, Chemical Engineering Department, Universidade Federal do Ceará, Fortaleza, Brazil

Francisco Murilo T. Luna, Chemical Engineering, Universidade Federal do Ceará, Fortaleza, Brazil

Marianna C. Aragao, Universidade Federal do Ceará, Fortaleza, CE, Brazil

Diana C. S. Azevedo, Chemical Engineering, Universidade Federal do Ceará, Fortaleza, Brazil

Celio L. Cavalcante Jr., Chemical Engineering, Universidade Federal do Ceará, Fortaleza, CE, Brazil

Biodiesel is an alternative diesel fuel consisting of alkyl monoesters of fatty acids prepared from vegetable oils. It has been the focus of a considerable amount of recent research because it is renewable, biodegradable and reduces the emission of some pollutants. However, biodiesel has lower oxidation stability as compared with petroleum diesel because biodiesel has high content of unsaturated methyl esters, especially poly-unsaturated methyl esters which can be easily oxidized such as methyl linoleate (C18:2) and methyl linoleate (C18:3), which lead to the formation of decomposed compounds such as acids, aldehydes, esters, ketones, peroxides and alcohols. These products not only affect the properties of biodiesel, but also bring problems for engine operation. Consequently, storage of biodiesel over extended periods may lead to degradation of fuel properties that can compromise fuel quality.

The objective of this study was to determine the oxidative stability of biodiesel produced by transesterification of castor oil. The evaluation was conducted by means of a rapid oxidation method (PetroOxy), at temperatures of 100,120 and 140°C at time intervals between 0 and 60min, and at pressure 700kPa. The Acidity Index (AI) of each oxidized sample from the accelerated tests was measured in order to evaluate the kinetics behavior of the oxidation process.

See more of this Session: Catalytic Biofuel Refining
See more of this Group/Topical: Fuels and Petrochemicals Division

Conversion of Vegetable and Algae Oils to Hydrocarbons Over Supported Metal Catalysts

Tonya Morgan, Center for Applied Energy Research, University of Kentucky, Lexington, KY

Daniel Grubb, Center for Applied Energy Research, University of Kentucky, Lexington, KY

Eduardo Santillan-Jimenez, Center for Applied Energy Research, University of Kentucky, Lexington, KY

Samuel A. Morton III,Center for Applied Energy Research, University of Kentucky, Lexington, KY

Mark Crocker, Center for Applied Energy Research, University of Kentucky, Lexington, KY

As an alternative to hydrotreating, we have investigated metal-catalyzed deoxygenation for the production of hydrocarbon fuels from triglycerides. In this approach, deoxygenation of the triglyceride is accomplished - in the absence of hydrogen - via removal of the carboxyl group in the fatty acid structure as CO2 and/or CO, thereby producing linear hydrocarbons. Upgrading of tristearin, containing saturated C18 fatty acid chains, over 20 wt% Ni/C at 350°C was found to result in near quantitative conversion of the triglyceride. Alkanes, corresponding to mainly n-heptadecane and n-pentadecane, constituted ca. 60% of the liquid product (by weight), the remainder consisting of fatty acids and unidentified heavy residues. Analysis of the gaseous products confirmed that oxygen was rejected as CO and CO2, while small amounts of light alkanes (C1-C4) and H2 were also formed. Similar results were obtained with triolein (a triglyceride containing only monounsaturated C18 chains); however, compared to tristearin, the selectivity to light hydrocarbons was slightly increased, reflecting the greater reactivity of the unsaturated fatty acid chains towards cracking. This trend was maintained in the deoxygenation of soybean oil; the greater unsaturation of the triglycerides present resulted in increased yields of C2-C10 hydrocarbons relative to triolein. 13C NMR spectroscopic analysis of the liquid product from soybean oil deoxygenation at intermediate reaction times suggested that one pathway for triglyceride deoxygenation involves liberation of fatty acids via C-O bond scission and concomitant H transfer, followed by elimination of CO2 from the acids in a later step. Compared to Ni, catalysts containing Pd or Pt supported on activated carbon showed lower activity for both triglyceride decarboxylation and for cracking of the fatty acid chains, although significant yields of linear alkanes and alkenes were obtained when soybean and algae oils were treated with these catalysts at 350 °C.

See more of this Session: Catalytic Biofuel Refining
See more of this Group/Topical: Fuels and Petrochemicals Division

Production of Biofuel From Biomass Via Fast Pyrolysis and Hydrotreatment Processes

Satya K. Jujjuri, Center for Sustainable Environmental Technologies, Iowa State University, Ames, IA

Justinus A. Satrio, Center for Sustainable Environmental Technologies, Iowa State University, Ames, IA

Robert C. Brown, Center for Sustainable Environmental Technologies, Iowa State University, Ames, IA

Depletion of fossil oil resources, global climate change and national security concerns have put the development of alternative energy to the central stage. Compared to the other alternative energy choices, the production of energy and chemicals from biorenewable resources, such as biomass, has attracted increasing research interests in recent years due to the vast potential availability of biomass materials and the availability of existing technologies that potentially enable the production of energy and chemicals in commercial scales in shorter time period in the future. Center for Sustainable Environmental Technologies (CSET) is a research center at Iowa State University (ISU) that has been working on promoting, developing and demonstrating the thermochemical technologies such as gasification and fast pyrolysis, and catalytic upgrading processes, for the production of fuels, chemicals, and energy from biomass. This presentation is to report the current work at CSET on a process of production of liquid fuels from biomass by using a combination of fast pyrolysis and catalytic upgrading technologies. In the process, cellulosic biomass is first fast-pyrolyzed to produce a liquid, typically called fast pyrolytic oil or bio-oil, as the primary product. Resembling petroleum crude oil in appearance, bio-oil is highly viscous reactive concoction of organic compounds, derived from biomass components (cellulose, hemicelluloses, and lignin) and is not only heterogeneous but also thermally and chemically unstable. Like petroleum crude, bio-oil needs to be further processed in order to be used for final applications (e.g. transportation fuels), however the chemical properties of bio-oil makes it more challenging to process, compared to the fossil-derived oil. The elemental composition of bio-oil contains significant amounts of oxygen as high as 35-40% with 54-58% carbon and about 5.5-7% hydrogen, which make bio-oil reactive and has lower energy content compared to petroleum-derived oil. Moreover bio-oil is represented by approximately more than 300 compounds of various classes. Many bio-oil components are highly polar compounds (such as acetic acid, furfural and hydroxyacetaldehyde). The presence of phenols, furans and carbohydrates (sugars) are also highly pronounced. The high oxygen content of bio-oil may be reduced by using the hydrodeoxygenation process, typically used in petroleum refineries; however not without significant modification of the process. The catalytic upgrading of bio-oil requires a multi-dimensional approach which involves both chemistry (including catalyst synthesis) and engineering approaches. At ISU, one engineering approach has been developed by the way of producing bio-oil, in which bio-oil produced from fast pyrolysis process is collected in fractions. This approach enables bio-oil components that cause bio-oil reactivity and instability, such as organic acids and water, to be separated out from the primary bio-oil components to be catalytically upgraded. At present, the catalytic upgrading work by using selected bio-oil fractions produced at Iowa State Unviersity is being performed . It is reported that the current work on hydrodeoxygenation of bio-oil focuses on the following four main steps. (i) Complete characterization of bio-oil employing complementary techniques such as GC-MS, HPLC and TGA. (ii) Synthesis and characterization of novel catalysts employing XRD, TGA, Chemisorption, BET and TEM. (iii) Study of model compounds such as acetic acid, furfural and phenol which are representative of bio-oil functionality. (iv) Fractionation of bio-oil and comparing each fraction and as a whole in evaluating the efficiency of the hydrodeoxygenation.

See more of this Session: Catalytic Biofuel Refining
See more of this Group/Topical: Fuels and Petrochemicals Division

Catalytic Conversion of Tall Oil to Biodiesel and Renewable Diesel

Kiran Pathapati, Chemical Engineering, University of Louisiana at Lafayette, Lafayette, LA

Stephen Dufreche, Chemical Engineering, University of Louisiana at Lafayette, Lafayette, LA

Rakesh Bajpai, Chemical Engineering, University of Louisiana at Lafayette, Lafayette, LA

Mark Zappi, Chemical Engineering, University of Louisiana at Lafayette, Lafayette, LA

The dependence of petroleum for nearly all of our transportation fuels fosters the importance and demand for renewable, biomass based transportation fuels. Biomass, the only current renewable source of liquid transportation fuel, has great potential to provide renewable energy. Biodiesel and Green Diesel can also be produced from non-conventional, biologically derived feedstock such as tall oil.

Crude tall oil is a by-product of the Kraft Sulfate processing of the softwood for pulp and paper manufacture. Unsaturated C18 fatty acids, resin acids and unsaponifiables make up the principal constituents of the tall oil. Esterification of the tall oil fatty acids using a heterogenous acid catalyst gives Biodiesel whereas catalytic cracking over a catalyst bed gives Green Diesel. Results from catalytic conversion of tall oil to biodiesel and renewable diesel will be shown, as well as comparisons to traditional feedstocks.

See more of this Session: Catalytic Biofuel Refining
See more of this Group/Topical: Fuels and Petrochemicals Division

BIOMASS Upgrading for the PRODUCTION of Biofuels FROM Biowastes

Michiel Van Der Stelt, Department of Chemical Engineering and Chemistry, Eindhoven University of Technology, Eindhoven, Netherlands

Keywords: Renewable Energy, Biomass, Torrefaction, Mild pyrolysis, Weight loss kinetics, Energy densification, Oxygen removal

Biomass resources play an important role in the energy transition. However, the use of raw biomass sources as fuel has various problems, such as high moisture content, low calorific value, hydrophilic nature and its tenacious and fibrous structure. The use of biomass is also subjected to limitation of land, water and competition with food production. The agricultural production of biomass is relatively land intensive and involves high logistics costs due to low energy density of biomass. For biomass based systems a key challenge is to develop efficient conversion technology which can also compete with fossil fuels.

Torrefaction is a technology which can improve biomass properties and therefore offers some solutions to above problems. Torrefaction is a thermal conversion pre-treatment technology of biomass resources to produce a high quality solid biofuel which can be applied in other energy conversion technologies such as combustion and gasification. Torrefaction operates in the temperature range of 200 - 300°C at atmospheric conditions in the absence of oxygen. In this temperature range the biomass decomposes and during this decomposition a gas is produced. Torrefied biomass has higher energy density, improved grindability and hydrophic nature.

Precise control over the process conditions temperature and residence time during biomass and large wood particles torrefaction provides a better reliability of the product quality. The key factor in torrefaction is the heat transfer phenomenon into and in the (large) wood particle. Reactivity of and volatile product formation from the biomass depend on the temperature at which torrefaction take place. The reactivity rate of the constituents of biomass hemicellulose, cellulose and lignin differ with temperature, residence time and matrix structure of the biomass material. Heat transfer limitations result in a temperature profile which leads to reactivity rates variance and heterogeneous material characteristics in the biomass particle itself. Endo and exothermal biomass decomposition reactions can enlarge the effect of the temperature profile and so influence the product quality, but also the process design and process operation regarding temperature set point.

In this work a causality model is derived that shows the causal effects between heating up the reactor temperature, biomass particle temperature and the reactivity of the biomass. The reactivity of small biomass particles and its relation with large biomass is important to describe the quality of the torrefied product. At powder scale biomass is torrefied using TGA to determine the reactivity model. Also cylindrical beech wood particles are torrefied in a fixed bed reactor setup under inert (argon) atmospheric conditions. Several thermocouples are used to determine the internal temperature profile.

It is shown that a causality model is a good model to describe the reactivity of biomass torrefaction. Also it is shown that biomass torrefaction is an exothermal process which can lead to temperatures in the biomass (wood) particle itself that are much higher than the reactor setpoint. This exothermal effect is quantified.

See more of this Session: Catalytic Biomass Pyrolysis and Gasification I
See more of this Group/Topical: Fuels and Petrochemicals Division

Low Temperature Catalytic GASIFICATION of BIOMASS for Syngas and Chemicals PRODUCTION

Foster A. Agblevor, Bilogical Systems Engineering, Virginia Polytechnic Institute and State University, Blacksburg, VA

Sedat H. Beis, Chemical and Biological Engineering, University of Maine, Orono, ME

Ofei Mante, Bilogical Systems Engineering, Virginia Polytechnic Institute and State University, Blacksburg, VA

Norredine Abdoulmoumine, Bilogical Systems Engineering, Virginia Polytechnic Institute and State University, Blacksburg, VA

Junia Pereira, Bilogical Systems Engineering, Virginia Polytechnic Institute and State University, Blacksburg, VA

Traditional gasification of biomass feedstocks occur a high temperatures because of the refractive tars that do no easily crack. We investigated the catalytic conversion of biomass feedstocks into gaseous products and phenolic liquids. Biomass feedstocks (pine, polar, switchgrass and corn stover) were pyrolyzed in a bubbling fluidized bed reactor using ZSM5 catalysts. The results showed that most of the carbohydrate fractions were gasified at 400-450 oC into CO, CO2, C1 to C4 hydrocarbons whereas the liquid fraction was mostly aromatic compounds. The liquid product yield was low, ranging from 25-35%, the char yield ranged from 15-25% and the difference was gas. The yield of various products was feedstock dependent.

The viscosity of the liquid fraction was much lower than that of conventional biomass pyrolysis oil and its pH was higher than that of conventional pyrolysis oil. The catalytic pyrolysis oil did not age as fast as conventional pyrolysis oil especially with respect to increase in viscosity with time. The higher heating value of the catalytic pyrolysis oil was also higher than that of the conventional pyrolysis oil. Fractionation of the liquid product showed that 54-70% was phenolic, 2-10% was aliphatic hydrocarbons, and 2-5% was polycyclic aromatic hydrocarbons (PAH). The relative yield of the fractions was also feedstock dependent. Switchgrass catalytic pyrolysis oil fraction was richer in hydrocarbon fraction compared to the woody biomass feedstocks.

See more of this Session: Catalytic Biomass Pyrolysis and Gasification I
See more of this Group/Topical: Fuels and Petrochemicals Division

1. Introduction

The research and development of biofuels has become a major focus of USA and other countries because of potential global warming concerns and the pollution emanating from the combustion of fossil fuels. Biofuels can be carbon neutral if produced in a sustainable manner. Rapid pyrolysis is one of the potential thermochemical processing technologies being developed to convert biomass feedstocks into liquid fuels and chemicals. Biomass pyrolysis oils also known as biooils, biocrude oils are typically acidic and tend to be unstable because of secondary post-pyrolysis chemical reactions that lead to increases in viscosity. Because of its poor stability, it is difficult to process biomass pyrolysis using conventional processing techniques such as distillation. Further, because of their high oxygen content, they are immiscible with petroleum derived fuels and they have low energy densities. Hydrogenation and hydrodeoxygination have been investigated for both stability and improving energy density of the fuel. Other research works have involved the use of acidic catalysts to improve on the quality of the pyrolysis oil. In this paper we report the investigation of fluid cracking catalyst (FCC) as a pyrolysis medium to produce improved quality biooils. The overall objective of the project is to produce stable biomass pyrolysis oils that can be processed using conventional petrochemical processing techniques.

2. Materials and methods

Debarked hybrid poplar wood samples that were air-dried at ambient conditions were used for the studies. The air-dried samples were knife-milled to pass 1-mm screen and their moisture content determined. The following catalysts and sand were investigated: HZSM5 and seven different FCC catalysts of various zeolite and acidic content supplied by BASF (BASF Inc, Florham Park, NJ). The FCC catalyst had particles sizes < 63μm and the sand had a particle size distribution between 305-500 μm and that of the HZSM5 was 124-500 μm.

Catalytic fast pyrolysis was carried out in a bench-scale fluidized bed reactor unit (Fig 1) which comprised of a K-Tron volumetric feeder, 50-mm bubbling fluidized bed reactor equipped with a 100-μm porous metal gas distributor, hot gas filter, two ethylene glycol-cooled condensers, an electrostatic precipitator and a packed column. The reactor was externally heated with a three-zone electric furnace. 150 g of fresh catalyst was used each time as the fluidizing medium. For FCC catalyst, the bed was fluidized with 3 L/min nitrogen and 12 L/min of nitrogen was used for the sand and HZSM5. 5L/min of nitrogen gas was used to entrain the feed through a jacketed air-cooled feeder tube into the fluidized bed. The reactor was maintained at an average temperature of 450°C. During pyrolysis, the mixture of char, gases and vapors that exited from the reactor were separated by a hot gas filter maintained at 380°C. The vapors were condensed in two ethylene glycol-cooled condensers connected in series. The condensers were maintained at -8°C with a 50/50 cooling mixture of ethylene glycol and water from an 18-liter A82 refrigerated circulating bath. Any condensable gases and aerosols that escaped from the condensers were captured by an electrostatic precipitator (ESP) kept at 16-20 kV and a packed column of glass beads.

Products analyses

The pH of the pyrolysis oils were measured using an Accumet pH Meter with F-55500-10 Accumet pH probe (Cole-Parmer Instrument Company, Vernon Hills, IL). The dynamic viscosities of the freshly produced oils were measured with a Brookfield DV-II+ Pro viscometer (Brookfield Engineering laboratories, Inc. MA.) and a programmable thermosel temperature controller. The measurement was conducted at 50°C with Spindle SC4-18 and speed ranging from 1-60 rpm depending on the viscosity of the oil. 7-ml of sample was used and the reading of the instrument stabilized within 5-15min. A moisture content of the oils were determined by Karl Fischer titration using hydranal® composite 5 reagent. About 60-100-mg of oil sample was loaded into a hypodermic plastic syringe and weighed and injected into the titration vessel and titrated in a methanol solvent using Metrohm 701KF Titrino (Brinkman Instruments Inc, New York, NY). The water content was titrated volumetrically and the resulting mass was recorded. The densities of the oils were determined at 23 °C using a Mettler Toledo DA-110M density meter (Greifensee, Switzerland) according to ASTM D4052. Calibrations were done prior to measurements with distilled water free from bubbles.

Samples of the pyrolysis oils were characterized by ^{13}C-NMR spectroscopy. Thus, about 2.0 g of oil was dissolved in 2.0 ml of dimethyl sulfoxide (DMSO) and 0.3 ml of tetramethylsilane (TMS) in a 10 mm sample probe. The ^{13}C-NMR spectra were recorded on a Varian Unity 400 MHz NMR spectrometer.

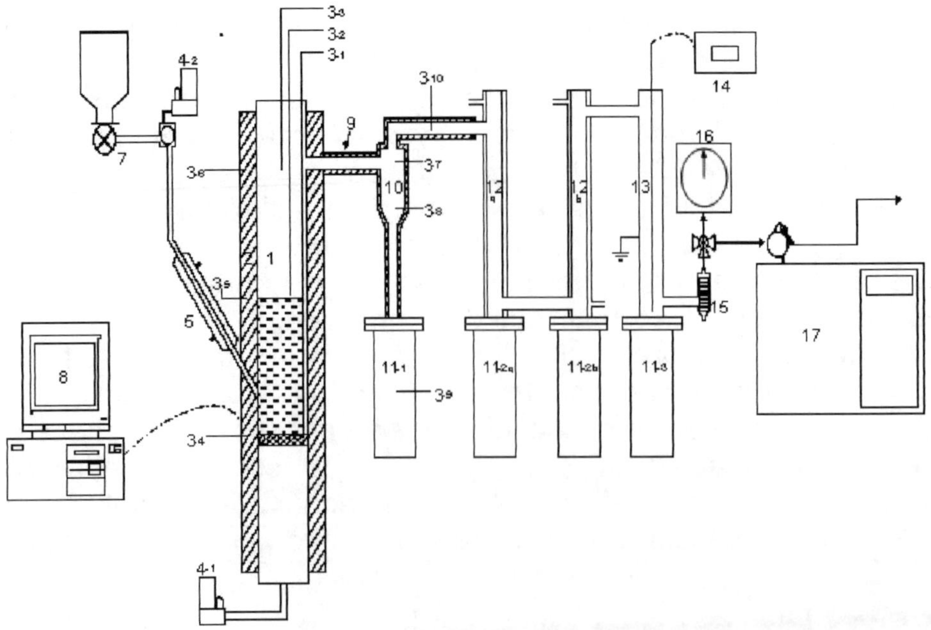

Fig 1. Schematic diagram of bench scale fluidized bed pyrolysis unit.

Results and Discussion

The pyrolysis products were segregated in the condensers and electrostatic precipitator. The first condenser which was immediately after the hot gas filtration unit collected mostly pyrolysis oils with small amounts water, while the second condenser collected almost 95 wt% water. The electrostatic precipitator which was located after the second condenser also collected mostly pyrolysis liquid with moisture content less than 5 wt%.

The yield of pyrolysis products using sand and various catalysts are shown in Table 1. It can be clearly seen from the bench scale studies that the FCC catalysts tended to crack the pyrolysis products extensively. Although the char yields were relatively low, the water content of the product was extremely high. The FCC VPI1 through 6 were FCC catalysts with different zeolite contents. Although the total liquid yields for all the catalysts were similar, the water contents varied considerably. FCC VPI1 had the lowest water content while FCC VPI4 had the highest water content. The char yields from all catalysts were similar. These results appear to suggest that the amount of zeolite in the catalyst had some influence on the degree of cracking of the pyrolysis products to produce water.

Table 1. Yields of pyrolysis products from sand and catalysts.

Catalyst/Medium	Product Yields and Distribution				
	Water (wt.%)	Organics (wt.%)	Total liquid (wt%)	Char (wt%)	Gas (wt%)
FCC- (<63 μm)	31.64	23.16	54.80	16.10	29.1
FCC VPI1(<63 μm)	20.53	38.17	58.70	14.13	27.17
FCC VPI2 (<63 μm)	27.27	31.47	58.74	13.72	27.54
FCC VPI3 (<63 μm)	28.99	29.36	58.35	14.87	26.78
FCC VPI4 (<63 μm)	42.67	27.62	70.29	16.67	13.04
FCC VPI5 (<63 μm)	31.13	34.64	64.77	14.54	19.69
FCC VPI6 (124 μm)	21.86	30.47	52.33	20.13	27.54
HZSM5	15.0	24.2	39.2	20.0	15.0
Sand	13.0	50.0	63.0	20.0	15.0

Table 2. Physical properties of pyrolysis oils from the ESP unit.

Catalyst	pH	Density (g/cm^3)	Dynamic viscosity at 50 °C (cP)	Kinematic viscosity (cSt)	Moisture content (wt%)
FCC- (<63 μm)	3.58	1.221	357.0	292.4	4.43
FCC VPI1(<63 μm)	2.98	1.184	197.0	166.4	3.10
FCC VPI2 (<63 μm)	3.01	1.210	325.0	268.6	2.50
FCC VPI3 (<63 μm)	3.24	1.194	276.0	231.2	3.60
FCC VPI4 (<63 μm)	3.29	1.211	365.4	301.7	3.50
FCC VPI5 (<63 μm)	3.14	1.190	286.0	240.3	3.56
FCC VPI6 (124 μm)	2.78	1.200	318.0	265.0	3.56
HZSM5	3.40	1.12	75.0	66.9	4.00
Sand	2.90	1.19	800.0	672.3	3.79

The sand medium produced less water than the FCC catalysts and additionally, it produced less char than the FCC catalysts. One major difference between the sand and the FCC catalysts was their particle size distributions. The sand (305-500 μm) and the HZSM5 size distributions were greater than those of the FCC catalysts. Thus, it appears the catalyst size distribution may be very important in the pyrolysis of the biomass feedstocks.

The physical properties of the pyrolysis oils are shown in Table 2. The viscosities of all the catalytic pyrolysis oils were lower than that pyrolyzed using sand. This clearly shows that the catalysts are able to reduce the viscosities of the oils because of the cracking of the heavy fractions. However, the pHs of the oils were similar to that of the sand and the densities were also similar. The most significant difference in the catalytic effect was that due to the HSZM5. The viscosity of the oil produced from this catalyst was several factors lower than those of the other catalysts and the sand (Table 2) although the pH and the densities were similar to the other oils.

The ^{13}C-nmr spectra show significant differences between the oils. The oils from the FCC and the sand appeared to be similar in chemical composition. They all contain large fractions of carbohydrate and lignin decomposition products (Fig 2, 3). The ^{13}C-nmr spectra can be divided into the following regions based on functional groups. The signals between 0-30 ppm were assigned to carbon atoms in aliphatic side chains. The peaks at 55-56 ppm were assigned to methoxyl carbons in lignin moities (either in the guaiacyl or syringyl groups). The peaks at 60-100 ppm were assigned to carbohydrate decomposition products such as levoglucosan. The peaks at 105-200 ppm derived from lignin decomposition products. Thus, it can be clearly seen that the carbohydrate decomposition region of the FCC catalysts were less intense compared to the sand pyrolysis oil sample. The strong peak at 56 ppm indicates that the methoxyl carbons were still present in the lignin decomposition products.

The ^{13}C-nmr spectrum of the HZSM5 poplar pyrolysis oil appeared significantly different from that of the other oils (Fig 4). In this spectrum, the signals in the carbohydrate decomposition products region appeared to be significantly reduced compared to those from the sand and the other catalysts whereas the lignin decomposition products signal appeared to have increased in intensity. This ^{13}C-nmr spectrum suggests that the biomass carbohydrate decomposition products were gasified into non-condensable gases whereas the lignin fraction was depolymerized into low molecular phenols. Consequently the gas fraction from this catalyst was much higher than those from the other catalysts. The methoxyl carbon signal also appeared to have reduced intensity suggesting that the HZSM5 also caused either demothoxylation or demethylation of the lignin moieties. It is also interesting to note that although there was significant gasification by the HZSM5 catalyst, it did not produce as much water as the FCC.

There appeared to be two different mechanisms in the catalytic activities of the FCC catalysts and the HZSM5 catalyst. In the case of the HZSM5 catalyst, the gasification was only effective on the carbohydrate decomposition products such as levoglucosan and not very effective on the lignin decomposition products. On the contrary, it appeared the FCC catalysts gasified the entire pyrolysis oils and therefore any residual oil left had similar composition as the original starting material.

These two different pathways have significant effect on the viscosity of the pyrolysis oils. The carbohydrate decomposition products, especially levoglucosan has significant influence on the viscosity of the pyrolysis oils. At room temperature, levoglucosan is partially crystallized and produces very viscous mother liquor. Thus, the presence of levoglucosan in the pyrolysis oils increases their viscosity. In the case of the HZSM5 pyrolysis oil, most of the levoglucosan was gasified and therefore it had minimal effect on the viscosity of the pyrolysis oil. Consequently the viscosity of this oil was several factors lower than those from the sand and the other catalyst. This observation suggests a potential route for lowering the viscosity of the poplar pyrolysis oils.

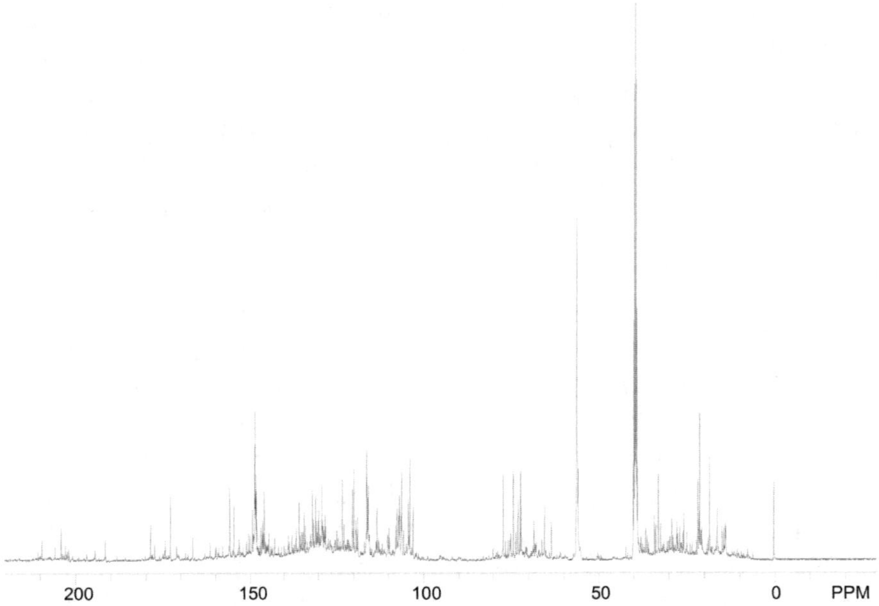

Fig 2. ^{13}C-nmr spectra of hybrid poplar wood pyrolysis oil produced using FCC catalyst. The oil sample was collected from the ESP unit.

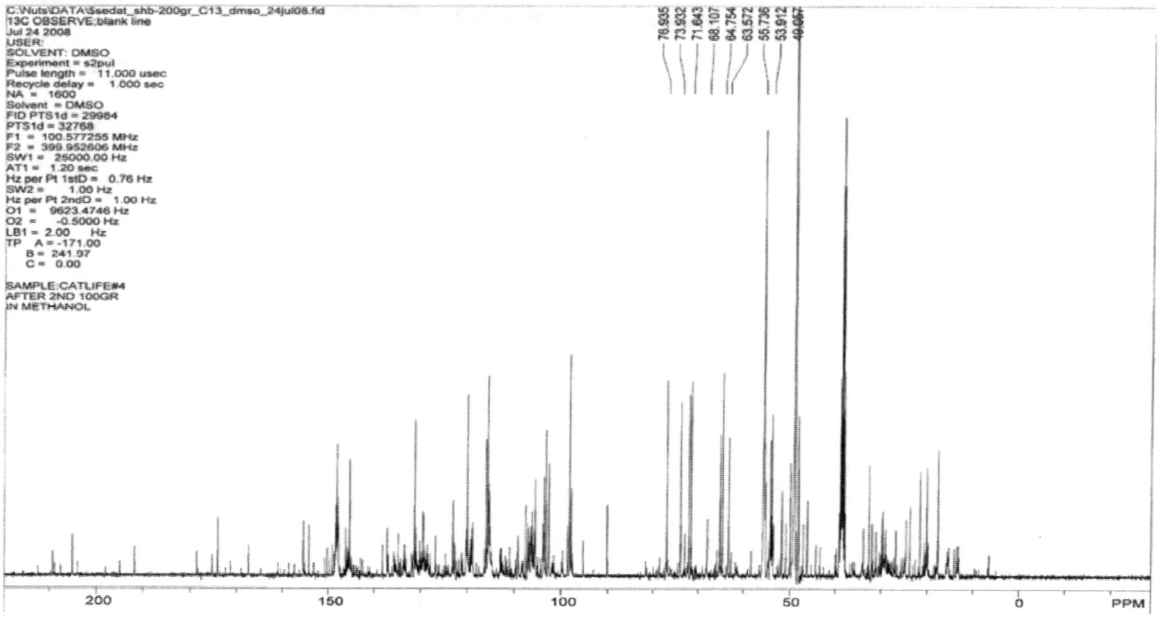

Fig 3. ^{13}C-nmr spectra of hybrid poplar wood pyrolysis oil produced using sand bed. The oil sample was collected from the ESP unit.

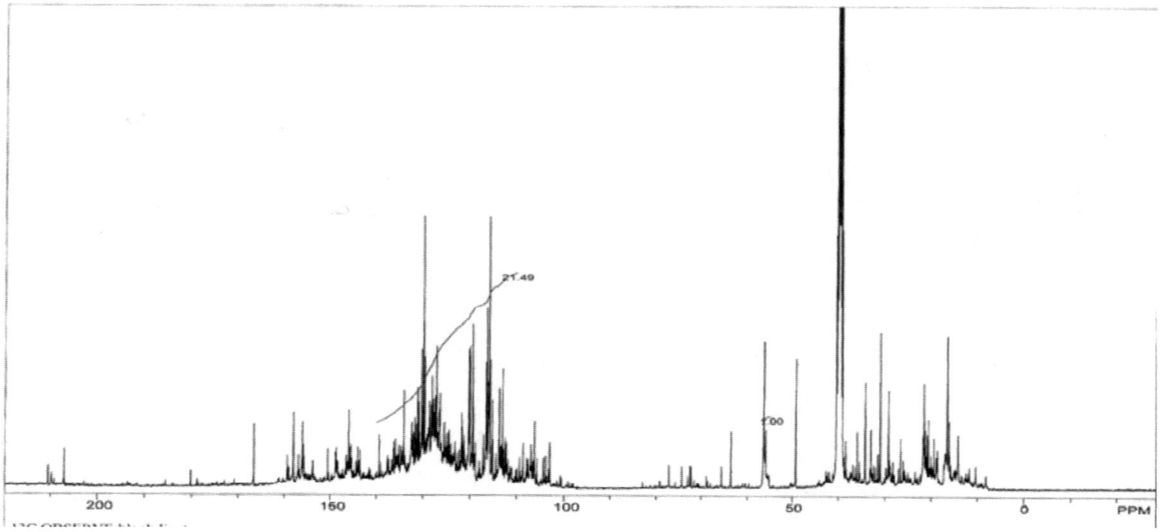

Fig 4. ^{13}C-nmr spectra of hybrid poplar wood pyrolysis oil produced using HZSM5 catalyst. The oil sample was collected from the ESP unit.

3. Conclusions

The FCC catalysts tended to extensively crack the biomass pyrolysis products resulting in the production of water. The pHs of the biooils were acidic, but the viscosities of the oils varied depending on the catalyst. The HZSM5 catalyst produced the oil with the lowest viscosity compared to the sand and the FCC catalysts. The mechanisms of catalysis of the HZSM5 appeared to be different from those of the FCC catalysts.

Liquid Phase Pyrolysis of Biogenic Feedstock

Verena Mertlitz, Department of Chemical Engineering and Environmental Technology, Graz University of Technology, Graz, Austria

Nikolaus Schwaiger, Department of Chemical Engineering and Environmental Technology, Graz University of Technology, Graz, Austria

Peter Pucher, BDI BioDiesel International AG, Grambach, Austria

Matthäus Siebenhofer, Department of Chemical Engineering and Environmental Technology, Graz University of Technology, Graz, Austria

Edgar Ahn, BDI BioDiesel International AG, Grambach, Austria

Pyrolytic conversion of biogenic feedstock has become a high priority task due to declining crude oil reserves.

This project focuses on pyrolytic conversion of CHO-feedstock such as fir chips under liquid phase operation conditions. From liquid phase pyrolysis improved convective heat transfer is expected, enabling fast pyrolysis conditions at ambient pressure. Investigations may be classified as isothermal. Operation temperature is limited to a range of less than 650 K.

The process is operated in a lab scale reactor in the semi-batch mode. The rate of degradation for nearly monodisperse feed is recorded. The feed quantity is limited to an amount which causes a deviation of the reactor temperature of less than 2 K each feed cycle because of the energy surplus of the reaction mass. The effect of temperature on product distribution is investigated. For comparison dry pyrolysis has been carried out for quasi isothermal conditions. Additionally, the influence of the heating rate on the mechanism and rate of pyrolytic degradation was analyzed.

The gaseous, liquid and solid products were characterized by GC-MS, NMR-spectroscopy and elemental analysis. Reproducibility was checked with replicates.

Figure 1 representatively shows the temperature dependency of water formation.

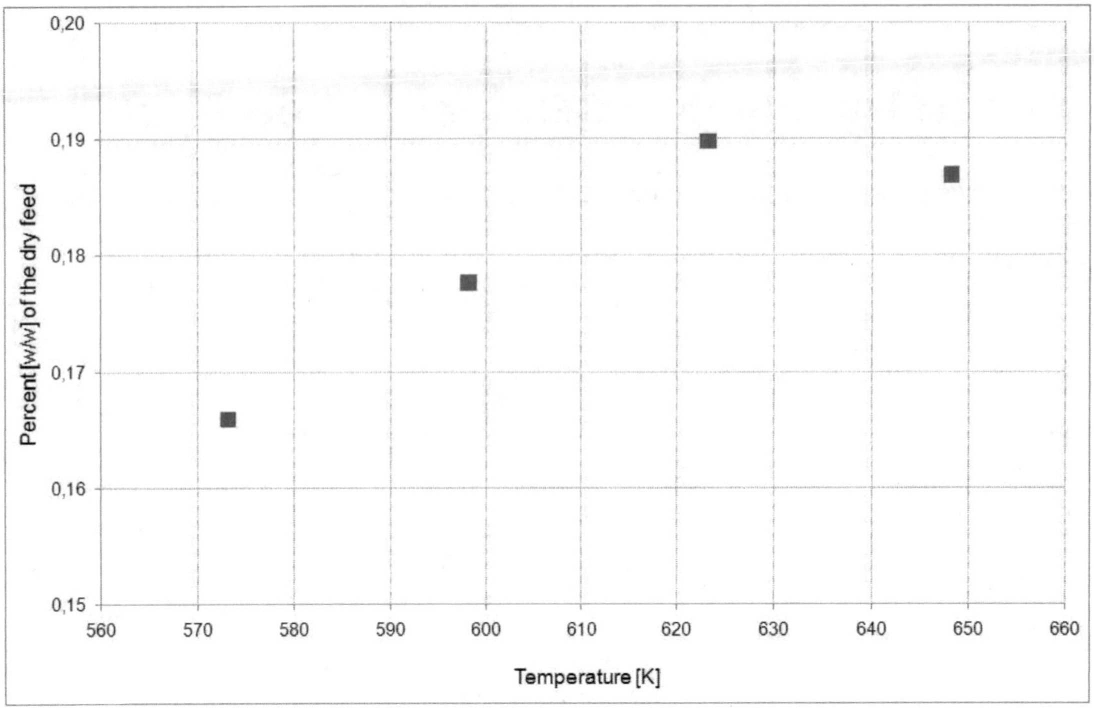

Figure 1: Temperature dependency of water formation

The ratio of water formed during pyrolytic degradation increases with increasing temperature, passing a maximum, while formation of CHO-products linearly correlates with the operation temperature. Carbon transfer from feed to gaseous and liquid product phase also linearly correlates with the temperature.

See more of this Session: Catalytic Biomass Pyrolysis and Gasification I
See more of this Group/Topical: Fuels and Petrochemicals Division

Decomposition of Heptylbenzene by Supercritical Water

Pradip C. Mandal[1], Tatsuya Shiraishi[1], Wahyu Diono[1], Mitsuru Sasaki[1] and Motonobu Goto[2]*

[1] Graduate School of Science and Technology, Kumamoto University

[2] Bioelectrics Research Center, Kumamoto University

2-39-1 Kurokami, Kumamoto 860-8555, Japan

* Corresponding author, TEL: +81-096-342-3664, e-mail: mgoto@kumamoto-u.ac.jp

Abstract

The oil sands are naturally occurring mixtures of sand or clay, water and an extremely dense and viscous form of petroleum called bitumen. Challenges of bitumen production including low mobility, difficulties in transportation to refineries, and limited bitumen refining capacity, make it necessary to upgrade bitumen. There are a number of upgrading processes including carbon rejection, hydrogen addition, and physical separation. The catalytic cracking is the traditional process to convert bitumen into lighter and high value fuels though the catalyst is deactivated in the presence of metals and asphaltenic molecules. The dramatic change of ion product and dielectric constant of water at supercritical condition makes it as acid catalyst. Long-chain n-alkylbenzenes are the simplest chemicals models of the alkyl aromatic moieties. So, in this study, Heptylbenzene (HPB) is used as a model compound and water is used instead of catalyst. The reaction was carried out in a batch reactor made by AKICO. The temperature profile at different molar ratio of HPB and water was observed during the study. The ability of supercritical water (SCW) to decompose alkylbenzene, like heptylbenzene, was also studied at 450 ^{0}C and 40 MPa. Heptylbenzene decomposed into light hydrocarbons (like, toluene, ethylbenzene, butylbenzene, propylbenzene etc.), heavy hydrocarbons (like, pyrene, m-terphenyl etc.) and gases under above mentioned conditions. Significant proportions of heptylbenzene were converted during the reaction with SCW. The obtained results suggest that water acts as a chemical reagent above its critical point (374 ^{0}C and 22.1 MPa).

Author Keywords: Batch reactor; supercritcal water; heptylbenzene.

Introduction

There is about 50-70% heavy compound in crude oil. So heavy oil has been the world`s largest petroleum resources for supplying energy and raw material. Long-chain n-alkylbenzenes, which exist in heavy crude oil or oil sands, are the simplest chemical models of the alkyl aromatic moieties. For pyrolysis of long-chain alkylbenzene, limited kinetic and mechanistic data exist (Savage and Klein, 1987). Cracking crude oil would convert heavy and low value oils into lighter and high value fuels, such as gasoline and light olefins. The fluid catalytic cracking (FCC) and thermal cracking is the typical process to produce lighter oil. The main obstacle to use catalytic process is the presence of metals and asphaltenic molecules in the heavy oil that might lead to deactivate the catalyst (Filho and Sugaya, 2001).

Supercritical water is highly diffusive steam. It can be miscible with gases and hydrocarbons to form a homogeneous mixture (Sato et al, 2003). The supercritical water oxidation processes are developed for treating organic waste (Savage et al, 1995) using solubility effect of supercritical water. The ion product of water increases significantly at high temperature and pressure. The dielectric constant of SCW ranges from 2 to 30 which are similar to the range from a non-polar solvent such as hexane, whose dielectric constant is 1.8, to a polar solvent such as methanol, whose dielectric constant is 32.6. This dramatic change of ion product and dielectric constant, SCW possesses the effect of an acid catalyst (Ogunsola, 2000). Supercritical water acts as a solvent and a reactant on hydrolysis based reaction and water molecules often take part as collision partners (Sato et al, 2003).

In this paper, heat transfer for batch reactor is examined to assess the impact of supercritical water. Reaction of HPB in the absence and in the presence of supercritical water is also described to better understand the reaction mechanism. This study provides some understanding of the reaction of similar alkylbenzene compounds in the presence of supercritical water.

Materials and experimental procedures

Materials

Commercially available HPB, 99.7% by GC analysis, was bought and used as received without further treatment. All experiments were carried out in an 8.8 ml batch reactor fabricated from hastelloy C-276. The both sides of the reactor were equipped with removable threaded cover containing facilities to inset temperature sensor, pressure sensor and collecting gaseous product. The reactor was designed by AKICO for maximum temperature 500 ^{0}C and maximum pressure 50 MPa.

Procedure

About 3 ml HPB and 2.5 ml distilled water were taken in the batch type reactor for pyrolysis and hydrolysis reaction. The reactor was reassembled and argon gas was used to purge the reactor to remove gaseous portion from the reactor. The electric furnace, made by AKICO, was heated to the expected temperature. Then the reactor was loaded into the furnace. During each run, the reactor was shaked by front and back motion approximately 70 cycles per minutes. For collecting temperature profile data a temperature sensor was connected in the middle of the reactor. After an hour, the furnace was stopped and the reactor was removed from the furnace and quenched in an ice bath. Then the reactor was opened and its contents collected by washing with acetone.

Analytical Chemistry

Identification of individual decomposed components in the reaction liquid mixture was done using the GC/MS method (qualitative analysis), while the composition of the reaction liquid mixture was determined by GC/FID (quantitative analysis). The GC-MS column was a non-polar HP-1MS capillary column, 30 m long by 0.32 mm diameter. The temperature programme was 45-285 ^{0}C heated at 5 ^{0}Cmin−1 heating rate, held at 285 ^{0}C for 25 min followed by heating to 295 ^{0}C at 10 ^{0}Cmin−1 heating rate and held at the temperature of 295 ^{0}C for 10 min, then, heating to 300 ^{0}C at 10 ^{0}Cmin−1 heating rate and held at the final temperature of

300 ^{0}C for 5 min. The GC-FID measurement was carried out using Shimadzu GC 2014 instrument using the same temperature program used in GC-MS.

Results and discussions

Heating system study

Figure 1 shows the furnace heating curve without loading reactor. After 31 minutes, the furnace was reached at temperature 450 ^{0}C. Then temperature was slightly increased and at 38 minutes the furnace was reached the fixed temperature 450 ^{0}C. When the reactor was inserted into the furnace, the reactor and reactants were started to consume heat and the furnace temperature was started to decrease (figure 2). Within a short time, the furnace temperature was started to rise and this time was called zero reaction time. After 10 minutes the furnace temperature was reached to a fixed value.

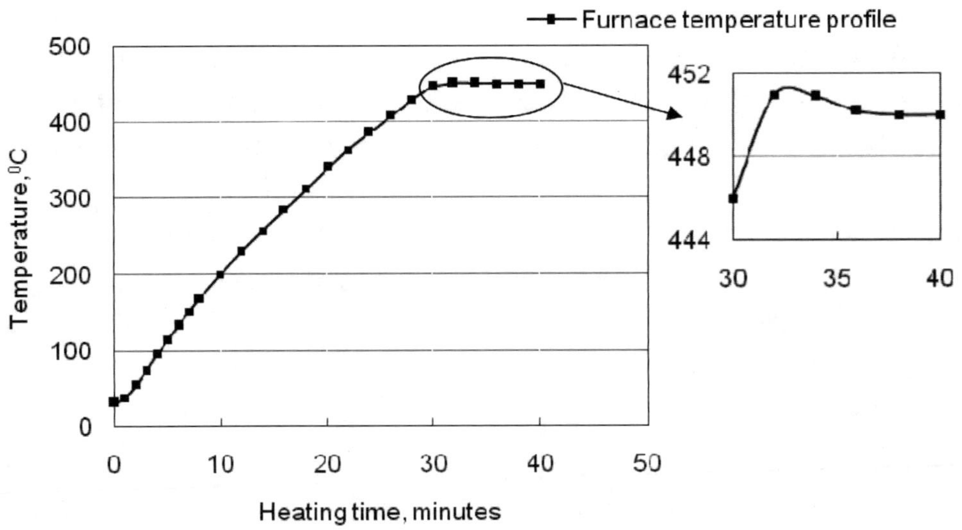

Figure 1: Furnace heating curve without loading reactor.

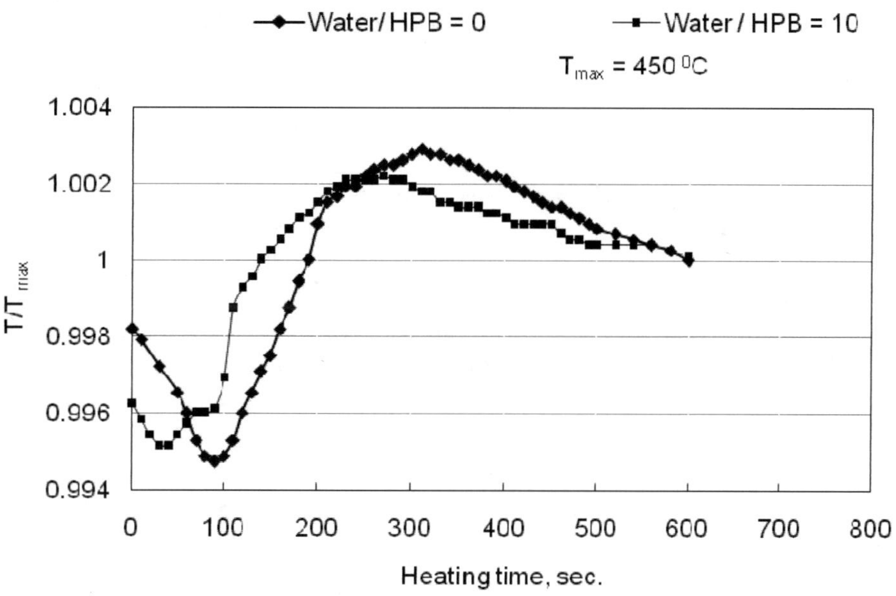

Figure 2: Furnace heating curve after loading reactor.

Figure 3 shows that during the heating of the reactants and products the mixture temperature evolves in non-linear way. The initial slope of temperature curve depends on the molar ratio of water to HPB and the slope is higher for the presence of water. Figure 4 shows that initial heat transfer rate is high for all cases. Among them, heat transfer rate is higher for presence of water. Figure 5 shows that reactant temperature increases quickly at shaking speed 70 cycles per minutes.

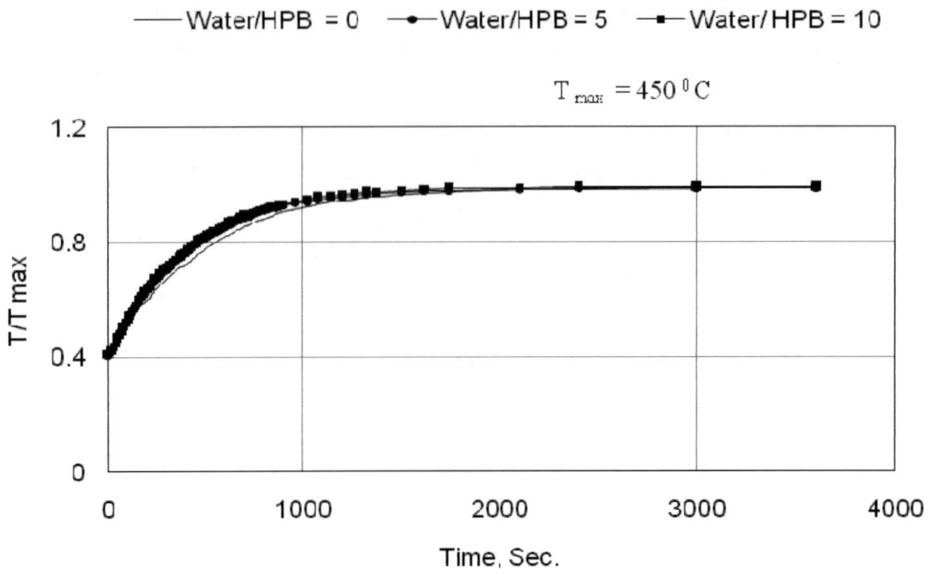

Figure 3: Temperature distribution, in dimensionless form, of fluid for batch reactor at different time and molar ratio.

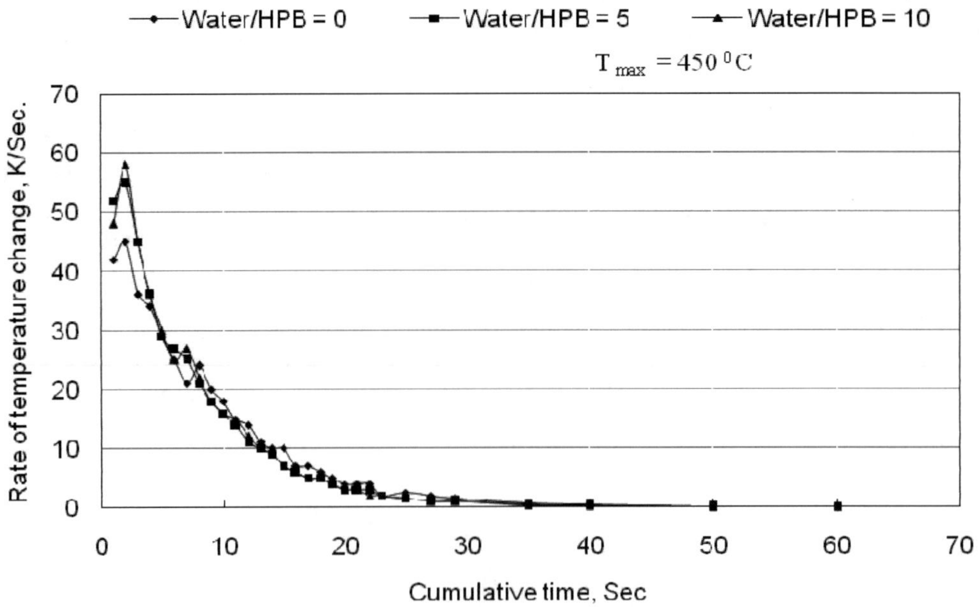

Figure 4: Rate of temperature change at different molar ratio of water to HPB.

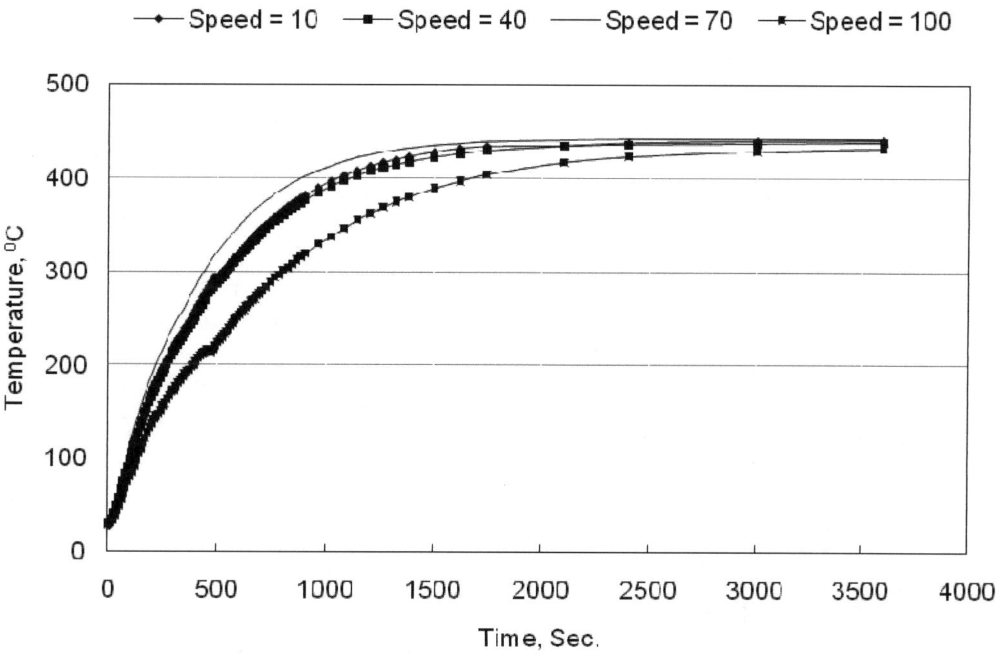

Figure 5: Temperature distribution of fluid for batch reactor at different shaking speed.

Decomposition of HPB at temperature 450 ^{0}C and pressure 40 MPa

Figure 6 shows a typical GC-MS chromatogram showing the reaction products during the pyrolysis of HPB at 450 ^{0}C. The major products of HPB pyrolysis were toluene, ethylbenzene, butylbenzene, propylbenzene, 1-methylethylbenzene, 1-methylhexylbenzene, pentylbenzene, hexylbenzene, naphthalene and pyrene. The reaction sequence in the presence of SCW is outlined in figure 7. The major products were toluene, p-xylene, butylbenzene, propylbenzene, pentylbenzene, 1-methylethylbenzene, naphthalene and 1-methylnaphthalene. The basic reaction mechanisms were shortening and rearranging side chain of benzene, ring formation and addition reaction in side chain of benzene.

Table 1 shows the GC-MS distribution of the dominant liquid products from pyrolysis and SCW interaction of heptylbenzene. Pyrolysis of HPB led to an ultimate HPB conversion of 88.81% at 60 minutes and 450 ^{0}C. But in the same conditions, the hydrothermal conversion was 84.67%. The product, pyrene is drastically reduced by the action of SCW. Large amount of p-xylene is also produced instead of ethylbenzene and 1-methylhexylbenzene during supercritical destruction of HPB.

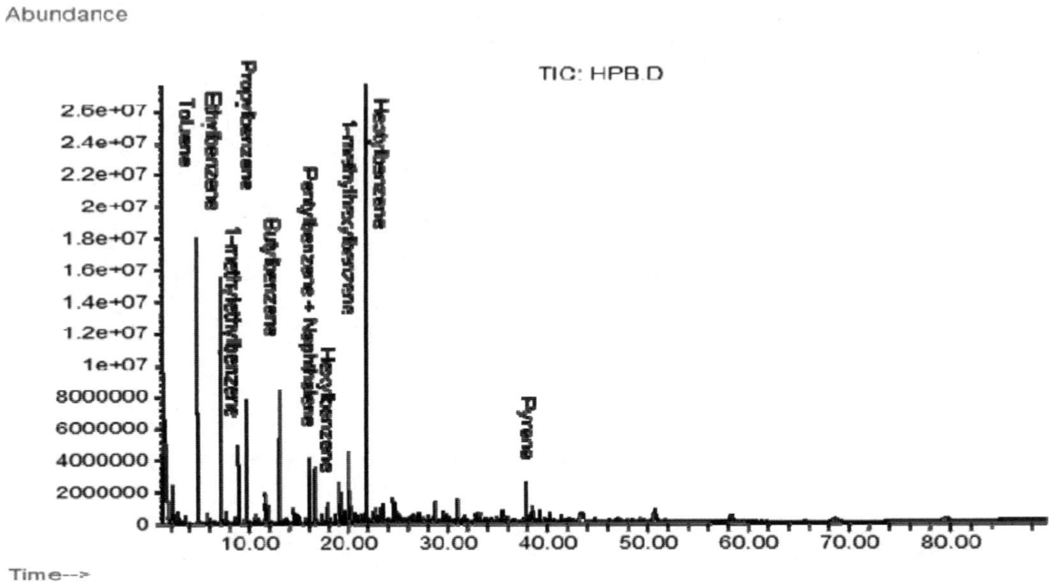

Figure 6: GC-MS result of HPB thermal cracking.

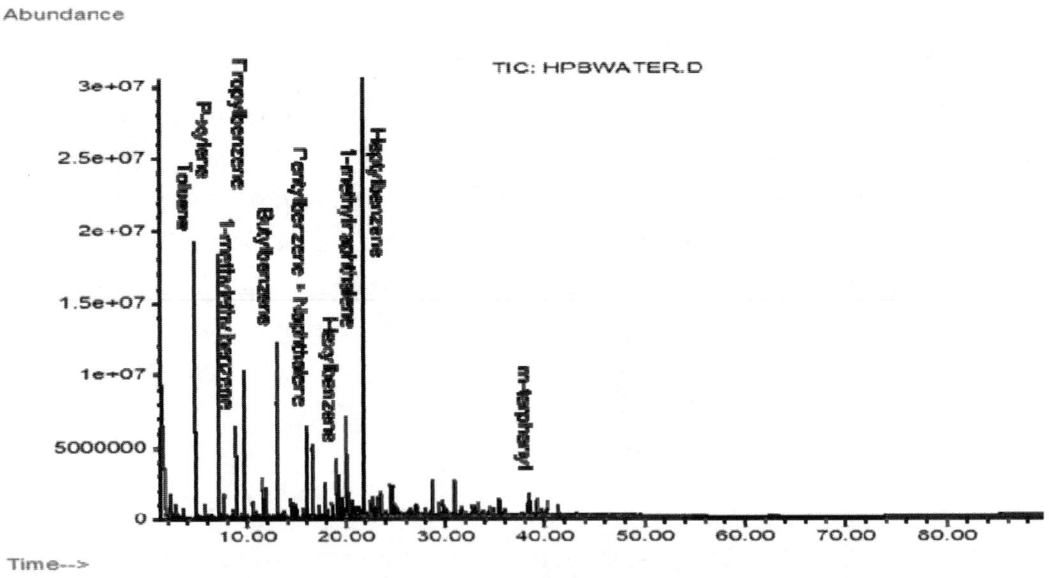

Figure 7: GC-MS result of HPB hydrothermal cracking.

Table 1. Reaction products from heptylbenzene pyrolysis and hydrolysis at temperature 450 ^{0} C and time 60 minutes.

Product	Formula	Molecular weight	Pyrolysis	(SCW) Hydrolysis
Yields (wt.%)				
HPB unreacted			11.19	15.33
Reaction liquid product [a]			63.34	54.4
Unaccounted for [b]			25.47	29.77
Composition of liquid products after reaction (Area% by GC-MS Analysis)				
		Alkylbenzene		
Toluene	C_7H_8	92.14	11.48	9.68
Ethylbenzene	C_8H_{10}	106.17	7.36	0
p-xylene	C_8H_{10}	106.17	0	6.83
Benzene, (1-methylethyl)-	C_9H_{12}	120.19	1.93	1.93
Benzene, propyl-	C_9H_{12}	120.19	3.15	3.19
Benzene, butyl-	$C_{10}H_{14}$	134.22	3.35	3.74
Benzene, pentyl-	$C_{11}H_{16}$	148.24	1.68	2.04
Benzene, hexyl-	$C_{12}H_{18}$	162.27	1.62	1.92
Benzene, (1-methylhexyl)-	$C_{13}H_{20}$	176.30	1.89	0
Heptylbenzene	$C_{13}H_{20}$	176.30	16.85	17.44
		Ring Formation		
Naphthalene	$C_{10}H_8$	128.17	1.42	1.57
Naphthalene, 1-methyl-	$C_{11}H_{10}$	142.20	0.33	1.45
Naphthalene, 1-ethyl-	$C_{12}H_{12}$	156.22	0.60	0.73
Pyrene	$C_{16}H_{10}$	202.25	1.10	0.13
		Addition		
Diphenylmethane	$C_{13}H_{12}$	168.23	0.62	0.73
Benzene, octyl-	$C_{14}H_{22}$	190.32	0.75	0.79
p-terphenyl	$C_{18}H_{14}$	230.30	0.45	0.61

[a] Not including gases.

[b] Possibly contained in product gases.

Conclusions

This study has shown that water is a good solvent and dominates the temperature profile in a batch reactor. The temperature of reactants increases quickly at shaking speed 70 cycles per minutes. Detailed studies of the decomposition behavior of HPB have been carried out under hydrothermal conditions. During reaction, low molecular weight products generated during interaction with supercritical water. Different products slates were also observed in the thermal cracking and hydrothermal cracking. So, supercritical water was active during the reaction in hydrothermal condition.

Acknowledgements

This work was supported by the Global COE program on Global Initiative Center for Pulsed Power Engineering.

References

1. Ogunsola, Olubunmi M. (2000), ``Decomposition of Isoquinoline and Quinoline by Supercritical Water,`` Journal of Hazardous Materials 74 (3), pp. 187-195.
2. Savage, P.E. and M.T. Klein (1987). Asphaltene Reaction Pathways. 2. Pyrolysis of n-pentadecylbenzene. Industrial and Engineering Chemistry Research, 26 (3), pp. 488-494.
3. Savage, Phillip E., Sudhama Gopalan, Tahmid I. Mizan, Christopher J. Martino and Eric E. Brock (1995), ``Reactions at Supercritical Conditions: Application and Fundamentals,`` AIChE Journal, July, pp. 1723-1725.
4. Sato, Takafumi, Tadafumi Adschiri, Kunio Arai (2003), ``Decomposition Kinetics of 2-propylphenol in Supercritical Water,`` Journal of Analytical and Applied Pyrolysis, 70, pp. 735-746.
5. Maciel Filho, R., and M.F. Sugaya (2001), ``A Computer Aided Tool for Heavy Oil Thermal Cracking Process Simulation,`` Computers and Chemical Engineering, 25, pp. 683-692.

Atomistic Simulations of Alternative Proton Exchange Membranes for Fuel Cell Applications

Chetan V. Mahajan, Chemical Engineering, University of Texas at Austin, Austin, TX

Venkat Ganesan, Chemical Engineering, University of Texas at Austin, Austin, TX

SPEEK (Sulfonated Poly(ether ether ketone)) membranes constitute promising alternatives to nafion for DMFC(Direct Methanol Fuel Cell) applications. In this talk, the results of extensive all-atom simulations of water and methanol solvated SPEEK membranes are reported. We focus on key features of changes in structure, transport and the corresponding coupling between the structure and transport. Our simulations suggest that the membrane swells appreciably with increasing water and methanol contents. The sizes of water clusters in the membrane are found to be appreciably lower than that reported for nafion in the literature, thus confirming experimental observations of narrower aqueous channels in SPEEK than nafion. The diffusivities of water, hydronium ions and methanol are also found to be lower than those published for nafion earlier, thus complementing experimental findings of lower penetrant mobility in SPEEK than nafion. Structural information pertaining to the relative locations of methanol and hydronium ions is used to derive insights into the behavior of the transport coefficients.

See more of this Session: Fuel Cell Technology II
See more of this Group/Topical: Fuels and Petrochemicals Division

Effects of the Micro-Porous Layer on Water Transport in a PEM Fuel Cell

Trung V. Nguyen & Xuhai Wang
Department of Chemical and Petroleum Engineering
The University of Kansas, Lawrence, KS 66044

Background

Experimental studies have shown that the performance of a PEMFC was improved by the application of a microporous layer (MPL) in the cathode [1, 2]. However, the actual effect of the MPL is not fully understood. One suggestion [3] is that the highly hydrophobic MPL forces the water that is generated at the cathode back to the anode resulting in higher membrane hydration and conductivity. In this study, a model was developed to evaluate the role of the MPL and the effects of its properties (capillary and permeability) on the liquid water saturation levels in the cathode porous media, the transport rate of oxygen to the reactive sites and the performance of a PEMFC. Three cases were selected for this study: a cathode 1) without an MPL and with 2) a less hydrophobic MPL and 3) a more hydrophobic MPL.

Results

The results validated that the fuel cell performance was improved by adding an MPL between the macro-porous gas diffusion layer and the catalyst layer in the cathode. The MPL, due to its high hydrophobicity, increases the liquid water pressure in the cathode to the levels much higher than those in the anode resulting in an increased back-transport rate of liquid water from the cathode to the anode. This increased back transport rate of liquid water reduces the liquid water transport rate from the catalyst layer to the macro-porous gas diffusion layer resulting in lower liquid water saturation levels in this layer and, consequently, higher oxygen gas transport to the catalyst layer in the cathode. The results also show that the more hydrophobic the MPL is, the higher liquid pressure increases, and, consequently, the higher driving force is created across the membrane to drive water from the cathode to the anode.

The benefit of lower liquid water saturation levels in the GDL is traded-off by the higher saturation levels in the CL. So, the impact of the MPL on the saturation levels in the GDL, MPL and CL must be considered systematically in order to obtain the best fuel cell performance. The interaction of these three layers will be discussed in this study.

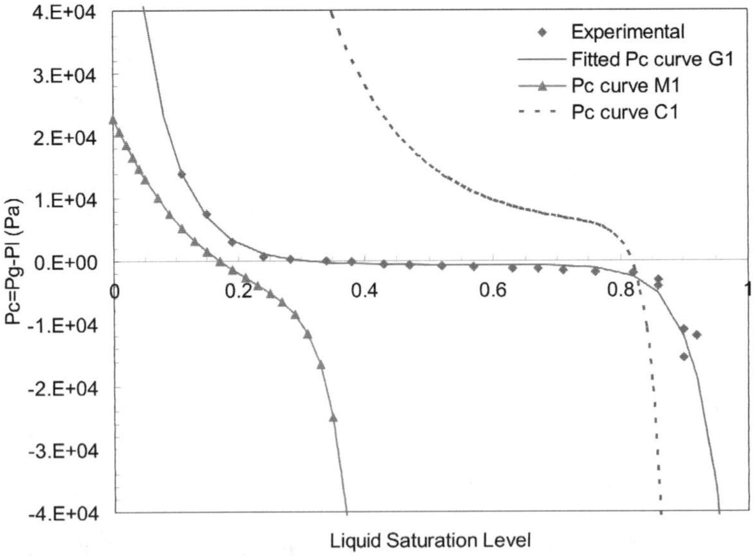

Figure 1. Capillary curves used in the case with and without the MPL.

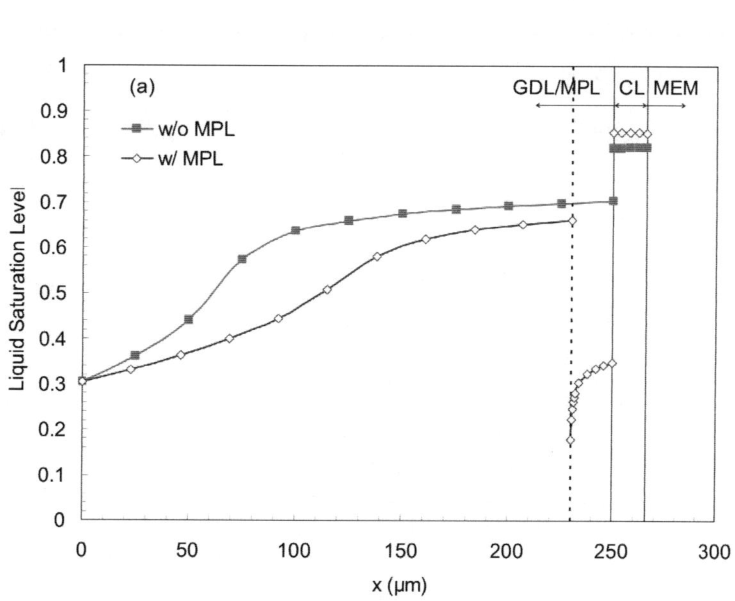

Figure 2. Liquid water saturation level distribution in the membrane-electrode-assembly at 0.6V

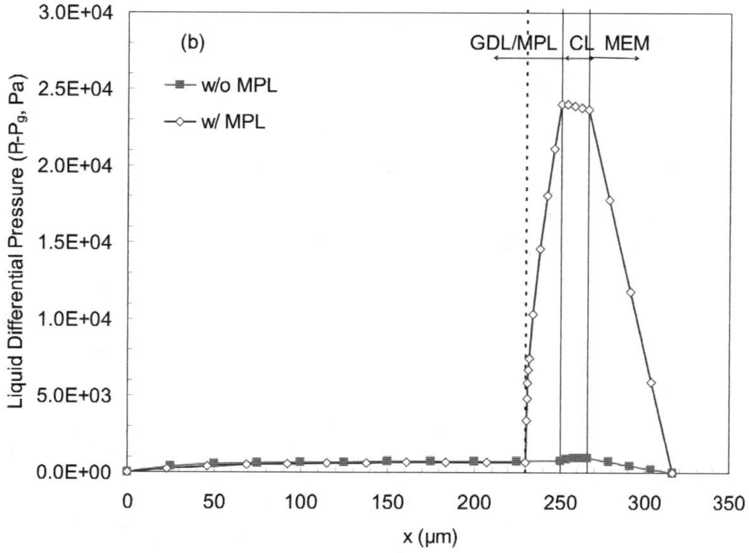

Figure 3. Liquid water pressure distribution in the membrane-electrode-assembly at 0.6V

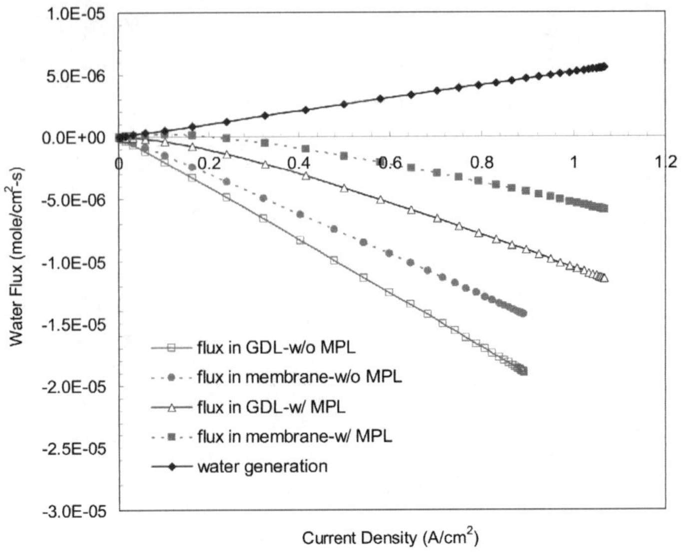

Figure 4. Water fluxes in the membrane-electrode-assembly with and without the MPL

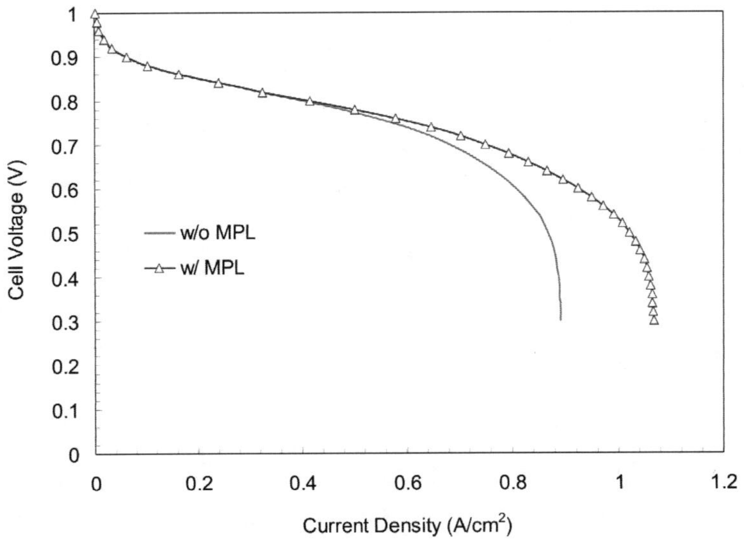

Figure 5. Fuel cell performance with and without the MPL in the cathode

Acknowledgements

This work was supported by the National Science Foundation through grant No. CBET- 0651758.

References

1. G. Lin and T. V. Nguyen, *J. Electrochem. Soc.*, **152** (10) A1942 (2005).
2. Z. Qi and A. Kaufman, *J. Power Sources*, **109,** 38 (2002).
3. A. Z. Weber and J. Newman, *J. Electrochem. Soc.*, **152** (4) A677 (2005)

Modelling of Cross-Over Effects in Direct Ethanol Fuel Cells (DEFCs)

Suresh N S, Department of Chemical Engineering, IIT Madras, Chennai, India

Sreenivas Jayanti, Department of Chemical Engineering, IIT Madras, Chennai, India

Alcohol-based fuel cells offer a possibility of using renewable fuels derived from biomass [1] to be used for power generation. In this context, direct ethanol fuel cells (DEFC), in which ethanol in liquid form is fed as the proton source in a regular proton exchange membrane-based fuel cell (PEMFC), offers the treble advantages of a renewable, environmentally friendly and easily transportable fuel to produce electrical power [2], while with respect to methanol it has higher energy density and is non-toxic . However, as in all alcohol-based PEMFCs, fuel permeation through the proton exchange is a problem. This "cross-over" of ethanol from the anode through the membrane electrode assembly (MEA) to the cathode side results in two disadvantages: it burns up chemically on the cathode side in the presence of platinum catalyst without producing any protons, and it also competes with electricity-producing protons for oxygen. The extent of ethanol cross-over can have an important bearing on the usefulness of DEFCs as a power provider [3].

The cross-over of ethanol is primarily a mass transfer phenomenon with add-on effects of electrochemistry. In the present study, a comprehensive mathematical model is presented to calculate the extent of fuel cross-over in a DEFC. The model is based on a similar model [4] for direct methanol fuel cell but differs from it in the treatment of the oxidation-reduction reaction kinetics on the anode side. The model takes account of all the important ethanol transport mechanisms in a DEFC. While ethanol permeation through the "gas diffusion layer" is by concentration-gradient-driven diffusion, ethanol cross-over across the PEM takes place by diffusion and convection [5]. Convective transport in the membrane is driven by electrostatic potential gradient and pressure gradient. All these effects are included in the present model. The electrochemical ethanol oxidation at the anode side of the DEFC is modelled using Tafel kinetics with a first order dependence.

Using the mathematical model, the effect of important cell operating parameters such as current density, ethanol feed concentration, cell temperature and cathode side pressure on the ethanol cross-over rate is investigated. The results obtained show that at low concentrations, the ethanol cross-over flux decreases as the current density increases. However, as the feed concentration increases, the ethanol cross-over flux remains constant or increases [6] with the current density and it poses serious problem to DEFC performance. At the same time, the permeated ethanol and its oxidation of intermediate products could poison the cathode catalyst and declines the active nature of platinum. For low ethanol feed concentrations, the potential gradient driven electro-osmotic cross-over flux is small, and as the current density increases, the ethanol concentration at the interface between anode catalyst layer and PEM decreases and becomes zero at limiting current density values. Thus, for low feed ethanol concentrations, the ethanol cross-over flux across PEM is purely by diffusion mode of transport. For higher ethanol concentrations, over a range of current densities, no sharp drop or any significant change in diffusive cross-over flux but the net cross-over rate increases because of the enhanced electro-osmotic cross-over flux with current density.

The effect of temperature on the ethanol cross-over is related to the diffusivity of the PEM. So, decreasing the cell temperature decreases the diffusivity of ethanol through PEM and hence the cross-over rate. It is also found that the ethanol cross-over flux can be slightly reduced by increasing the cathode side pressure which is attributable to the increased negative effect of the pressure gradient driven ethanol cross-over flux that takes place in the opposite direction i.e., from the cathode side to the anode side.

The model is thus able to capture the effect of important parameters on ethanol cross-over and it can be used for optimizing the design and operating parameters of a DEFC.

References

[1] Guangchun Li, Peter G.Pickup, Analysis of performance losses of direct ethanol fuel cells with the aid of

a reference electrode, Journal of power sources,2006 (161):256-263.

[2] Lamy et al, Recent progress in the direct ethanol fuel cell: development of new platinum-tin electrocatalysts, Electrochemical Acta,2004 (49):3901-3908.

[3] Song et al, Ethanol crossover phenomena and its influence on the performance of DEFC, Journal of power sources,2005(145):266-271.

[4] D. Kareemulla, S.Jayanti, Comprehensive one-dimensional semi-analytical mathematical model for liquid-feed polymer electrolyte membrane direct methanol fuel cells, Journal of Power Sources,2009(188):367-378.

[5] Bernardi, D.M., and M.W.Verbrugge, A mathematical model of the solid -ploymer-electrolyte fuel cell, Journal of Electrochemical society, 1992(139), 2477-2491.

[6] S.Kontou, V.Stergiopoulos, S.Song, P.Tsiakaras, Ethanol/Water mixture permeation through a Nafion based membrane electrode assembly, Journal of Power Sources, 2007(171):1-7

See more of this Session: Fuel Cell Technology II
See more of this Group/Topical: Fuels and Petrochemicals Division

Steady State Modeling and Optimization Studies of a Tubular Proton Exchange Membrane Fuel Cell (PEMFC) Validated with Experimental Studies

Brian Bullecks, Chemical Engg. Dept., Clarkson University, Potsdam, NY

Debangsu Bhattacharyya, Department of Chemical Engg., West Virginia University, Morgantown, WV

Raghunathan Rengaswamy, Department of Chemical Engineering, Texas Tech University, Lubbock, TX

A significant reduction in cost is required to make the PEMFC (Proton Exchange Membrane Fuel Cell) technology commercially viable. Several options are being explored for reducing the cost. In this work, a tubular architecture is considered for reducing the cost. The design also significantly improves the power/weight ratio of the cell compared to the commercial planar PEMFCs. It can be mentioned that a significant increase in the power/weight ratio can reduce the parasitic losses in the system especially for automotive applications of the cell.

The tubular cell is designed and manufactured for both air breathing and pressurized cell applications. Commercially available hardware components are used for manufacturing the cells. Testing of the cell is done with a Scribner Associates 850E HT system. The performance of the pressurized cell is found to be better than the state-of-the-art planar PEMFC available commercially. The power/weight ratio of the tubular cell is found to be higher by an order of magnitude than its commercial counterpart. The design also results in a simplification of the complicated flow-field design typically used for enhancing the performance of planar PEMFCs. In addition, the design improves the water management, particularly at a high current density.

A theoretical model of the cell is also developed. In this isothermal half cell model, mass and momentum conservation equations are considered inside the cathode gas flow channel. Mass/species conservation equations are written for the gas diffusion layer and the reaction layer. The reaction layer is characterized as both macro homogenous and spherical agglomerate models. Effects of these characterizations on the model predictions will be presented. In order to reduce the time for computations, the method of Thiele Modulus and effectiveness factor are adopted for the spherical agglomerate characterization. The model of both the air-breathing and the pressurized cells are developed. Both the models are validated by the experimental data collected over a wide range of air flow rate and operating pressure.

A dimensional optimization study is done with the developed steady state model. A nonlinear constrained multiobjective optimization problem is solved to generate the optimum dimensions for the tubular cell that results in an enhanced performance of the cell compared to the base case. The optimized dimensions are used to fabricate a new cell. The performance of the optimized cell in comparison with the base case cell and that of the theoretical prediction will also be presented.

See more of this Session: Fuel Cell Technology II
See more of this Group/Topical: Fuels and Petrochemicals Division

AC Impedance in Characterization of Solid Oxide Fuel Cell and an Interpretation of the Low Frequency Inductive Loop

Ying Zhu, Wenhua H. Zhu, and Bruce J. Tatarchuk
Center for Microfibrous Materials Manufacturing
Department of Chemical Engineering
207 Ross Hall, Auburn University, AL
Email: brucet@eng.auburn.edu

Abstract

Impedance spectroscopy was applied on the small button cell at various conditions, which was fabricated with anode and cathode reference electrodes to facilitate separate testing of anode and cathode. Equivalent circuit models were derived from the impedance data and validated by changing cell temperature, load current and reactant concentration. The impedance spectrum of total cell was analyzed, in comparison with the sum of half cell spectra, to evaluate the propriety for separation anodic and cathodic losses by using reference electrodes. A small but significant inductive loop at low frequency was found in the anode impedance spectrum, and its physical meaning was interpreted by an equivalent circuit corresponding to the possible electrode process.

Introduction

Fuel cells have been widely recognized as the environmentally benign power generation devices. Fuel cell technology has been undergoing large progress during past years. The Solid Oxide Fuel Cells (SOFCs) are now catching more and more attentions for their higher efficiencies than other power generation systems, with minimal air pollutant emissions and low greenhouse gas emissions.

Impedance spectroscopy is a novel and powerful method to assess the dynamic response characteristics of an electrochemical system. An AC signal is superimposed on the DC output of an electrochemical cell to perform the impedance spectroscopy measurement. A set of impedance data is collected over a spectrum of frequencies and fitted with an equivalent circuit model, which is a circuit of electrical elements producing a similar load response to the device under investigation. Different circuit elements in equivalent circuits theoretically represent different relaxation processes occurring in the electrochemical cell(s), and values of circuit elements can be used to distinguish to what extent the processes are limiting cell behavior.

Most AC impedance has been performed near open circuit voltage (OCV) due to equipment limitations. In the following work, impedance spectroscopy has been applied to test the button cell at load under various operating conditions. Fitted equivalent circuits have also been derived.

Experimental

The solid oxide fuel cell tested here is a disc like button cell (Ceramatec, Inc.). The button cell measurements are essential to perform before testing the full-sized stack cells, because conditions are stable and the electrodes can be measured separately. The picayune nature of the button cell minimized temperature and current distribution over the active surface. Of even greater significance is the ability to use a reference electrode on the button cell in order to separately probe anode from cathode, because installing reference electrodes in a stack proves difficult. The ultimate goal is to apply the individual button cell electrode analyses to data obtained from a commercial fuel cell stack.

The button cell consisted of an 8 cm^2 disc of zirconia based electrolyte (~170 μm thick) onto either side of which a 2 cm^2 nickel based anode and a 2 cm^2 manganite based cathode had been brushed. A small reference electrode of the same material as the corresponding electrode was brushed on both sides of the electrolyte near the electrode, as depicted in Figure 1. Mixtures of hydrogen and nitrogen, at different hydrogen content, were bubbled through water on the anode side of the button cell, and air was fed at ~15 psig on the cathode side. Cell temperature was maintained with furnaces controlled by a LabVIEW™ (National Instruments) program. Furnaces were cooled down to 500°C during long intervals between tests to minimize cell aging.

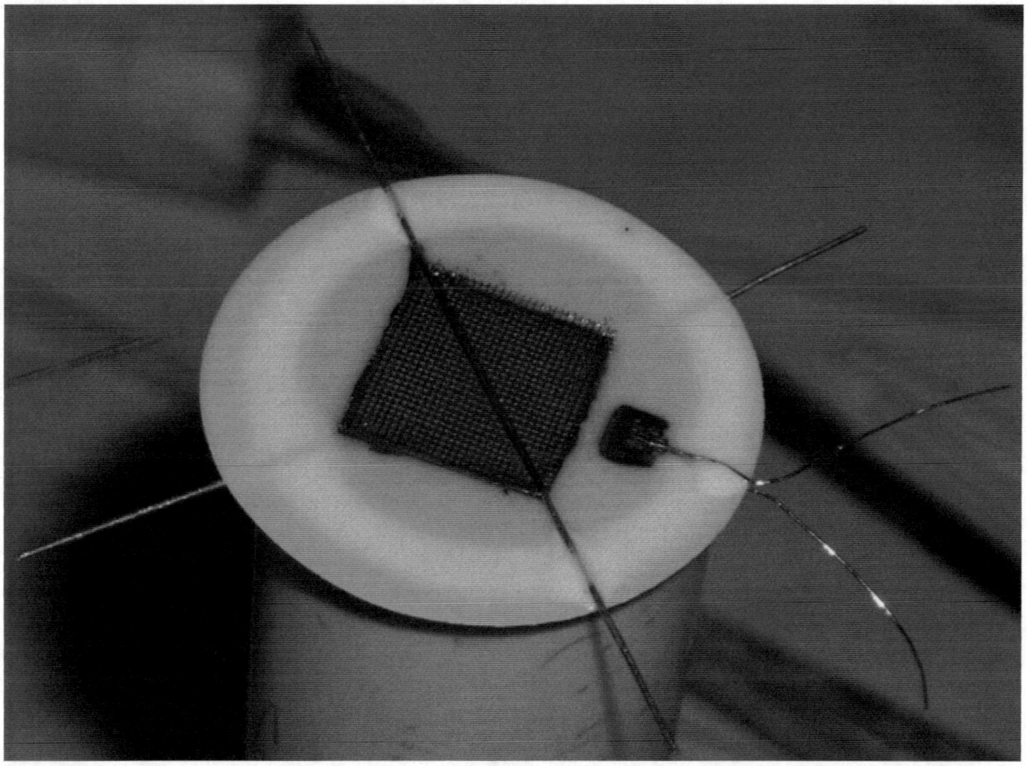

Figure 1. Tested button cell, where the cathode is the large electrode and the reference, the smaller. The opposite face of the electrolyte appears the same, with an anode and anode reference electrode.

The button cell was tested using a Solartron Analytical 1255A frequency response analyzer in conjunction with a Schlumberger SI-1286 potentiostat/galvanostat. The test was controlled and data were recorded through a LabVIEW™ program. Impedance spectra were measured with an FC350™ potentiostat (by Gamry Instruments) in conjunction with a TDI-Dynaload® RBL488 programmable load. Data were acquired galvanostatically for the anode, cathode, and total-cell at a given current. In each case, the system was perturbed with a sinusoidal wave with amplitude of 5 mV over a spectrum of 1.0 MHz to 10 mHz. All spectra collected from the anode, the cathode and the total cell were fitted to equivalent circuits with Gamry's Echem Analyst software using the Levenberg-Marquardt algorithm.

The independent variables studied were cell current, cell temperature, and anode feed composition (the inlet percentage of hydrogen). Gas mixtures were altered using rotameters upstream from the water bubbler maintaining a total flow of 39 SCCM. Outlet gas compositions were measured with a molecular sieve column in an Agilent 3000A gas chromatograph and analyzed with Agilent's Cerity Network Data System software. Inlet compositions were assumed to be equal to those of the outlet at open circuit voltage (OCV).

Results and Discussion

Cathode

AC impedance was performed on the button cell cathode with varying load current and cell temperature. The nyquist plots for the cathode impedance consist of a depressed semicircle and a relatively small capacitive tail, as shown in Figure 2. Each loop in the spectrum corresponds to a relaxation time associated with a physical process.

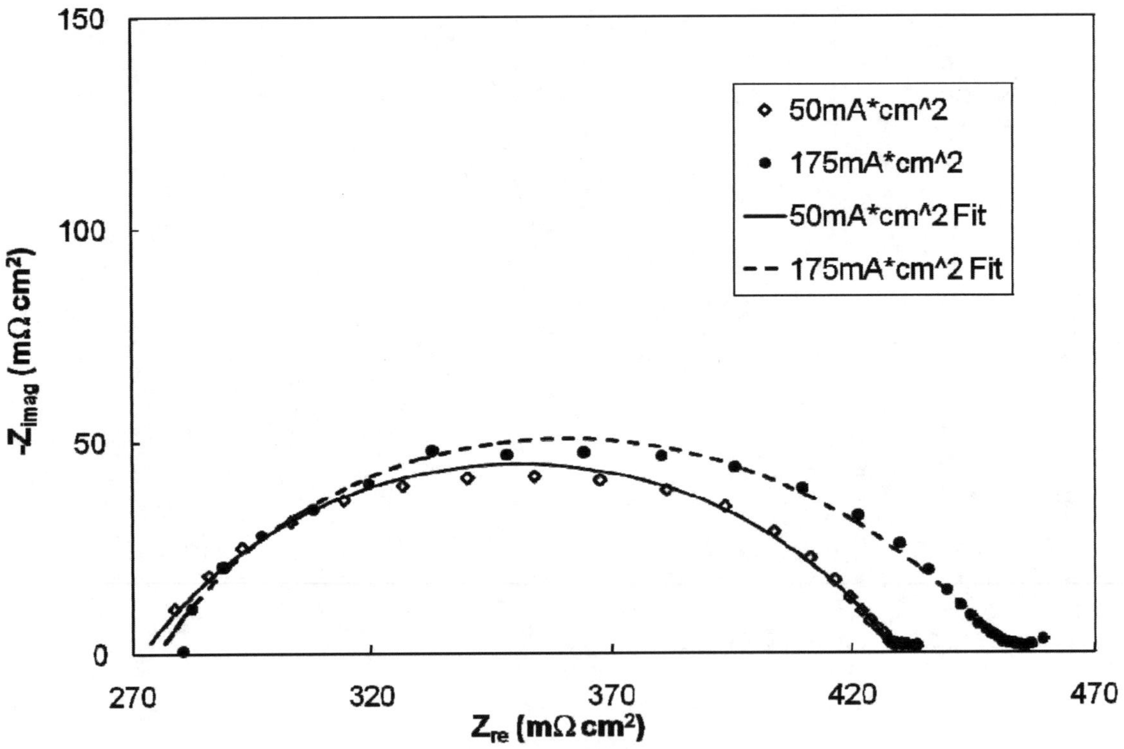

Figure 2. Nyquist plots for the button cell cathode at 800°C with varying load current. (The data fits were obtained with the cathode equivalent circuit model shown in Figure 3.)

The half reaction occurring at the cathode of a SOFC is the oxygen reduction reaction (ORR), and the overall reaction can been expressed as:

$$\frac{1}{2}O_2 + 2e^- \leftrightarrow O^{2-} \qquad (1)$$

Several different mechanisms have been proposed for this reaction; however, it is unknown exactly what steps are involved. Following one of the general proposed mechanisms consisting of two reaction processes, an equivalent circuit (Figure 3) was derived to fit the cathode impedance data having two time constants; one related to charge transfer and the other, to adsorption. The circuit is typical for the electrode reaction involving one adsorbed intermediate.

The higher frequency depressed capacitive semicircle in the cathode nyquist plot (Figure 2), ascribed to charge transfer, obviously shows a constant phase element (CPE) behavior. The impedance of a CPE is expressed as:

$$Z = Y_0 (j\omega)^{-\alpha} \tag{2}$$

When $\alpha=1$, it reduces to a pure capacitor. The general cause for CPE behavior is either surface inhomogeneity or roughness, which derivates the double layer from an ideal capacitor. After replacing the C_d in Figure 3 with a CPE, the one adsorbed intermediate equivalent circuit model was fitted well to the cathode impedance data (Figure 2).

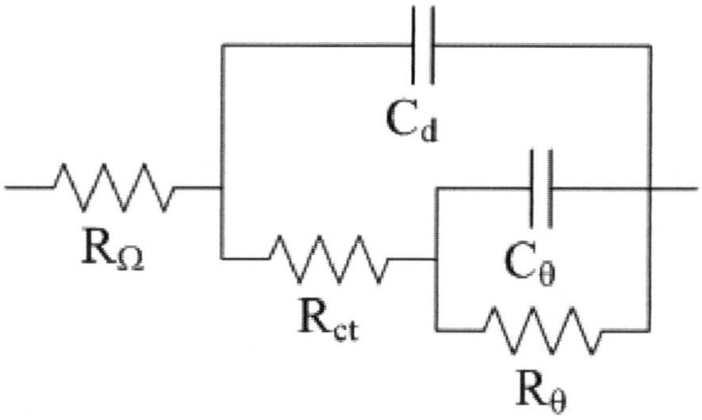

Figure 3. The equivalent circuit for the button cell cathode. (C_d was replaced by a CPE to fit the data in Figure 2.)

Anode

Impedance spectroscopy was also applied to the button cell anode at different DC current, cell temperature and hydrogen concentration. Figure 4 shows the variations of impedance spectra with varying hydrogen concentration at 800°C and 175 mA•cm^{-2}.

The equivalent circuit proposed for the anode half cell (Figure 5) follows the hydrogen oxidation reaction (HOR), involving one charge transfer process and one adsorption process associated with two adsorbed intermediates. The overall reaction of HOR is expressed as:

$$H_2 + O^{2-} \leftrightarrow H_2O + 2e^- \tag{3}$$

The higher frequency capacitive semicircle in the anode nyquist plot (Figure 4) exhibits purely capacitive behavior, and could be well modeled by a double layer sub-circuit, shown as R_{ct} paralleled with C_d in Figure 5.

Adler[1], who studied on the distortions and limitations of the measurements with reference electrodes, had found small low frequency inductive arcs on SOFC half cells. He ascribed this inductive effect to be an artifact. However, he did not give specific explanations on this conclusion.

Certain tendency of the low frequency inductive arcs in Figure 4 obviously shows the change of impedance with varying inlet hydrogen composition on the anode side. Following Epelboin[2] and Gabrielli's[3] theory, the low frequency capacitive arc combined with the lower frequency inductive arc was explained as two adsorbed species competing for adsorption sites. The two $R_\theta C_\theta$ sub-circuits in Figure 5 were used to model this two species adsorption process, and fitted with the low frequency inductive loops very well.

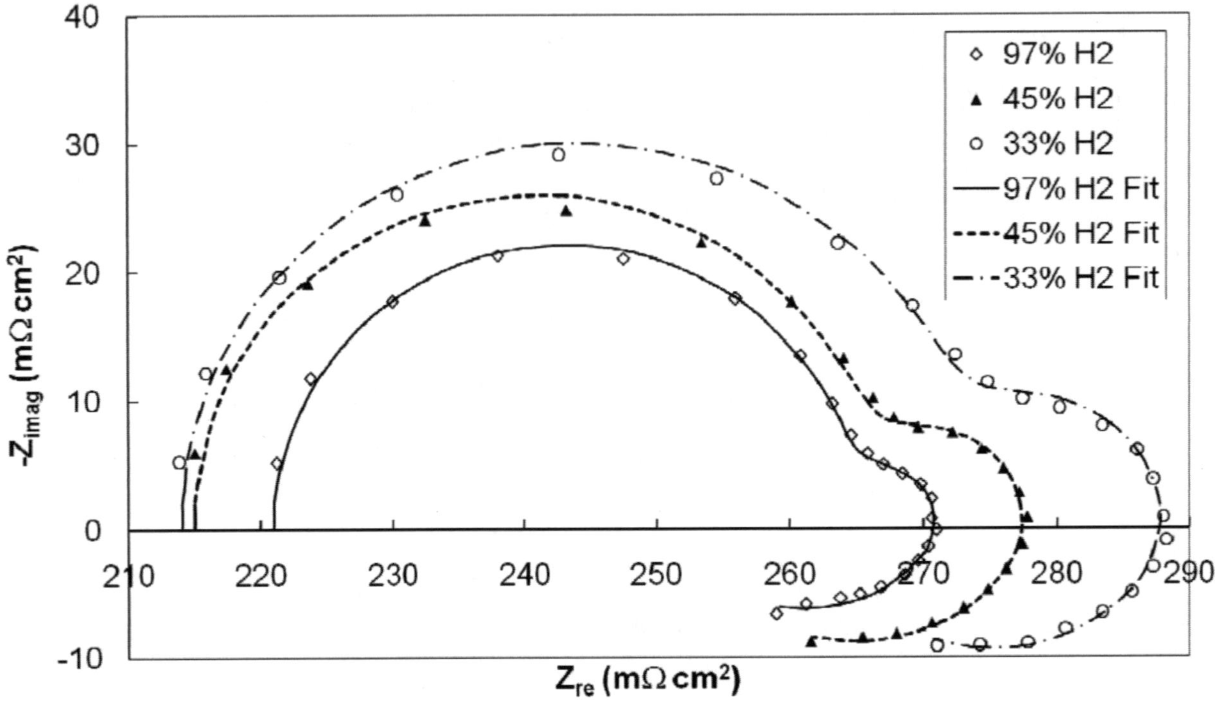

Figure 4. Nyquist plot for the button cell anode operating at 800°C and 175 mA•cm^{-2} and fed with 3% H_2O and H_2 diluted with varying amounts of N_2. (The data fits were obtained with the anode equivalent circuit model shown in Figure 5.)

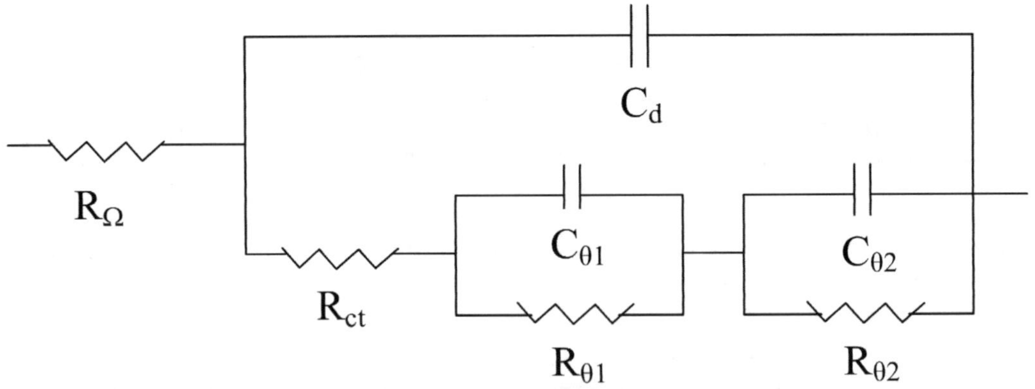

Figure 5. Equivalent circuit model for the button cell anode.

Total Cell

The total cell impedance was measured and compared with the sum of half cell impedances at the same operating conditions (Figure 6). A low frequency tail exists in the impedance spectrum of the total cell, which is absent from the sum of the anode and cathode impedances.

Winkler[4] and his group built up numerical models to simulate current density and potential distributions in three-electrode SOFCs of various electrode geometries and reference electrode locations, coming up with that distortions of electrode polarization resistance measurements would inevitably be introduced by using reference electrodes. Later, Adler[1][5] examined potential distributions in solid electrolytes with several different geometries. Combined with Winkler's work, he provided validated models and data to explain how reference electrodes distort the polarization resistances.

In our experiments, two reference electrodes were established to separate anodic and cathodic characterizations. The distortions of polarization resistances introduced by inhomogeneous current density and potential distributions, though exist (comparing polarization resistances shown in Figure 2, Figure 4 and Figure 6), are not the most significant features in Figure 6.

During the separate measurement on the anode and cathode electrodes, the use of two reference electrodes might cancel the effects of some certain physical process which occurred on both the anode and cathode electrodes to the same extent. Thus, the impedance spectrum corresponding to that certain process, i.e. the low frequency tail (Figure 6), exists only in the total cell nyquist plot.

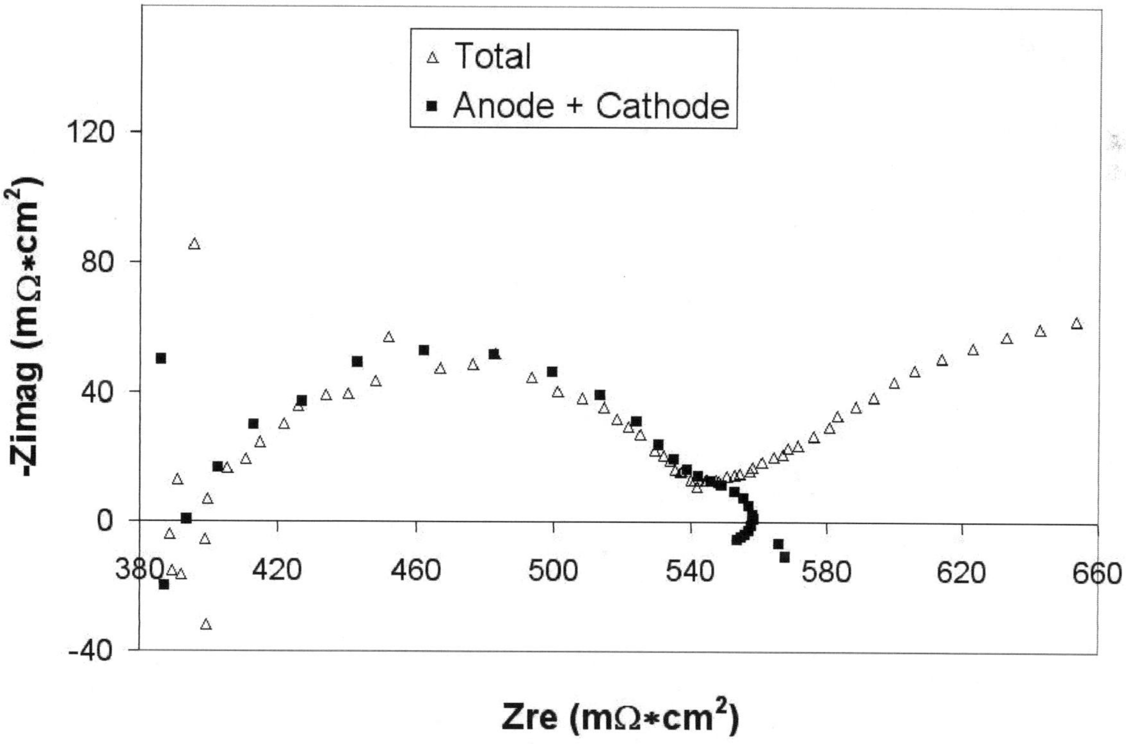

Figure 6. Nyquist plot of the total cell compared with the sum of the individual anode and cathode spectra for a button cell operating at 800°C and 175 mA•cm^{-2} and fed with 3% H_2O and balance H_2.

Conclusion

The characterization of SOFC had been studied on a small disc like button cell with reference electrodes on both sides. Impedance spectroscopy was applied to the anode and cathode of the button cell separately. Equivalent circuits were derived to fit experimental impedance data in order to determine the physical processes associated with each circuit elements. The significant low frequency inductive arc had been ascribed to an adsorption process having two adsorbed species competing for the adsorption sites.

The impedance of total button cell had been measured and compared to the sum of the anode and cathode spectra. The inconsistent between the impedance spectra illustrated possible distortions and limitations introduced by reference electrodes.

References

1. S. B. Adler, Reference Electrode Placement in Thin Solid Electrolytes, J. Electrochem. Soc., 149(2002) E166-E172.
2. C. Gabrielli, Identification of Electrochemical Processes by Frequency Response Analysis, Solartron Instrumentation Group, Ref.No. 004/83 (Farnborough, England, 1983) 16-20.
3. I. Epelboin, M. Keddam, J. C. Lestrade, Faradaic Impedances and Intermediates in Electrochemical Reactions, Dis. Faraday Soc., 56(1973) 264-275.
4. J. Winkler, P. V. Hendriksen, N. Bonanos, M. Mogensen, Geometric Requirements of Solid Electrolyte Cells with a Reference Electrode, J. Electrochem. Soc., 145(1998) 1184-1192.
5. S. B. Adler, B. T. Henderson, M. A. Wilson, D. M. Taylor, R.E. Richards, Reference Electrode Placement and Seals in Electrochemical Oxygen Generators, Solid State Ionics, 134(2000) 35-42.

Topological Modeling of Reaction Networks for Engineering of Microbial Fuel Cells

Adam C. Baughman, Chemical & Biological Engineering, Rensselaer Polytechnic Institute, Troy, NY

Lealon L. Martin, Chemical and Biological Engineering, Rensselaer Polytechnic Institute, Troy, NY

We discuss the application of a novel framework for the topological modeling of complex reaction networks (TMA) for the rational redesign of microbial fuel cell devices. We demonstrate the ability of our approach to quantitatively capture all metabolic activities of the chosen microbial reducing agent (Chlamydomonas Reinhardtii, in this case), as well as all electrochemical processes external to the microbe, using a compartmentalized modeling approach within a single framework. We show that a rigorous examination of microbial metabolism and its participation within the fuel cellcan reveal targets for genetic redesign of the microbe improving both the overall bioremediation and power generation performance of the fuel cell device. We close by discussing aspects of practically constructing the redesigned fuel cell devices identified using our modeling approach.

See more of this Session: Fuel Cell Technology II
See more of this Group/Topical: Fuels and Petrochemicals Division

Understanding Long-Term Changes in Microbial Fuel Cell Internal Resistances Using Electrochemical Impedance Spectroscopy

Doug S. Aaron, Civil and Environmental Engineering, Georgia Institute of Technology, Atlanta, GA

Abhijeet P. Borole, Oak Ridge National Laboratory, Oak Ridge, TN

Choo Y. Hamilton, University of Tennessee, Knoxville, Oak Ridge, TN

Costas Tsouris, Oak Ridge National Laboratory, Oak Ridge, TN

Microbial fuel cells (MFCs) exhibit a promising synergy between wastewater treatment and power production. Exoelectrogenic bacteria simultaneously produce an electric current, while degrading carbon compounds in wastewater. The performance of MFCs is controlled via internal resistance mechanisms that limit reaction rates and material transport. Electrochemical impedance spectroscopy (EIS) was used to study how these internal resistances changed over time. The biocatalyst was enriched in a compact, flow-through, porous anode chamber; the passive air cathode was separated from the bioanode with a Nafion polymer electrolyte membrane (PEM). The anode charge-transfer resistance was observed to decrease during the establishment of the microbial consortium in the anode. The solution resistance, which included the PEM, was observed to stay relatively stable throughout the experimental period, while the cathode resistance increased slightly. This study integrates power density and EIS studies to provide a better understanding of how an MFC changes over time and also to provide insight into operational parameters that facilitate long-term, high-power density operation of MFCs.

See more of this Session: Fuel Cell Technology II
See more of this Group/Topical: Fuels and Petrochemicals Division

Soy-Based Oligomeric Polyols with Reduce Isocyanate Loadings

Galen J. Suppes, University of Missouri-Columbia, Columbia, MO

Zuleica Lozada, University of Missouri-Columbia, Columbia, MO

Arnold Lubguban, University of Missouri-Columbia, Columbia, MO

The soy-based polyols (urethanes) industry is one of the fastest growing of the renewable chemical industries in the U.S. In particular, urethane rigid foams are valued as construction and insulation materials. Soy-based polyols directly replace petroleum-based oligomers of ethylene oxide and propylene oxide.

The primary applications of polyols are as B-side monomers in combination with A-side diisocyanates to produce rigid, flexible, and elastomer urethanes. Many of the formulations for these renewable materials require more of the A-side diisocyanates in the final formulation (not good) and none reduce the diisocyanate loading of the formulation. This project is on the engineering of polyol molecules and urethane polymers to use reduced diisocyanate loadings.

See more of this Session: Catalytic Biomass Conversion to Chemicals
See more of this Group/Topical: Fuels and Petrochemicals Division

Conversion of Biomass Derived Sugars to 2,5-Dihydroxymethyltetrahydrofuran

Mark H. Tucker, Chemical and Biological Engineering, University of Wisconsin, Madison, Madison, WI

Yomaira Pagán-Torres, Chemical and Biological Engineering, University of Wisconsin, Madison, Madison, WI

Anthony Crisci, Department of Chemistry, University of California, Santa Barbara, Santa Barbara, CA

Susannah L. Scott, Departments of Chemical Engineering and Chemistry, University of California, Santa Barbara, Santa Barbara, CA

James A. Dumesic, Chemical and Biological Engineering, University of Wisconsin, Madison, Madison, WI

The key obstacle in the production of chemicals from biomass derived sugars is the selective removal of excess oxygen functionality. We have studied the solid-acid catalyzed dehydration of six-carbon sugars to 5-hydroxymethylfurfural (HMF) and the subsequent hydrogenation of HMF to 2,5-dihydroxymethyltetrahydrofuran (DHMTHF). DHMTHF can be used as a monomer for the production of biopolymers or can be further converted to 1,6-hexanediol, a specialty intermediate chemical.

HMF undergoes undesirable degradation reactions in the presence of an acid catalyst, which has led researchers to explore the use of biphasic systems to extract the product. To further increase the yield of furan species we have explored the conversion of HMF to the more stable molecule DHMTHF. A major challenge in this process is the hydrogenation of HMF in the presence of fructose, the latter of which can be hydrogenated to polyols, eliminating the ability to recycle unreacted fructose. To limit the hydrogenation of fructose, we have synthesized a series of hydrophobic heterogeneous catalysts composed of ruthenium on functionalized supports. Through this functionalization, we have been able to decrease the rate of fructose hydrogenation in the absence of HMF. Furthermore, kinetic studies of the hydrogenation of fructose and HMF in a biphasic system have shown that fructose hydrogenation is inhibited by HMF site blocking. This leads to slower rates of fructose hydrogenation while HMF is in the system, followed by faster rates once the HMF has been consumed.

The observation that fructose hydrogenation is inhibited by HMF site blocking has led to the design of a double reactor system for the production of DHMTHF. The first reactor would consist of a solid acid catalyst to dehydrate fructose to HMF. The effluent of this reactor would be fed to a second reactor containing a metal catalyst to hydrogenate HMF, while preserving the fructose for recycle. To test the feasibility of recycle, we have explored the effect of DHMTHF as a phase modifier in the dehydration reaction. This approach allows for the selective production of target molecules from biomass derived sugars through reactor design and novel functionalized hydrogenation catalysts.

See more of this Session: Catalytic Biomass Conversion to Chemicals
See more of this Group/Topical: Fuels and Petrochemicals Division

Etherification of Glycerol and Other Biomass-Derived Polyols: New Routes to Valuable Bulk Chemicals

Pieter C. A. Bruijnincx, Inorganic Chemistry & Catalysis, Debye Institute for Nanomaterials Science, Utrecht University, Utrecht, Netherlands

Agnieszka M. Ruppert, Inorganic Chemistry & Catalysis, Debye Institute for Nanomaterials Science, Utrecht University, Utrecht, Netherlands

Andrei N. Parvulescu, Inorganic Chemistry & Catalysis, Debye Institute for Nanomaterials Science, Utrecht University, Utrecht, Netherlands

Maria Arias, Inorganic Chemistry & Catalysis, Debye Institute for Nanomaterials Science, Utrecht University, Utrecht, Netherlands

Peter J. C. Hausoul, Inorganic Chemistry & Catalysis, Debye Institute for Nanomaterials Science, Utrecht University, Utrecht, Netherlands

Monica Calatayud, Laboratoire de Chimie Theorique, Universite P. M. Curie-Paris 06, Paris, France

Robertus J. M. Klein Gebbink, Chemical Biology & Organic Chemistry, Debye Institute for Nanomaterials Science, Utrecht University, Utrecht, Netherlands

Bert M. Weckhuysen, Inorganic Chemistry & Catalysis, Debye Institute for Nanomaterials Science, Utrecht University, Utrecht, Netherlands

New catalytic routes have to be developed to transform renewable highly oxygenated platform molecules to valuable bulk and fine chemicals. The polyols constitute an important class of these biomass-derived oxygenates, with glycerol being a prime example. Glycerol is produced in large quantities as a by-product of biodiesel production, i.e. for every 1000 kg of fatty acid esters around 100 kg of glycerol is formed. Although glycerol itself has a lot of commercial applications, e.g. in the cosmetic and pharmaceutical industries, new attractive applications should be introduced to improve the economics of the biodiesel process. Furthermore, glycerol is regarded as a potential platform molecule since it can be directly produced from sugars or sugar alcohols, which are considered as the cornerstones of future biorefinery schemes. The development of new catalyst systems for glycerol valorization is therefore of great importance. In this respect, etherification of glycerol or, indeed, other biomass-derived polyols represents an important application as the products can be used as fuel additives, intermediates in the pharmaceutical industry, agrochemicals or as non-ionic surfactants. In this contribution we will discuss the recent efforts of our group which have focused on two distinct approaches for glycerol/polyol valorization [1-3] through etherification. We have developed efficient heterogeneous catalyst systems for both routes.

The first approach aims at the synthesis of di- and triglycerols, which have numerous applications in the cosmetic and pharmaceutical industries. We have shown that the alkaline earth metal oxides CaO, SrO and BaO are highly active and selective as solid base catalysts for glycerol oligomerization (> 90 % selective at 60 % conversion) [1]. Importantly, glycerol oligomerization was not accompanied by any significant acrolein formation. Catalyst surface area and number of basic sites, but also Lewis acid strength turned out to be key parameters. Catalyst stability tests revealed the spontaneous formation of colloidal CaO particles which, when isolated, turned out to have superior etherification activity. A DFT study provided additional insight on glycerol absorption and reactivity in the oligomerization mechanism and the role of surface basicity and Lewis acidity therein [2]. The theoretical results were found to be in good agreement with the experimental data on catalyst reactivity. In addition, theoretical evidence was found for the possible direct involvement of

Lewis acid sites in the catalytic etherification of bio-derived alcohols.

In our second approach, we have extended these studies to the etherification of glycerol with long linear alkenes, e.g. 1-octene, using solid acid catalysts [3]. Remarkably, the catalytic etherification route of direct nucleophilic addition of glycerol to linear, long-chain olefins has been little explored. We have successfully developed efficient and selective solid acid catalysts systems for the etherification of glycerol and other biomass-derived polyols with 1-octene over solid acid catalysts. This provides a direct route to long alkyl chain ethers with potential surfactant application. To this extent, various heterogeneous acid catalysts were screened in the etherification of neat glycerol with 1-octene. Zeolites showed modest conversions compared to Amberlyst 70 or pTSA, but superior selectivities towards the most valuable monoether products. The highest activity was obtained with H-Beta zeolites, which gave excellent selectivities for the mono and di-octyl ethers of glycerol of always > 85 %. H-Beta with a Si/Al ratio of 12.5 gave the highest conversion (15 %) and the highest selectivity to the mono-octyl ether (94 %). Several factors were found to influence both the etherification activity and the selectivity. Hydrophilic properties and porous structure of the catalyst turned out to be the critical parameters. Other parameters like reaction time, 1-octene:glycerol molar ratio, reaction temperature and the addition of an inert gas were investigated in order to improve the etherification activity of H-Beta zeolite. Catalysts deactivation due coke formation was investigated as well. In addition, the catalyst was recovered and re-used in 3 reaction cycles without any loss of activity or selectivity. The scope of the catalyst system could be successfully extended to other bio-based polyols. Very promising results were obtained in the etherification of, for instance, ethyleneglycol and 1,2-propanediol with 1-octene, as good conversions (around 60 %) and excellent selectivities (around 85 %) were obtained.

We would like to thank the ASPECT-ACTS Program for financial support.

[1] Ruppert, A. M.; Meeldijk, J. D.; Kuipers, B. W. M.; Erne, B. H.; Weckhuysen, B. M. Chem. Eur. J. 2008, 14, 2016-2024

[2] Calatayud, M.; Ruppert, A. M.; Weckhuysen, B. M. Chem. Eur. J. 2009, in press

[3] Ruppert, A. M.; Parvulescu A. N.; Arias, M.; Hausoul P. C.; Bruijnincx P. C. A.; Klein Gebbink, R. J. M.; Weckhuysen, B. M. J. Catal. 2009, submitted for publication

See more of this Session: Catalytic Biomass Conversion to Chemicals
See more of this Group/Topical: Fuels and Petrochemicals Division

Reactivity and Carbon Deposition Study On Steam Reforming of Bio-Oil

Pedro J. Ortiz-Toral, Department of Chemical and Biological Engineering, Iowa State University, Ames, IA

Justinus A. Satrio, Center for Sustainable and Environmental Technologies, Iowa State University, Ames, IA

Robert C. Brown, Department of Mechanical Engineering, Iowa State University, Ames, IA

Brent H. Shanks, Department of Chemical and Biological Engineering, Iowa State University, Ames, IA

Two major concerns with the utilization of fast-pyrolytic oil (bio-oil) as a biofuel are its high oxygen and water contents, which makes it unstable and difficult to process. Improving the stability of bio-oil and upgrading bio-oil into bio-fuels requires a large amount of hydrogen. Commercially, hydrogen is synthesized from natural gas or petroleum oil. For upgrading bio-oil, the hydrogen required can be generated from steam reforming of specific fractions of bio-oil. It has been shown that bio-oil fractions composed mainly of low molecular weight compounds, which are soluble in water, are more favorable to the production of hydrogen compared to the higher molecular weight components, which are typically insoluble in water. However, even with bio-oil fractions containing low molecular weight compound in water, the production of hydrogen by steam reforming is still challenging. Catalyst deactivation by coking and a high degree of thermal decomposition are the major known limitations, although the responsible components in the water soluble fractions are not yet identified due to the complexity of the mixture. As an effort to identify the key components that contribute most significantly to coking in reforming bio-oil, a study using selected model compounds representing different types of functionalities in bio-oil has been performed. The goal of the study is to determine relative reactivities and tendencies to generate surface carbon. Acetic acid, furfural and levoglucosan were chosen to model carboxylic acids, aldehydes, and anhydrosugars respectively. These feedstocks were tested in a small fixed bed catalytic reactor in the presence of excess steam under kinetically controlled conditions. These results will help determine what types of bio-oil fractions are more desirable for reforming while minimizing conversion limitations.

See more of this Session: Catalytic Biomass Conversion to Chemicals
See more of this Group/Topical: Fuels and Petrochemicals Division

Conversion of Short Aldehydes From Bio-Mass Over $Ce_xZr_{1-x}O_2$ and Basic Zeolites

Anirudhan Gangadharan, Chemical, Biological and Materials Engineering, University of Oklahoma, Norman, OK

Q. V. Amen, Chemical, Biological, and Materials Engineering, University of Oklahoma, Norman, OK

T. Sooknoi, Department of Chemistry, King Mongkut's Institute of Technology Ladkrabang, Ladkrabang, Bangkok, Thailand

L.L. Lobban, Chemical, Biological, and Materials Engineering, University of Oklahoma, Norman, OK

Daniel Resasco, School of Chemical, Biological and Materials Engineering and Carbon Nanotube Technology Center, University of Oklahoma, Norman, OK

R.G. Mallinson, Chemical, Biological, and Materials Engineering, University of Oklahoma, Norman, OK

Coupling and condensation reactions are an important route for converting small oxygenates found in bio-oil into transportation fuels or fuels additives. Aldehydes and carboxylic acids possess reactive carbonyl moieties which can be leveraged to carry out carbon-carbon bond forming reactions, increasing the molecular weight of the oxygenate stream.

In this study, condensation and coupling reactions of propanal were investigated over $Ce_xZr_{1-x}O_2$ mixed oxide and solid base catalysts. The latter includes basic zeolites (NaX, KNaX, KX, and CsNaX) and traditional solid base oxide (MgO and MgO-Al_2O_3 hydrotalcites). The reactions were tested in an integral flow reactor at atmospheric pressure. Over $Ce_xZr_{1-x}O_2$ mixed oxide, 3-pentanone, 2-methyl-2-pentenal and other higher oxygenated compounds are produced. The highest selectivity to 3-pentanone was observed with He, followed by CO_2 and H_2. The addition of potassium as a promoter suppressed the formation of 2-methyl-2-pentenal and increased selectivity to 3-pentanone. It was found that oxygen vacancies plays an important role in creation of Lewis acid sites, which catalyze aldol condensation to 2-methyl-2-pentenal as well as promote ketonization to 3-pentanone. Over basic zeolites and solid base oxides, 2-methyl-2-pentenal and 2-ethyl-3,5-dimethylcyclo-2-pentenone are mainly produced. Minor products include small amounts of 2,4,6-trimethylphenol and mesitylene. Activity and selectivity were largely depended on the catalyst surface area and its Na+ content. Kinetic studies verify the proposed reaction pathway that proceed primarily via aldol condensation of propanal to 2-methyl-2-pentenal and subsequent cyclization to 2-ethyl-3,5-dimethylcyclo-2-pentenone. The later is largely promoted in the presence of Na+. KX shows the highest stability and selectivity to the primary product.

See more of this Session: Catalytic Biomass Conversion to Chemicals
See more of this Group/Topical: Fuels and Petrochemicals Division

ETHERIFICATION OF 2-METHYL-PENTANAL ON SUPPORTED PALLADIUM

Trung T. Pham[1], Steven. P. Crossley[1], Tawan Sooknoi[2], Lance L. Lobban[1], Daniel E. Resasco[1], and Richard G. Mallinson[1]

[1]Center for Biomass Refining, Chemical Biological and Material Engineering,University of Oklahoma, Norman, OK 73019

[2]King Mongkut Institute of Technology Ladkrabang, Bangkok, Thailand

Abstract

Dialkyl ethers can be selectively produced from etherification of 2-methyl-pentanal and other oxygenates on supported Pd catalysts. Selectivities of 90% of ether were observed when feeding 2-methyl-pentanal and 2-methyl-pentanol at a molar ratio 1:1. Yields of up to 79 % have also been achieved. The reaction appears to be sensitive to metal particle features and molar feed ratios.

Introduction

Biofuels production has attracted substantial attention in recent years [1-4]. There is strong motivation to produce fuels from biomass derived oxygenates and much research has been geared toward production of fungible fuel molecules in a sustainable manner. Ethers are one of the potential fuel components that may be used in either gasoline or diesel blends. Ethers were first commercially introduced in Italy in 1973 as a blending component in gasoline [5]. For the diesel market, Olah [6] has patented the use of C_2-C_{24} alkyl ethers as a blending component, demonstrating cetane number improvement from 2 to 20 points.

Etherification is well known for alcohols on acid catalysts [5]. There are only a few studies of ether formation on metal catalysts [7-9]. These studies looked at alcohol and aldehyde condensation. The need for ensembles of sites to accomplish the condensation was also identified [8,9].

This study investigates the etherification of 2-methyl-pentanal on Pd catalysts. From previous studies, 2-methyl-pentanal may be selectively produced from hydrogenation of 2-methyl-2-pentenal, itself produced from aldol condensation of glycerol or propanal. The objective of the study is to understand the pathways of etherification of smaller bio-derived oxygenates by metal catalysis in order to maximize ether production to high carbon chain ethers (here a C_{12} dialkyl ether) that have potential as fuel components for gasoline or diesel.

Experimental

Catalyst synthesis. Different loadings (3, 10 and 16 wt.%) of Pd catalysts were prepared by incipient wetness impregnation of the metal precursor on precipitated silica (HiSil 210 obtained from Pittsburg Plate Glass Co.) having a BET surface area of 124 m^2/g. The metal loading in each catalyst was calculated based on the amount of metal precursor incorporated in the catalyst by impregnation. The liquid/solid ratio used to achieve incipient wetness was 1.26 ml/g. The precursor was $Pd(NO_3)_2$ purchased from Sigma Aldrich. After impregnation, the catalysts were dried overnight in an oven, and then calcined in air at 400°C for 4 hours. For the activity measurements, catalysts (0.1-0.15 g) were loaded in a tubular reactor and reduced in-situ in flowing H_2 at 150°C (or other temperature) for 1 h, then cooled down to the reaction temperature at 125°C.

Characterization. CO chemisorption was used to determine dispersion of the Pd. CO adsorption was conducted in a dynamic chemisorption unit, including a pulse system with an SRI Model 310C Gas Chromatograph equipped with a Thermal Conductivity Detector (TCD). Temperature Program Reduction (TPR) of 3% Pd/SiO_2 shows one peak profile at 50°C.

Table 1. CO chemisorption uptake on 3, 10, and 16 wt.% Pd/SiO_2. Reduction temperature: 150°C.

Weight loading	Pd 3%	Pd 10%	Pd 16%
CO/Pd	0.071	0.053	0.039

Catalyst testing. Catalytic activity measurements were carried out in a continuous flow system that includes a syringe pump delivering the organic feed to a heated line into a stream of H_2 that feeds to a reactor within an electric furnace. 100 to 150 mg of catalyst was used, with the appropriate flow rate to achieve the desired W/F. Operating conditions were 200°C and atmospheric pressure. Experiments were continued for 2-3 h. All the lines were heated with heating tapes to keep the feed and products in the gas phase. The carrier gas, H_2, was used to reduce the catalyst and also flowed at a relative molar ratio of 12:1 during reaction with 2-methyl-pentanal (MPAL) and/or 2-methyl-pentanol (MPOL). Quantitative product analysis was by online gas chromatography using a HP 6890A with a flame ionization detector. Organic mole fractions are reported. A Shimadzu GCMS-QP2010S provided product identification. The light gas product CO was detected online via gas chromatography using a Carle series 400 AGC with TCD.

Results and Discussion

Reaction of 2-methyl-pentanal on 3 wt.% Pt/SiO_2 vs. 3 wt. % Pd/SiO_2. 2-methyl-pentanal conversion was compared on both 3 wt.% Pt/SiO_2 and 3 wt.% Pd/SiO_2 in Fig. 1 and Fig. 2. The overall activity is similar on both catalysts. However, the selectivity toward ether is much higher on the Pd catalyst while the highest selectivity on Pt is alcohol. No reaction was observed on the SiO_2 support alone in H_2 carrier gas.

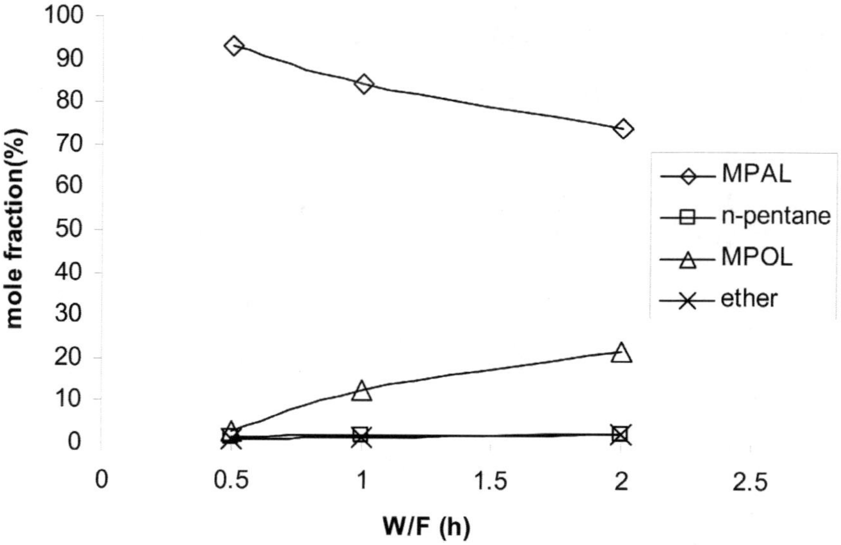

Figure 1. MPAL conversion on 3% Pt/SiO_2 reduced at 250°C and run at 125°C, 1 atm, H_2: MPAL = 12:1

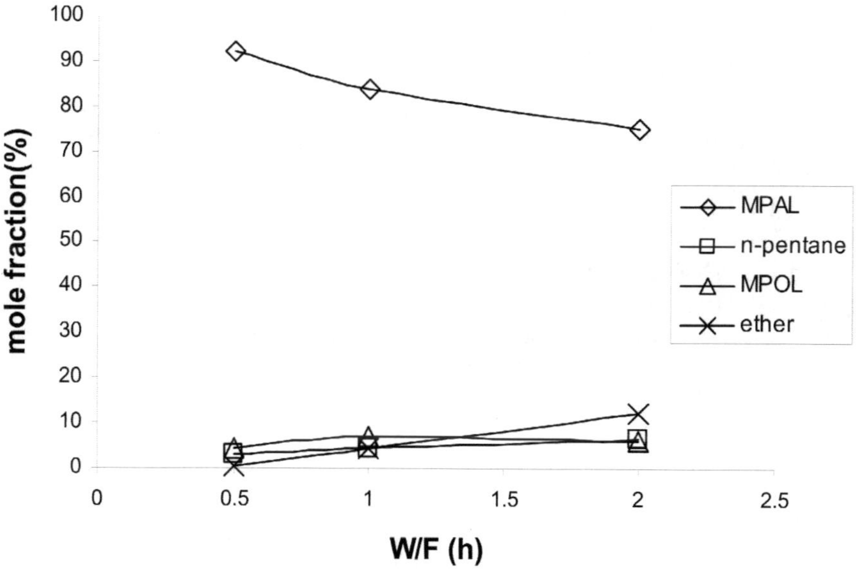

Figure 2. MPAL conversion on 3% Pd/SiO_2 reduced at 250°C and run at 125°C, 1 atm, H_2: MPAL =12:1

Reaction of 2-methyl-pentanal on 3 wt.% Pd/SiO_2, 10 wt.% Pd/SiO_2 and 16 wt.% Pd/SiO_2. Different metal loadings were compared for the condensation of MPAL on Pd catalysts at similar conversions. Increasing loading from 3 to 16 wt % increases the particle size based on the CO chemisorption measurements shown in Table 1. The ether yield and selectivity are highest at 16 wt % Pd. The estimated conversion rate per surface atom are similar on the three catalysts. The 16wt% Pd, with the largest particle size, may have a higher density of large ensembles necessary for condensation to form the ether or a larger fraction of low index planes (e.g. Pd(111)).

Table 2. Activity and yields of 3, 10, and 16 wt% Pd/SiO_2 reduced at 150°C with product yields from MPAL conversion at 125°C, H_2: MPAL =12:1

Wt loading	3% Pd	10% Pd	16% Pd
Dispersion	7.1%	5.3%	3.9%
W/F (h)	2	1	1
Conversion (%)	20.9	20.4	23.8
n-pentane (%)	7.0	5.3	5.5
MPOL (%)	6.7	7.4	6.4
Ether (%)	7.2	7.7	11.9

Table 3. Product Selectivities for MPAL conversion on 3, 10, and 16wt % Pd/SiO_2 reduced at 150°C and run at 125°C, H_2: MPAL=12:1

W/F (h)	2	1	1
n-pentane (%)	33.5	26.0	23.1
MPOL (%)	32.0	36.4	26.8
Ether (%)	34.5	37.6	50.1

Effect of reduction temperature: comparing 3 wt.% Pd reduced at 250°C vs. 16% wt. % Pd reduced at 150°C. When the 3 wt.% Pd was reduced at 150°C, the yield of ether was low (lowest compared to 10% and 16% Pd). However, when 3 wt.% Pd was reduced at 250°C, the yield and selectivity to ether were much improved compared to 3% Pd reduced at 150°C. The overall activity at W/F = 2h, shown in Fig. 2, is similar to that of 16 wt.% Pd at W/F =1h, shown in Fig. 3. This suggests that at higher reduction temperature, the Pd surface may anneal, resulting in a different surface structure more similar to the higher loading catalysts, having more sites and ensembles that are selective for ether formation.

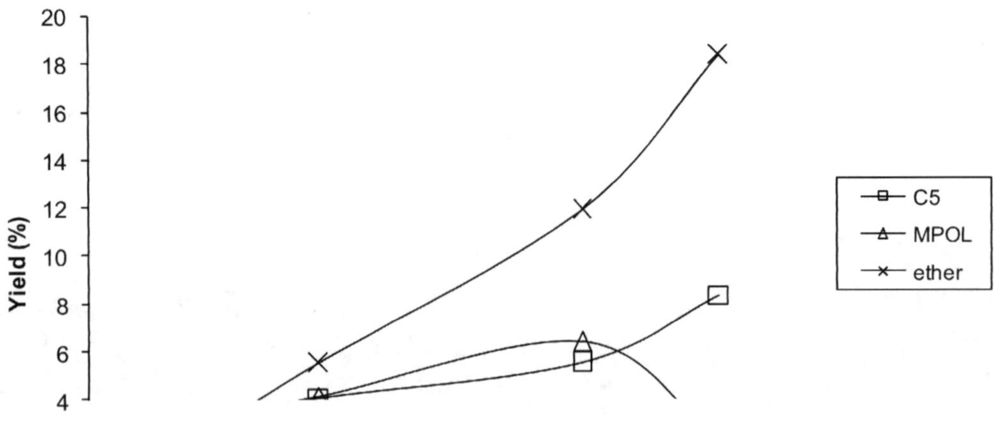

Figure 3. MPAL conversion on 16% Pd/SiO_2 reduced at 150°C and run at 125°C, 1 atm, H_2: MPAL =12:1

Effect of feed: co-feed mixture of 2-methyl-pentanal and 2-methyl-pentanol with different aldehyde concentrations. In Fig.4, instead of feeding MPAL as in earlier experiments, pure alcohol 2-methyl-pentanol was fed on 16% Pd. The ether yield was very low. The main product is the aldehyde, 2-methyl-pentanal. The yield of the product of decarbonylation, n-pentane, is also low. This suggests that the aldehyde is more active for ether formation than alcohol.

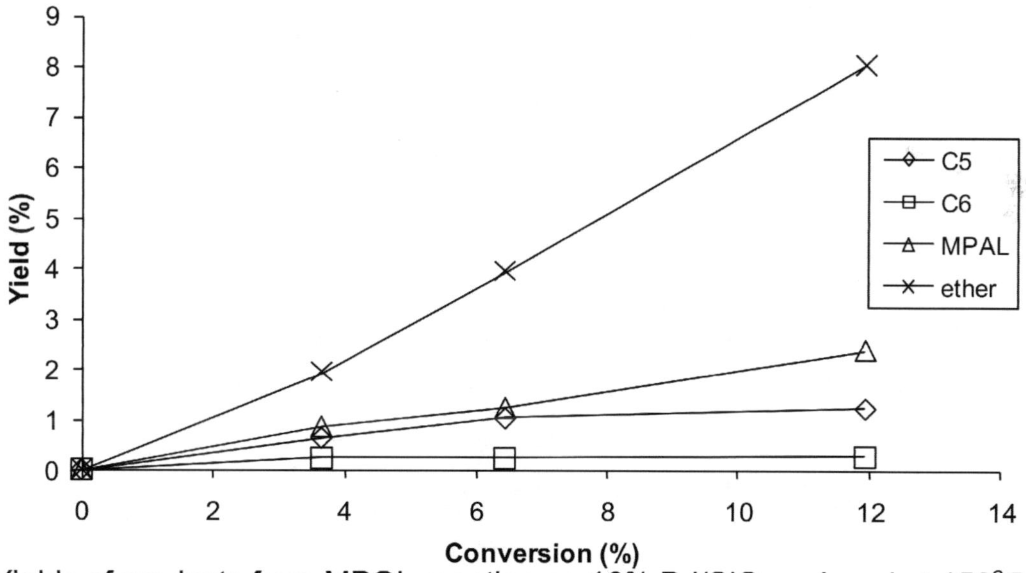

Figure 4. Yields of products from MPOL reaction on 16% Pd/SiO_2 reduced at 150°C and run at 125°C, 1 atm, H_2: MPOL =12:1

Further mixture experiments were carried out with different ratios of MPAL and MPOL, shown in Fig. 5. The mixed feed results show much higher ether yield compared to pure feeds of ether MPAL or MPOL. The highest selectivity of ether achieved is 90% at 1:1 feed ratio of MPAL to MPOL (y_{MPAL} =0.5).

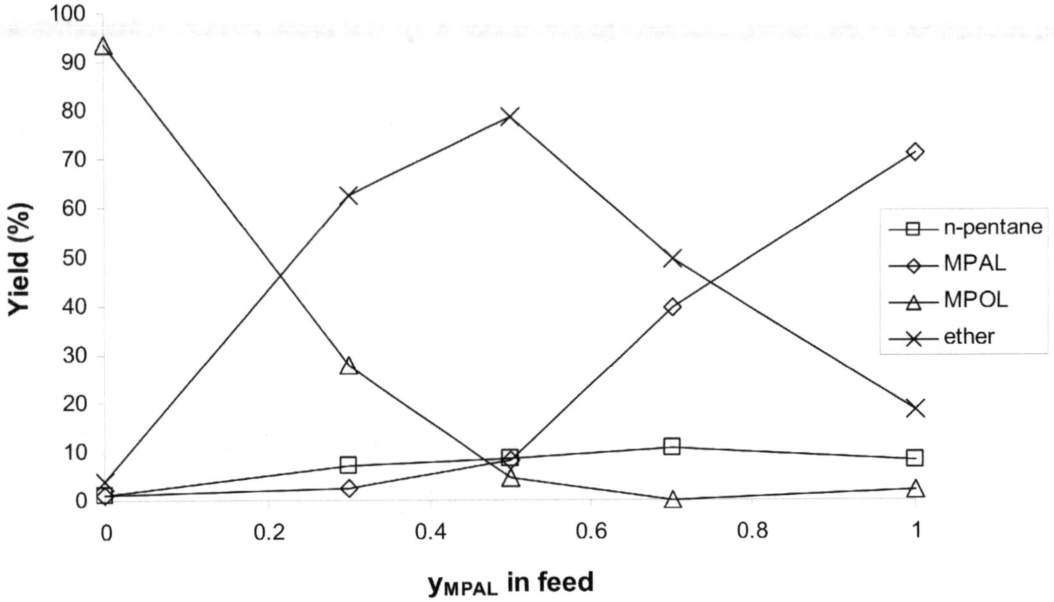

Figure 5. Co-feed mixtures of MPAL and MPOL on 16% Pd/SiO_2

Reaction scheme of 2-methyl-pentanal on supported Pd. The proposed reaction pathway for MPAL on Pd is summarized in Fig. 6. Both, aldehyde and alcohol can condense to form ether, most probably by first inter-converting to have both species present. However, the highest yield was observed when both aldehyde and alcohol were co-fed. Cross etherification will be investigated with different aldehydes and alcohols, for example, MPAL and n-butanol, to better determine the mechanism of the surface reaction.

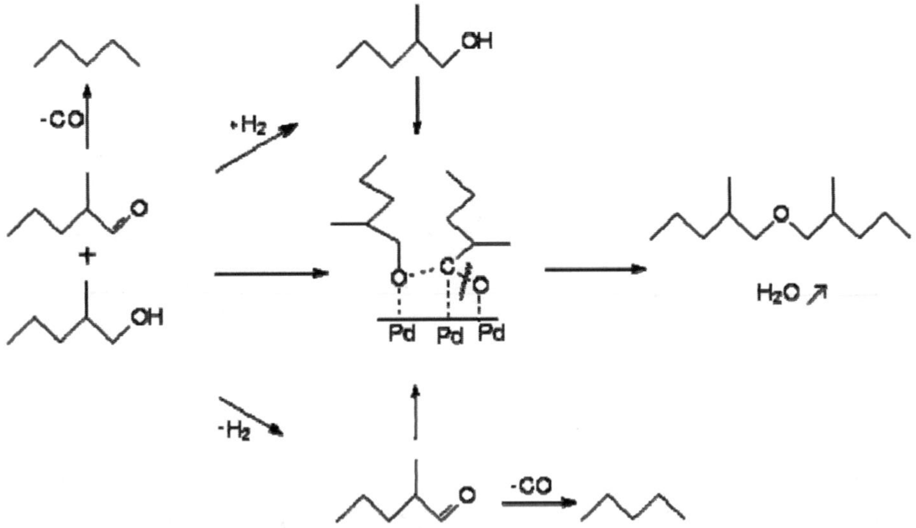

Figure 6. Schematic reaction pathway of 2-methyl-pentanal on Pd catalyst.

Acknowledgement. The support of the Oklahoma Secretary of Energy and the Oklahoma Bioenergy Center are gratefully acknowledged.

References

(1) J. N. Chheda, J. A. Dumesic, *Catal. Today*, **2007**, 123, 59.
(2) R. M. West, Z. Y. Liu, M. Peter, C. A. Gärtner, J. A. Dumesic, *J. Mole. Catal. A* **2008**, 296, 18.
(3) T. Sooknoi, T. Danuthai, L. Lobban, R. G. Mallinson, D. E. Resasco, *J. Catal.* **2008**, 258, 199.
(4) M. Snare, I. Kubickova, P. Maki-Arvela, D. Chichova, K. Eranen, D. Y. Murzin, *Fuel* **2008**, 87, 933.
(5) L. Karas, W. J. Piel, Ethers, Kirk-Othmer Encyclopedia of Chemical Technology. John Wiley & Sons, Inc. **2004**.
(6) G. A. Olah, US Patent 5 520 710 **1996**.
(7) X. Wang, R. Y. Saleh, U. S. Ozkan, *J. Catal.* **2005**, 231, 20.
(8) A. Van der Burg, J. Doornbos, N. J. Kos, W. J. Ultee, V. Ponec, *J. Catal.* **1978**, 54, 243.
(9) G. M. R. van Druten and V. Ponec, *Appl. Catal. A.* **2000**, 191, 153.

A Microfluidic Fuel Cell as a Platform for Individual Fuel Cell Component Analysis

Fikile R. Brushett, Department of Chemical & Biomolecular Engineering, University of Illinois at Urbana-Champaign, Urbana, IL

Paul J. A. Kenis, Department of Chemical & Biomolecular Engineering, University of Illinois at Urbana-Champaign, Urbana, IL

Fuel cell-based systems are promising alternative power sources for a range of energy conversion applications due to their high efficiency, high energy density and low emissions. For portable and transportation applications polymer electrolyte membrane (PEM)-based fuel cells, utilizing either hydrogen or small organics as fuels, are considered the most promising configuration [1]. Unfortunately, for widespread commercialization of fuel cell technologies to be realized, significant reductions in system costs and improvements in durability are required. For example, the cathodic oxygen reduction reaction (ORR) requires high loading of expensive platinum (Pt)-based catalysts to maintain acceptable power densities. While alternative Pt-transition metal alloys are being investigated to improve catalytic activity and lower costs, such alloys exhibit insufficient stability under acidic conditions [2]. As recently highlighted in C&E News, the processes that govern individual component (i.e., catalyst layer, Nafion membranes and gas diffusion layers) performance and degradation within an operating fuel cell remain poorly understood [3]. At present, analytical tools for the in-situ investigation of these complex electrochemical and transport phenomena do not exist.

To address this need, we have developed a microfluidic fuel cell with a flowing aqueous electrolyte instead of a stationary PEM. The constantly refreshing electrolyte stream eliminates water management issues, facilitates by-product removal (i.e., carbonates) and enables electrolyte flexibility (i.e., no pH restrictions). By placing an external reference electrode at the electrolyte stream outlet, in-situ electrochemical analyses (i.e., CA, CV, EIS) of individual electrode characteristics can be conducted to identify performance-limiting factors. This microfluidic fuel cell enables the experimental versatility of a three-electrode electrochemical cell within an operating fuel cell without disassembly. Both the performance of individual components as well as the effects of operating conditions can be evaluated. We have demonstrated the utility of such an analytical tool by investigating the performance and durability of several novel electrodes and catalysts under acidic and alkaline conditions [4-6]. Here, we will present our ongoing work further investigating electrode degradation processes and optimizing membrane-electrode assemblies (MEAs) for use in conventional PEM-based fuel cells.

References

[1] L. Carrette, K. A. Friedrich and U. Stimming, ChemPhysChem, 1, 2000, 162.

[2] H.A. Gasteiger, S.S. Kocha, B. Sompalli, F.T. Wagner, Appl. Cat., B, 56, 2005, 9.

[3] M. Jacoby, Chem. Eng. News, 87(13) 2009, 39.

[4] R.S. Jayashree, M. Mitchell, D. Natarjan, P.J.A. Kenis, Langmuir, 23, 2007, 6871.

[5] F.R. Brushett, W.P. Zhou, R.S. Jayashree, P.J.A. Kenis, J. Electrochem Soc., 156, 2009, B565.

[6] F.R. Brushett, H.T. Duong, J.W.D. Ng, A. Wieckowski, P.J.A. Kenis, in preparation.

See more of this Session: Fuel Cell Technology
See more of this Group/Topical: Fuels and Petrochemicals Division

Operation of Polymer Electrolyte Membrane Fuel Cells with Dry Feeds. Design and Operating Strategies

Jay B. Benziger, Department of Chemical Engineering, Princeton University, Princeton, NJ

The operation of polymer electrolyte membrane fuel cells (PEMFCs) with dry feeds has been examined with different fuel cell flow channel designs as functions of pressure, temperature and flow rate. Auto-humidified (or self humidifying) PEMFC operation is improved at higher pressures and low gas velocities where axial dispersion enhances "back-mixing" of the product water with the dry feed. We demonstrate auto-humidified operation of the channel-less, self-draining fuel cell operation from 25 to 115°C with dry feeds. Design and operating requirements are derived for the auto-humidified operation. We show why the traditional serpentine flow design is incompatible with autohumidified operation. New engineering designs which improve mixing and exploit gravity offer substantial benefits for simplicity of operation and control including: the ability to self drain reducing flooding, the ability to uniformly disperse water removing current gradients and the ability to operate on dry feeds eliminating the need for humidifiers.

See more of this Session: Fuel Cell Technology
See more of this Group/Topical: Fuels and Petrochemicals Division

Layer by Layer Assembly of Multi-Functional Graphite/Silica Nanoparticle Coatings for PEM Fuel Cell Applications

Feng Wang, Chemical and Materials Engineering, University of Cincinnati, Cincinnati, OH

Mubarak Alazemi, Chemical and Materials Engineering, University of Cincinnati, Cincinnati, OH

Indrajit Dutta, Chemical and Materials Engineering, University of Cincinnati, Cincinnati, OH

Anastasios Angelopoulos, Chemical and Materials Engineering, University of Cincinnati, Cincinnati, OH

Layer by layer (LBL) assembly of chemically heterogeneous nanoparticles is investigated as a method to fabricate proton exchange (PEM) fuel cell bipolar plate composite coatings which are both hydrophilic and electrically conductive. The strong polyelectrolyte, acrylamide/β-methacryl-oxyethyl-trimethyl-ammonium copolymer, is used to adhere ~10 nm thick graphite platelets which are electrically conductive yet hydrophobic and ~20 nm diameter silica nanoparticles which are hydrophilic yet electrically insulating onto gold-coated bipolar plate surfaces. Various LBL assembly schemes are explored to obtain a detailed understanding of multilayer formation and functional performance. The optimum assembly scheme employs cationic binder between the graphite and silica nanoparticles to yield porous silica nanoparticle layers deposited directly onto the graphite platelets. For an ~100 nm thick coating, the electrical contact resistance is ~4 mΩ-cm^2 (comparable to that of a pure graphite platelet coating) while maintaining an advancing contact angle of 20° (comparable to that of the pure silica nanoparticle coating). Such behavior represents an order of magnitude reduction in the power loss in state-of-the art proton exchange membrane (PEM) bipolar plate fuel cell applications relative to the use of comparably thick pure silica nanoparticle coatings. In addition to such performance gains, LBL assembly of this heterogeneous graphite/silica nanoparticle system is shown capable of de-convoluting the effects of surface chemistry from those of morphology on dynamic surface wetting.

See more of this Session: Fuel Cell Technology
See more of this Group/Topical: Fuels and Petrochemicals Division

Improvements to the Miniature Direct Formic Acid Fuel Cell: Catalyst Layer Optimization

Robert D. Morgan, Chemical & Biomolecular Engr, University of Illinois at Urbana-Champaign, Urbana, IL

John L. Haan, Department of Chemistry, University of Illinois at Urbana-Champaign, Urbana, IL

Richard I. Masel, Chemical and Biomolecular Engineering, University of Illinois at Urbana-Champaign, Urbana, IL

Small fuel cells have been proposed to replace batteries for such applications as small hand-held electronic devices, on chip sensors, implantable devices, as well as for communications [1-3]. Formic acid fuel cells are ideal, because they have a lower crossover than DMFCs [4]. This allows one to use more concentrated fuels [5-6], and practically no balance of plant.

A brief overview of previous optimization of membrane thickness and fuel concentration will be given for our 4.9 mm2 fuel cell. Currently, the fuel cell can obtain 150 W-hr/l as well as a peak power density of 43 mW/cm^2. Also, work that will be presented will focus on the optimization of the catalyst ink layer. Nafion is an important binder in fuel cell catalyst inks. The proper amount needs to be added to the ink to ensure both good adhesion between the catalyst layer and the membrane, but also enough to allow for good proton and electron conduction[7-11]. Using a design of experiments approach, the optimum ink composition for the miniature formic acid fuel cell will be presented.

1. S. Hsieh, C. Huang, J. Kno, H. Tsai and S. Yang, J. Solid State Electrochem. 9 (2005), pp. 121-131

2. S.H. Chan, N. Nguyen, Z. Xia and Z. Wu, J. Micromech. Microeng. 15 (2005), pp. 231-236.

3. Meyers, J.P. and H.L. Maynard, Journal of Power Sources, 2002. 109(1): p. 76-88.

4. Y.-W. Rhee, S. Ha and R.I. Masel. J. Power Sources 117 (2003), pp. 35-38

5. C. Rice, S. Ha, R.I. Masel, P. Waszczuk, A. Wieckowski and T. Barnard. J. Power Sources 111 (2002), p. 83

6. C. Rice, S. Ha, R.I. Masel and A. Wieckowski. J. Power Sources 115 (2003), p. 229

7. McGovern, M.S., et al., Effects of Nafion as a binding agent for unsupported nanoparticle catalysts. Journal of Power Sources, 2003. 115(1): p. 35-39.

8. O'Hayre, R. and F.B. Prinz, The Air/Platinum/Nafion Triple-Phase Boundary: Characteristics, Scaling, and Implications for Fuel Cells. Journal of the Electrochemical Society, 2004. 151(5): p. A756-A762.

9. Song, D., et al., A method for optimizing distributions of Nafion and Pt in cathode catalyst layers of PEM fuel cells. Electrochimica Acta, 2005. 50(16-17): p. 3347-3358.

10. Wagner, N., T. Kaz, and K.A. Friedrich, Investigation of electrode composition of polymer fuel cells by electrochemical impedance spectroscopy. Electrochimica Acta, 2008. 53(25): p. 7475-7482.

11. Wang, S., et al., Effect of Nafion® ionomer aggregation on the structure of the cathode catalyst layer of a DMFC. Journal of Power Sources, 2007. 165(1): p. 128-133.

See more of this Session: Fuel Cell Technology
See more of this Group/Topical: Fuels and Petrochemicals Division

Nano-Materials to Control Two-Phase Transport in PEM Fuel Cell Electrodes

Trung V. Nguyen
Department of Chemical and Petroleum Engineering
The University of Kansas, Lawrence, KS 66044

Background

To achieve good performance and high durability polymer electrolyte membrane fuel cells (PEMFC) are operated at high water saturation levels. However, excessive liquid water generated at the cathode side by electro-osmosis and the oxygen reduction reaction can block pores within the catalyst layer (CL) and thus obstruct the transport of reactant gases to the active catalyst sites. To prevent electrode flooding, an ideal CL structure has to provide continuous hydrophobic pathways for reactant gas transport and continuous ionic and electronic pathways for proton and electron conduction.

The idea of adding Teflon® as hydrophobic phase for reactant gas transport was rejected by Wilson and Gottesfeld in 1992 because of the concern that it might block active catalyst sites [1]. Uchida et al. did a study on both the Nafion® and the PTFE content within the CL [2] and found that while Nafion® affected the whole current density range, Teflon® showed its effect only in the high current density regions. Since then, however, very little research in this area has not been done.

Results

In this work, the Teflon® and Nafion® content and its distribution in the cathode CL were optimized by means of a novel 2-step preparation method of the catalytic ink [3, 4]. With this method it was possible to control the distribution of Nafion® and Teflon® within the CL. Figure 1a and 1b show schematics of controlled and uncontrolled distribution of Nafion® and Teflon® in a catalyst layer, respectively.

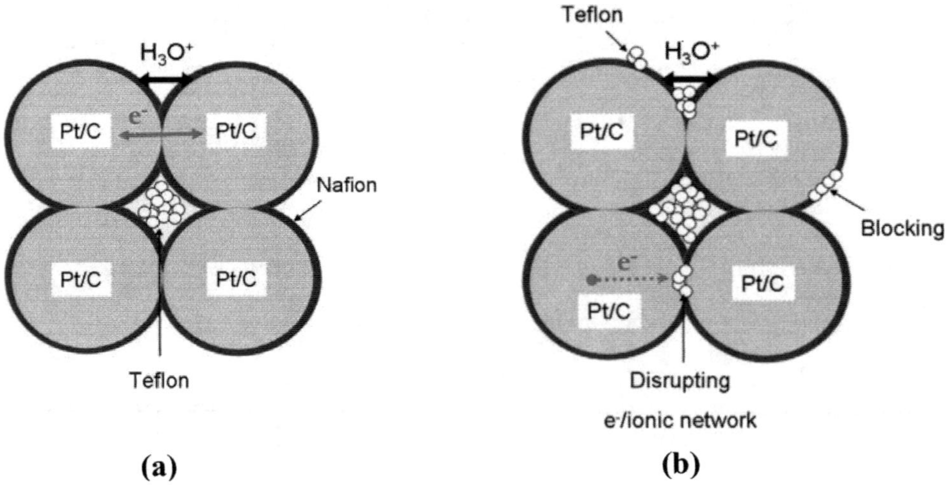

Figure 1. (a) Controlled and (b) uncontrolled distributions of Nafion® and Teflon® in the catalyst layer

As seen in Figure 1b, Teflon® can place itself on the surface of the catalyst particles and block active catalyst sites. The area blocked by Teflon® becomes unavailable for electrochemical reaction resulting to lower catalyst utilization. In the second place, Teflon® can place itself in between the catalyst agglomerates and disrupt the continuity of the ionic and also electronic network and thus lead to a complete loss of activity for those agglomerates.

A desirable structure for a CCL is presented in Figure 1a. Here the whole catalyst area is covered by Nafion® and thus electrochemically active. Teflon® doesn't block any active catalyst sites or disrupt the ionic/electronic network, but rather places itself within the void space to create hydrophobic pathways for gas transport. The region between the cluster of Teflon® particles and the Nafion® phase is available for water transport because of the hydrophilic property of the ionomer phase. Hence this optimal CL structure addresses both the problem of low catalyst utilization and water transport within the CL. To achieve such a CL structure our group had developed a novel 2-step preparation for the catalyst ink [3,4]. In this method, Pt/C and Nafion® are mixed together in the first step to form a slurry that then undergoes a heat treatment in a steam environment at a temperature above the glass transition temperature of Nafion® to form a material consisting of ionomer-coated Pt/C catalyst.

By this heat treatment the ionomer phase is steam-annealed onto the catalyst particles to cover all accessible catalytically active surface area. This annealed coated catalyst powder is then used as the starting material for the second step, where additional Nafion® ionomer, needed for particle to particle linkage, and Teflon® nano particles are added. Since the catalyst particles are already covered by the ionomer, Teflon® can no longer block the active catalyst sites and thus places itself either into the voids or over the ionomer film.

Figure 2 shows the performance curves for a PEM fuel cell with four different cathode catalyst layers under the same operating conditions, two using the 1-step catalyst preparation method (CTRL) with C:Nafion®:Teflon® weight ratios of 1:1:0.25 and 1:1:1 and two using the new 2-step method with similar C:Nafion®:Teflon® weight ratios. The catalyst loadings in these cathodes were about 0.35 mg Pt/cm^2. First, note that at 0.25 Teflon® level both catalyst layers (1-step and 2-step) performed quite well with that of the 2-step method slightly better than that of the 1-step method. However, when the Teflon® level was increased to 1 there was a significant vertical drop in the voltage of the whole polarization curve of the fuel cell with the 1-step cathode catalyst layer showing that there was a significant loss in the catalyst active area. On the other hand, the fuel cell with the 2-step cathode catalyst layer showed no loss in catalyst active surface area with the low current density portion of the polarization curve being almost identical to those with 0.25 Teflon® level. The rapid loss in performance at high current density could be attributed to the fact that these high level Teflon® catalyst layers were much thicker than those with lower Teflon® content causing more severe mass transport resistance. The results in Figure 2 validated the hypothesis that the 2-step method would eliminate the problem of Teflon blocking the catalyst surface area of the 1-step method.

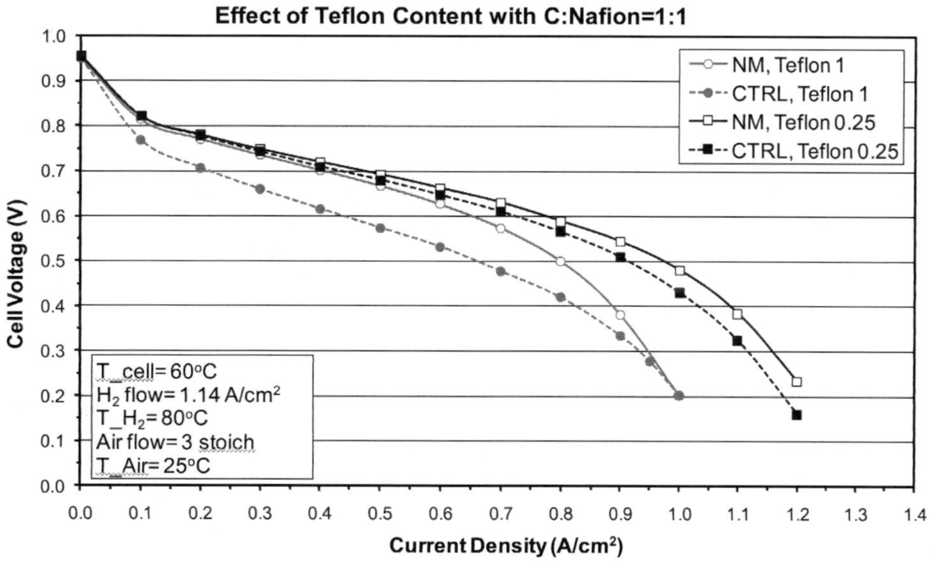

Figure 2. Effect of Teflon® content on performance for the 1-step and 2-step methods

Figure 3 shows the scanning electron micrographs for the catalyst layers with Teflon® level of 1 with 3(a) for the 1-step method and 3(b) for the 2-step method. Note the excessive localized Teflon® amount in the micrograph for the 1-step case, and since with this preparation method Teflon® can cover active surface area this could be the cause for the vertical shift in the polarization curve. On the other hand, Teflon®

appeared be more uniformly distributed and occupy mainly the void space between the particles in the 2-step method as highlighted in Figure 1(a).

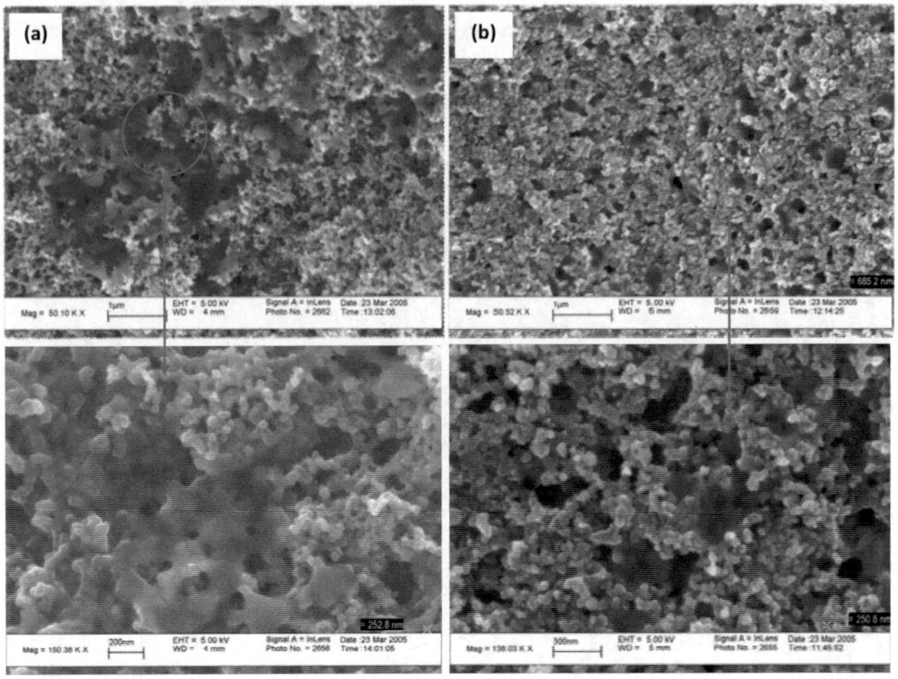

Figure 3. Teflon® distribution in the catalyst layer prepared by 1-step and 2-step methods

Acknowledgements

This work was supported by the National Science Foundation through grant No. CBET- 0651758.

References

[1] M.S. Wilson, S. Gottesfeld, *J. Appl. Electrochem.*, **22**, 1 (1992).
[2] M. Uchida, Y. Aoyama, N. Eda, A. Ohta, *J. Electrochem. Soc.*, **142**, 4143 (1995).
[3] T.V. Nguyen, D. Natarajan, R. Jain, *ECS Transactions*, **1**, 501 (2006).
[4] R. Friedmann, T.V. Nguyen, *ECS Transactions*, **16**, 2021 (2008).

Micro Computed Tomography Investigation of Water Distribution and Transport in a Gas Diffusion Media

Wei Zhu, Chemical and Biomolecular Engineering, University of Illinois, Urbana-Champaign, Urbana, IL

Richard I. Masel, Chemical and Biomolecular Engineering, University of Illinois at Urbana-Champaign, Urbana, IL

Proper water management has long been identified in the literature as a key component for optimal fuel cell performance. Current literature suggests that the lack of in-situ water data is slowing the progress of water management technology. Water motion in an operating polymer electrolyte membrane (PEM) fuel cell has been examined with Micro computed tomography (CT) in our previous research. In this work, Micro CT was used to look through the gas diffusion layer (GDL) to provide accurate two-dimensional (2D) and three-dimensional (3D) pictures of water inside two types of carbon paper, both plain and standard wet proof.

The results demonstrated many details of the water motion in the GDL. 2D pictures showed that water was diffused perpendicularly to the GDL surface. However, once water entered the carbon paper, it diffused parallel to the surface as well. The diffuse rate in the plain carbon paper was much quicker than the diffuse rate in the wet proof treated carbon paper. 3D images further confirmed recent numerical and experiment studies of the co-existence of hydrophilic and hydrophobic areas of the GDL. Although previous numerical studies demonstrated that the more hydrophilic the GDL was, the more water accumulated inside, our 3D images indicated deviations from those results. We found less water distribution on the surface of the more hydrophobic carbon paper, however, inside the carbon paper, the water distribution was almost the same as the plain carbon paper. A potentiostatic test was also conducted in order to find the relationship between characteristics of the GDL and of PEM fuel cell performance. Better fuel cell performance was observed with the carbon paper with standard wet proof. This was believed to be due to the different water distribution pattern inside these two types of carbon paper, which caused different mass transfer barriers.

See more of this Session: Fuel Cell Technology
See more of this Group/Topical: Fuels and Petrochemicals Division

Effect of Wall Adhesion On Liquid Water Transport through the Gas Diffusion Layer of a PEM Fuel Cell

P. V. Suresh, Department of Chemical Engineering, IIT Madras, Chennai, India

Jayanti Sreenivas, Department of Chemical Engineering, IIT Madras, Chennai, India

Prathap Haridoss, Metallurgical & Materials Engineering, IIT Madras, Chennai, India

Abhijit P. Deshpande, Chemical Engineering, IIT Madras, Chennai, India

The gas diffusion layer (GDL) is one of the key components which play a crucial role in the successful operation of PEM fuel cell. It assists in the transportation of the reactants from the flow channels uniformly over the catalyst layers as well as in draining out the liquid water from the catalyst layer back to the channels in addition to conducting the electrons with low resistance. Generally it is a carbon-based porous medium that allows for the transport of electrons through the fibers and allows the reactants and the liquid water to flow through the void space to/from the catalyst layer in a PEM fuel cell. This carbon fiber paper is often coated with PTFE to give a nature of hydrophobicity to the medium, and this is expected to facilitate the liquid water transport. Liquid water transport through the porous structure is found to be greatly affected by the nature of the porous medium, specifically, its hydrophobicity which is indicated by the wall adhesion angle or contact angle made by water on the porous medium. Contact angle is a quantitative measure of the wetting of a solid by a liquid. The actual mechanism of the flow of liquid water through the porous structure is not very well understood. And it is essential to understand the liquid water transport through the porous gas diffusion layer which helps in taking good measures to mitigate flooding of the cell, thereby improving the cell performance. In this paper, the dynamic behaviour of the liquid water transport through an idealized GDL is simulated numerically using computational fluid dynamics (CFD) techniques along with Volume of Fluid (VOF) method [1] and Continuum Surface Force (CSF) formulation [2], subject to hydrophilic/hydrophobic boundary condition at the fibre-fluid interface. VOF method has been used in the area of fuel cells recently to simulate liquid water emergence in the flow channels [3-5]. A simple three-dimensional structure with three layers each containing cylindrical fibres of 8 micron diameter and 40 micron length with a layer to layer gap of 1 micron is used to represent a small portion of the GDL. These dimensions are typical of the values observed in the carbon paper which is a commonly-used GDL in PEM fuel cells. Liquid water is injected at one place in the bottom surface with a very low velocity ($Re \sim 0.4$) to simulate the situation of liquid water entering into the GDL from the catalyst layer. The Volume of fluid (VOF) method is then used to track the air-water interface subject to a specified wall adhesion angle at fibre-fluid interface. Comparative simulations have been performed for two cases with different contact angles, each representing a hydrophilic and a hydrophobic fibre. The transport of the liquid water is tracked in the fibrous domain and the water evolution is shown at different time intervals in the Fig. 1 for both hydrophilic and hydrophobic GDLs. It is observed that the action of the interfacial tension in hydrophilic GDL is to spread the liquid water along the fibre and fills the bottom surface, thus blocking the flow of reactants to the catalyst layer. In case of hydrophobic GDL, liquid water is prevented from spreading laterally. Instead of spreading along the fibre or bottom surface, the water droplet grows bigger and bigger in size and touches the fibre in the next layer. The growth of the water droplet in the hydrophobic GDL with time is shown in figure 2. From there, the water may channel through the gap between the fibres and get evacuated from the GDL. This prevention of lateral spreading allows the pathway for the reactants to enter in the counter direction. References: 1. C.W. Hirt, B.D. Nichols. Volume of Fluid (VOF) method for the dynamics of free boundaries, J. Comput. Phys. 39 (1981) 201-225. 2. J.U. Brackbill, D.B. Kothe, C. Zemach, A continuum method for modeling surface tension, J. Comput. Phys. 100 (1992) 335-354. 3. A. Theodorakakos, T. Ous, M. Gavaises, J.M. Nouri, N.

Nikolopoulos, H. Yangaihara. Dynamics of water droplets detached from porous surfaces of relevance to PEM fuel cells, J. Colloid Interface Sci. 300 (2006) 673-687. 4. X. Zhu, P.C. Sui, N. Djilali. Dynamic behavior of liquid water emerging from a GDL pore into a PEMFC gas flow channel, J. Power Sources 172 (2007) 287-295. 5. X. Zhu, P.C. Sui, N. Djilali. Three-dimensional numerical simulations of water droplet dynamics in a PEMFC gas channel, J. Power Sources 181 (2008) 101-115.

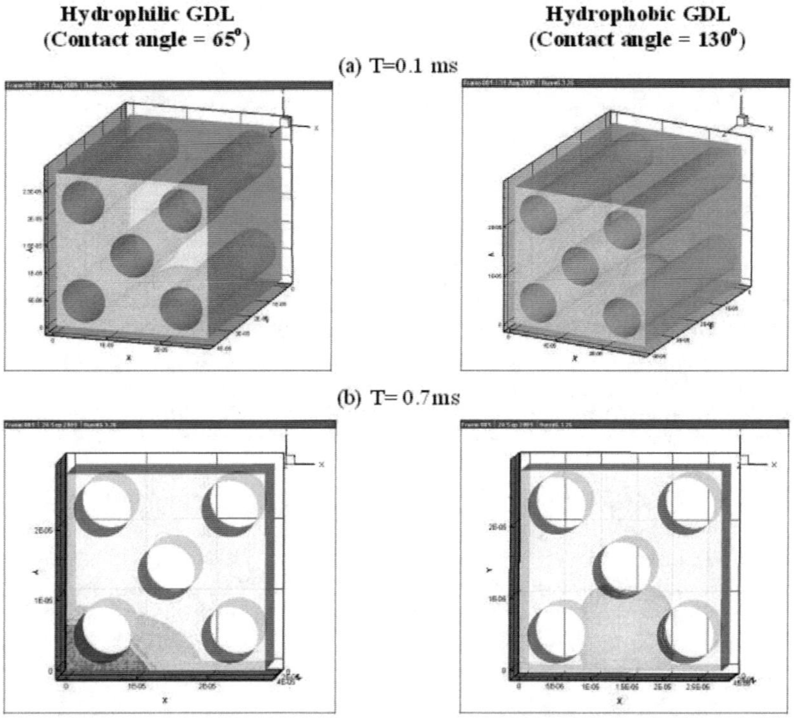

Figure 1. Time evolution of water transport through the hydrophobic (Contact angle of 130°) and hydrophilic (Contact angle of 65°) fibrous structures.

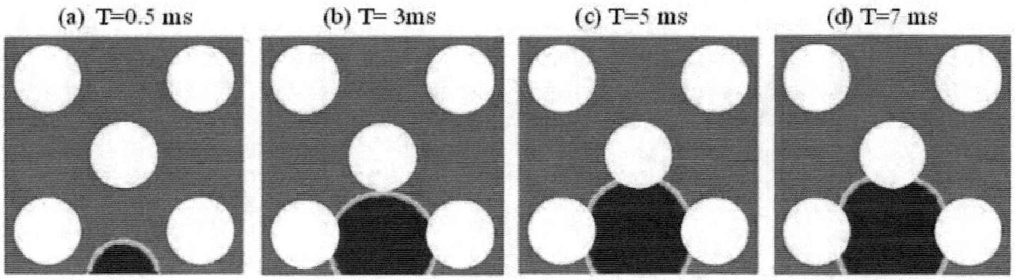

Figure 2. Water droplet growth with time in a hydrophobic fibrous structure (Contact angle of 130°)

See more of this Session: Fuel Cell Technology
See more of this Group/Topical: Fuels and Petrochemicals Division

A Comparative Study of the Hydrated Morphologies of Perfluorinated Ionomers with Distinct Protogenic Groups

Dongsheng Wu, Chemical and Biomolecular Engineering, The University of Tennessee in Knoxville, Knoxville, TN

Stephen J. Paddison, Chemical and Biomolecular Engineering, The University of Tennessee in Knoxville, Knoxville, TN

The hydrated morphologies of a perfluorosulfonic acid (PFSA) ionomer and a bis[(perfluoroalkyl)sulfonyl]imide-based ionomer have been investigated through dissipative particle dynamics (DPD) simulations as a function of ionomer equivalent weight (EW) and degree of hydration. The PFSA ionomers with EWs of 589, 651 and 801 g/mol were chosen to determine the effect of EW on their hydrated morphologies. With a common hydrophobic polytetrafluoroethylene backbone, the PFSA ionomer and imide-based ionomer were further compared to investigate the effect of the different protogenic group on the morphologies.

Coarse-grained mesoscale models were constructed by dividing the hydrated ionomer into components consisting of a polytetrafluoroethylene backbone bead, ionomer specific backbone bead, terminal side chain beads and a water bead consisting of a cluster of six water molecules. Flory-Huggins χ-parameters and repulsion parameters between DPD beads were calculated based on the optimized structures of the beads. Equilibrated morphologies of the ionomers containing various water contents were then determined in a simulation box with dimension of about 32.4 nm. The hydration level (λ) was varied in both systems with water contents corresponding to 6, 12, 18 and 24 H2O per side chain.

Water contour plots revealed that as the hydration level was increased, the isolated water clusters present at the lower water contents increase in size eventually resemble channels or pores in the PFSA ionomer. While for the imide-base ionomer, the water clusters have stronger segregation and only form continuous water region at relatively high hydration level (e.g. $\lambda = 24$). This illustrate why the imide-base ionomer shows especially low conductivity at low to intermediate hydration levels and gain conductivity closed to other PFSA ionomers again at a specific higher hydration level. The simulations of the PFSA at different EWs suggest that the high EW ionomers exhibit significantly greater dispersion of water regions that the low EW ionomers. The high EW ionomers stimulate the segregation of hydrophobic and hydrophilic phases in the PFSA membranes. Finally, water-water particle radial distribution functions (RDFs) were evaluated for all of the PFSA ionomers and the average size of water clusters estimated.

See more of this Session: Fuel Cell Technology III
See more of this Group/Topical: Fuels and Petrochemicals Division

A Highly Conductive Quaternary-Phosphonium Functionalized Membrane for High-Performance Hydroxide Exchange Membrane Fuel Cells

Shuang Gu, Chemical and Environmental Engineering, University of California, Riverside, Riverside, CA

Rui Cai, Chemical and Environmental Engineering, University of California, Riverside, Riverside, CA

Ting Luo, Chemical and Environmental Engineering, University of California, Riverside, Riverside, CA

Kurt Jensen, Chemical and Environmental Engineering, University of California, Riverside, Riverside, CA

Yushan Yan, Chemical and Environmental Engineering, University of California, Riverside, Riverside, CA

As a novel fuel cell technology, the hydroxide (OH-) exchange membrane fuel cells (HEMFCs) have been demonstrated to have the potential to solve the proton exchange membrane fuel cells' (PEMFC's) crippling dependence on precious metals as electrode [1]. In addition, HEMFCs can completely avoid the electrolyte leakage and potentially eliminate the metal (bi)carbonate formation problems of the traditional liquid alkaline fuel cells (AFCs) [2].

As one of the most crucial parts of HEMFCs, the hydroxide exchange membrane (HEM) usually directly decided the performance of HEMFCs through its hydroxide-conductivity and membrane-thickness. Current HEMs are based on the quaternary-ammonium-hydroxide (QAOH) functional groups, and have low performance due to its hydroxide conductivity. By switching from the QAOH to the quaternary-phosphonium-hydroxide (QPOH), the HEM may have the potential to solve the low hydroxide-conductivity problem and thus improve the HEMFC performance drastically.

In this work, by optimizing the degree of chloromethylation and the preparation-procedure of membrane, we synthesized and prepared a quaternary-phosphonium functionalized hydroxide exchange membrane (TPQPOH152) with typical thickness of 50-100 μm. TPQPOH showed a 45 mS/cm (20 oC) of hydroxide-conductivity that is substantially higher than those of the commercial and currently experimental QAOH functionalized membranes. In addition, the HEMFC single cell with a 50 μm TPQPOH152 as HEM, at 70 oC, exhibited substantially high peak power density of 258 mW/cm2 and significantly low internal resistance of 0.21 Ωcm2. The high hydroxide-conductivity and high-performance of HEMFC indicate that the quaternary-phosphonium functionalized TPQPOH152 HEM is highly promising for use in HEMFC.

References

[1] S. Gu, R. Cai, T. Luo, Z. Chen, M. Sun, Y. Liu, G. He, Y. Yan, Angewandte Chemie International Edition 2009, in press.

[2] J. R. Varcoe, R. C. T. Slade, Fuel Cells 2005, 5, 187.

See more of this Session: Fuel Cell Technology III
See more of this Group/Topical: Fuels and Petrochemicals Division

PVDF-Acrylic Semi-Interpenetrating Network Proton Exchange Membranes

Carson Meredith, School of Chemical and Biomolecular Engineering, Georgia Institute of Technology, Atlanta, GA

Pedro J. Zapata, School of Chemical and Biomolecular Engineering, Georgia Institute of Technology, Atlanta, GA

The proton exchange membrane (PEM) is a fundamental component of the polymer electrolyte membrane fuel cell (PEMFC); one of the most promising alternative energy conversion approaches for low to medium power applications. Existing commercially available PEMs are predominantly based on perfluorosulfonic acid polymers (PFSA) (e.g., Nafion®), which are characterized by elevated costs owing to convoluted manufacturing processes. In this study the viability of semi-interpenetrating ionomer-polymer networks from blends of poly(vinylidene fluoride) (PVDF) and covalently-crosslinked sulfonated acrylic polyelectrolytes (PE) as potential PEMs is examined. A total of 80 dissimilar PVDF/PE membranes, prepared from five different grades of Kynar® PVDF (homo- and copolymers) and two types of PE, were characterized in terms of proton conductivity and mechanical properties using custom-developed high-throughput screening tools. In addition to PE type and content, the crystalline characteristics (i.e., crystallinity and crystallite size) and melt viscosity of the inert PVDF phase were found to play a major role on proton conductivity. Particularly, membranes based on highly crystalline and viscous PVDF homopolymers exhibited the lowest proton conductivity due to precluded segmental motion of the PE chains during crosslinking. Mechanical properties of the membranes were dominated by the PVDF grade incorporated. Membranes based on stiffer PVDF homopolymers exhibited higher elastic modulus and tensile strength; while those based on the more flexible copolymers were tougher. In general PVDF/PE membranes compared favorably to Nafion®, having conductivities in excess of 130 mS/cm (vs. 84.5 mS/cm for Nafion®) and mechanical properties 2 to 5 times higher in some cases. The selection of a particular PVDF/PE membrane, however, implies a tradeoff between conductivity and mechanical properties and is determined by the requirements of the PEMFC final application.

See more of this Session: Fuel Cell Technology III
See more of this Group/Topical: Fuels and Petrochemicals Division

Sodium Silicate Based Sol-Gel Structures as Proton Exchange Membranes for Microfluidic Fuel Cells

Debashis Dutta, University of Wyoming, Laramie, WY

Chandelle Wadsworth, Chemistry, University of Wyoming, Laramie, WY

In this work, we report a new approach to fabricating ultra-thin proton exchange membranes (PEM) that can be readily integrated to a microfluidic fuel cell. The proposed approach takes advantage of the larger capillary forces within sub-micrometer scale channels to retain nanoliter sized volumes of a precursor material between two compartments carrying the fuel and oxidant. This material is then provided a suitable physical/chemical treatment to transform it into a PEM with thicknesses of the order of 10μm. In this design, the mechanical stability of the membrane is not compromised however, as this structure is embedded between two thick glass substrates. The proposed approach to fabricating integrated PEM in microfluidic fuels cells has been demonstrated in this work by creating a sodium silicate based sol-gel structure within an array of sub-micrometer scale channels (nanochannels) that bridge two micrometer sized conduits (microchannels). The device is then operated by filling up one of the microchannels with a fuel e.g., 1M formic acid, and the other with an oxidant e.g., 0.15M $KMnO_4$ in 0.5M H_2SO_4. Upon bringing in contact two platinum electrodes to these solutions, a voltage is generated. In this circuitry, the porous silicate structure acts as the PEM that preferentially allows protons to pass through its pores. With our current design and the choice of fuel/ oxidant, we have been able to generate an open circuit potential of 1.3V and a maximum current of about 40μA. Interestingly, the power output of our device is an order of magnitude greater than that reported for other microfluidic fuel cells with integrated PEMs and the same choice of fuel/oxidant. Moreover, a detailed characterization of our current device shows that its power output is limited by the Ohmic resistance of the microchannels rather than that of the PEM. In this situation, a further enhancement in the maximum current deliverable by our device is possible through optimization of the microchannel design.

See more of this Session: Fuel Cell Technology III
See more of this Group/Topical: Fuels and Petrochemicals Division

Unraveling the Problem of Carbonate Formation in Alkaline Fuel Cells

Matthew S. Naughton, Department of Chemical & Biomolecular Engineering, University of Illinois at Urbana-Champaign, Urbana, IL

Fikile R. Brushett, Department of Chemical & Biomolecular Engineering, University of Illinois at Urbana-Champaign, Urbana, IL

Paul J. A. Kenis, Chemical and Biomolecular Engineering, University of Illinois at Urbana-Champaign, Urbana, IL

Section: Alternate Fuels and New Technology (16d)

Session: 16D07 Fuel Cell Technology

Fuel cells are promising power sources due to their ability to bypass Carnot efficiency limitations by directly converting chemical energy into electrical energy. However, the high costs of catalysts and membranes, as well as component durability issues, have barred widespread implementation. Alkaline fuel cells (AFCs) have gained increasing attention due to superior electrode kinetics in alkaline media, compared to acidic media, which enables the use of non-noble metal catalysts [1]. Unfortunately, carbonate formation is a major barrier to mass commercialization of AFC technologies. Hydroxide anions in the electrolyte react with carbon dioxide from organic fuel oxidation and/or air to form carbonates. This reaction reduces electrolyte conductivity and can deactivate electrodes through carbonate precipitation with an available cation [2].

The effects of carbonate formation are largely determined by the fuel cell electrolyte. The electrolyte can be a solid polymeric membrane or a liquid electrolyte, either of which have an initial concentration of hydroxide anions. With a stagnant liquid electrolyte, carbonates can easily precipitate onto electrodes when the nearby solution is saturated, while a flowing electrolyte remains dissolved in solution as long as the entire solution volume remains unsaturated [3]. Metal cation-free membrane systems do not suffer from carbonate issues since no mobile cations are present to facilitate precipitation onto an electrode or within the membrane [4]. Nevertheless, overall fuel cell performance is still hampered by conductivity losses and the precise effects of carbonate formation have yet to be researched in detail.

We present research quantifying the effect and rate of carbonate formation in a microfluidic fuel cell operating with a flowing alkaline electrolyte and utilizing either hydrogen or methanol as the fuel. Precise control over electrolyte composition enables the study of carbonate formation effects in an operating fuel cell by varying carbonate-electrolyte ratios. The membraneless design allows for individual analyses of anode and cathode performances via an external reference electrode.

[1] Tewari et al., Journal of Power Sources, 2006, 153, 1-10

[2] Rolla et al., Journal of Power Sources, 1980, 5, 189-196

[3] Brushett et al., Journal of the Electrochemical Society, 2009, 156, B565

[4] Lu et al., Proceedings of the National Academy of Sciences, 2008,105 52, 20611-20614

See more of this Session: Fuel Cell Technology III
See more of this Group/Topical: Fuels and Petrochemicals Division

Development of Cathode Air Filters for PEM Fuel Cell Using Microfibrous Entrapped Sorbents

Abhijeet Phalle, Center for Microfibrous Materials Manufacturing, Department of Chemical Engineering, Auburn University, Auburn, AL

Bruce Tatarchuk, Center for Microfibrous Materials Manufacturing, Department of Chemical Engineering, Auburn University, Auburn, AL

The performance of Proton Exchange Membrane Fuel Cell (PEMFC) is degraded significantly as a result of poisoning of cathode catalyst by airborne contaminants, in particular $SO_2$and NO_x in trace levels. In the literature, it has been suggested that the best method for addressing fuel cell contamination is by the inclusion of adsorptive filtration with a cathode air filter. This work thus aims at the development of adsorptive cathode air filters, mainly to remove SO_2 because of its irreversible effects on PEMFC. Two types of sorbents were prepared; an alumina support impregnated with different transition metal (Mn, Fe, Ni, Cu and Zn) oxides and inorganic chemicals (KOH and $KMnO_4$). Metal nitrates of respective transition metals were used as precursors for impregnation of alumina with metal oxides; except for Manganese; where in addition to $Mn(NO_3)_2$, $MnSO_4$ was also used as precursor. The adsorption capacity of these sorbents was investigated in a tubular packed column under different relative humidity conditions (RH of 15%, 50% and 75%) and at room temperature. At identical loadings, among all metal oxide based sorbents, MnO_x/Al_2O_3 prepared using $MnSO_4$ as precursor showed the highest saturation and breakthrough capacity.A bi-component polymer fibers-entrapped MnO_x/Al_2O_3 sorbent (PFES) is designed for SO_2 removal from air. Composite bed consisting of packed bed with large size particles (~1mm) of sorbent followed by polishing layer of PFES media was prepared and its performance was compared with packed bed with similar loading of sorbent. Due to the microfibrous media and supported sorbent design, composite bed demonstrated superior SO2 removal performance as compared with a packed bed. At equivalent loading of ($KMnO_4$+$NaHCO_3$), the composite bed yielded ~100% higher breakthrough capacity than that for same sorbent in the packed bed. This is attributed to the high contact efficiency of small particulates, and decreased intrabed channeling observed in microfibrous media. Comparison of different configurations of adsorbent bed based on their pressure drop as well as adsorption performance and optimal design of sorebnt bed for cathode air filter will be discussed.

See more of this Session: Fuel Cell Technology III
See more of this Group/Topical: Fuels and Petrochemicals Division

Ethanol Production From the Biological Fermentation of Synthesis Gas
16D00 Alternative Fuels and Enabling Technologies
Richard E. Tobey, Coskata, Warrenville, IL

Summary

This paper illustrates the process of feedstock-flexible ethanol production from various carbonaceous sources via the biological fermentation of synthesis gas. Specific focus is spent on the unique capabilities of Coskata's syngas-to-ethanol platform which holds potential for contributing valuable technological innovation to the ethanol industry.

The paper also reviews the progress Coskata has made in commercializing the syngas-to-ethanol platform. A briefing on Coskata's semi-commercial operation, based in Madison, Pa. will be included in the presentation at the AIChE Annual Conference in November.

The Cellulosic Ethanol Production Challenge

As next generation ethanol technologies continue to be developed, several production pathways have emerged. However, feedstock flexibility and cost remain as important hurdles, preventing the wide-spread adoption of these advanced technologies.

Research and development in the cellulosic ethanol industry remains focused on bringing costs down while improving conversion efficiencies. Other challenges facing the industry include the need for innovation to produce on a large commercial scale in order to contribute to the Renewable Fuels Standard (RFS) first signed into law in the U.S. under the Energy Policy Act of 2005, and revised under the 2007 Energy Independence and Security Act (EISA).

Ethanol derived from corn can contribute to the RFS, but a feedstock flexible platform must emerge for the wide scale adoption of cellulosic biofuels around the globe. The 2007 EISA sets the mandate at 21 billion gal./yr. of production coming from cellulosic and advanced biofuel sources. The way to meet this mandate, and extend it even further to as much as 90 billion gallons in the future, will come from technology providers that have feedstock flexibility. An affordable, flexible, efficient and scalable cellulosic ethanol production path must be deployed for making an impact on the RFS requirements.

The final remaining hurdle to achieving the widespread adoption of cellulosic biofuels is the project financing barriers that exist in the renewable energy market today. Banks are not lending to one another, let alone to wind, power, or biofuel projects. Commercial lending must catch up to the technology being ready for commercial deployment.

Solution: Producing Ethanol From The Biological Fermentation of Synthesis Gas

The biological fermentation of synthesis gas offers several advantages over traditional ethanol production. This paper argues that the biological fermentation of synthesis gas can yield improvements in efficiency and affordability, which makes it the leading pathway available today.

The biological fermentation process starts with the thermological conversion of carbonaceous feedstock into synthesis gas, where the chemical energy is contained in the hydrogen and carbon monoxide gasses. One gasification approach to create synthesis gas is plasma gasification. This technology allows feedstock flexibility, while contributing to the geographic versatility of the system, enabling ethanol production near locally abundant feedstocks. Steam that is recovered from the cooling of the hot synthesis gas can be utilized in a co-located process such as a pulp and paper mill, sugar mill, or even an existing corn ethanol plant.

The biological fermentation of synthesis gas is achieved as bacteria ingest synthesis gas, and excrete only fuel-grade ethanol via their chemical pathways. By using anaerobic bacteria, the biological fermentation process does not require expensive enzymes or pretreatment. With the process yielding only fuel-grade ethanol, unlike chemical catalysis which produces mixed alcohols, higher yields of the desired product are able to be obtained per dry ton of input material.

Coskata's work in facilitating the biological fermentation of synthesis gas into ethanol includes the development of patented bioreactor systems, which increase the mass transfer of the synthesis gas being converted directly into fuel-grade ethanol. The resulting yield is approximately 100 gallons of ethanol per dry ton of wood biomass input material. The bioreactors also operate at lower pressure and lower temperatures, delivering a cost and energy advantage vs. chemical catalysis.

The microorganisms feed on the syngas and nutrients in the bioreactor, and produce ethanol in an aqueous solution, which is then separated from the water using either traditional distillation or vapor permeation technology, resulting in saleable fuel-grade ethanol. The bioreactor continues to recycle the process water, delivering one of the lowest net water use numbers in the industry per gallon of ethanol produced.

Coskata's research and development efforts have also focused on systematically breeding the microorganisms for performance and ethanol yield. This ongoing efficiency and performance work, done at the company's Warrenville, Illinois headquarters and pilot plant, matched with the optimization of bioreactor design, has enabled the company's biological fermentation process to become one of the most energy efficient, affordable and geographically flexible ethanol production technologies available today, achieving one of the lowest production costs in the industry.

Argonne National Laboratory performed a GREET analysis in 2008 that showed Coskata's process can produce up to 7.7 times more energy output than the fossil energy inputs used in making the ethanol, whereas corn-derived ethanol produces approximately 1.3-1.5 times the fossil energy required. Additionally, Argonne's Well-to-Wheel analysis showed that the Coskata process, compared to conventional gasoline, can reduce carbon dioxide emissions by as much as 96% when wood biomass is used as an input material.

The efficiency, flexibility and affordability advantages that biological fermentation hold over other ethanol production pathways allows the process to make a meaningful impact on the RFS standards while contributing to the growth of local and national economies.

In addition to growing local economies and creating green jobs, large-scale production of cost-efficient biofuels will help ensure energy security by offsetting national dependency on petroleum, helping to localize ethanol production by utilizing a wide variety of feedstocks including municipal waste and trash, agricultural waste, bagasse and other biomass to eliminate food-for-fuel concerns.

The unlocking of feedstock flexible ethanol production can be very economically competitive against gasoline – ethanol produced through this pathway can compete in the long-term unsubsidized against gasoline – helping to move ethanol past the point of a transitional fuel and into the realm of a truly renewable biofuel. This process can also significantly reduce the industry's risk of exposure to the feedstock pricing fluctuations which have been demonstrated with tradition ethanol and biodiesel.

Coskata's Commercialization Progress

The flexibility, efficiency and affordability advantages of Coskata's process has drawn funding from world class investors, including The Blackstone Group, Khosla Ventures and General Motors in efforts to commercialize this promising technology.

GreatPoint Ventures incorporated Coskata in July 2006, as the result of research Aaron Mandell and Andrew Perlman, two GreatPoint Ventures partners, had been doing since 2001 into alternative concepts for low cost cellulosic ethanol production.

Coskata has successfully produced fuel grade ethanol utilizing the biological fermentation process at the company's Warrenville, Ill. pilot plant, ensuring the feasibility of Coskata's biological fermentation process. Coskata's semi-scale facility, Project Lighthouse, is located in Madison, Pennsylvania, near Pittsburgh. Project Lighthouse is a showcase facility for the world's first commercially-viable flex ethanol process. It represents the successful scale-up of Coskata's technology and is a clear demonstration that its efficient, affordable and flexible conversion technology is ready today to help reduce our country's dependency on oil in an economically and environmentally sustainable way.

The plant demonstrates that Coskata's technology is ready today to reduce our dependence on oil, provide an environmentally sustainable solution to fossil fuels, and fuel economic growth through innovation and jobs. Project Lighthouse also demonstrates Coskata's feedstock flexibility and industry leading ethanol productivity that will help to meet the 36 billion gallon mandate for biofuels contained in the new Renewable Fuels Standard. Coskata has concluded the front end engineering design on a 55 Million gallon commercial facility, as well as started work with licensees to begin the design and construction of additional commercial FlexEthanol™ facilities.

Design and Operation of Novel Photobioreactors for the Cultivation of Microalgae for the Biofuels Sector

Nadia Abunasser, New Markets, Michigan Economic Development Corporation, Livonia, MI

Haiying Tang, Chemical Engineering and Materials Science, Wayne State University, Detroit, MI

Mario Enrique Danton Garica Perez, Alternative Energy Technology, Wayne State University, Detroit, MI

Meng Chen, Chemical Engineering, Wayne State University, Detroit, MI

Ansheng Li, Chemical Engineering, Wayne State University, Detroit, MI

Hongzhi Ma, Chemical Engineering, Wayne State University, Detroit, MI

Mahbuba Ara, Chemical Engineering and Materials Science, Wayne State University, Detroit, MI

Steven Salley, Chemical Engineering and Materials Science, Wayne State University, Detroit, MI

K. Y. Simon Ng, Chemical Engineering and Materials Science, Wayne State University, Detroit, MI

A major hurdle to the use of microalgae based biofuels is the lack of an economical method for the microalgae's cultivation. This problem is addressed by the development of two novel photobioreactors for algal growth. The first reactor is an external loop airlift reactor built with the capability of using either internal or external illumination. The riser diameter in this reactor is 6 inches while the downcomer is 2 inches. The second reactor is a 55 gallon bubble column designed with a novel mixing/ internal illumination mechanism. Both reactors were designed with automated control and monitoring systems. Dunaliella tertiolecta UTEX LB 999 grown in Erdschreiber's Medium and Chlorella minutissima UTEX 2219 in Modified Bold 3N Medium were used to study the operating parameters of both reactors. The effects of light intensity, photoperiod, and carbon dioxide concentration in the feed gas, gas flowrate and mixing on the microalgal growth were studied. The effects of the varying nutrient concentrations were also studied, and a procedure was developed to maintain the concentrations of the essential nutrients for algal growth. The microalgae were harvested at the end of each experiment and the algal oil was extracted and analyzed for biofuels potential.

See more of this Session: Alternative Fuels and Enabling Technologies
See more of this Group/Topical: Fuels and Petrochemicals Division

Biodiesel Oxidative Stability Study: Factors and Improvement

Rhet Joseph De Guzman, Department of Chemical Engineering and Materials Science, Wayne State University, Detroit, MI

Haiying Tang, Department of Chemical Engineering and Materials Science, Wayne State University, Detroit, MI

Steven Salley, Department of Chemical Engineering and Materials Science, Wayne State University, Detroit, MI

K. Y. Simon Ng, Department of Chemical Engineering and Materials Science, Wayne State University, Detroit, MI

Biodiesel, which is produced from animal and plant based-feedstocks, is more prone to oxidation than petroleum diesel. Fatty acid profile and natural antioxidant content are two of the inherent factors that affect the stability. Likewise, external factors such as exposure to air, heat, light and metals can initiate the onset of autoxidation and the subsequent degradation of fuel properties such as the kinematic viscosity and total acid number. This study examines the influence of the biodiesel composition, especially the minor components on the oxidative stability of biodiesel. Additionally, binary combinations of butylated hydroxyanisol (BHA), propyllgallate (PG), pyrogallol (PY) and t-butyl hydroxyquinone (TBHQ) were evaluated for their effect on oxidative stability under indoor and outdoor storage conditions as a function of time. Finally, the catalytic effect of metallic contaminants (Al, Cu, Fe and Zn) on biodiesel oxidation and possible antioxidant deactivation by the metals was investigated.

Results of this study provide a better understanding why biodiesel from different feedstocks has different degrees of oxidative stability and different responses to antioxidants. Additionally, permissible levels of metallic contaminants, with and without the presence of antioxidants, are described.

See more of this Session: Alternative Fuels and Enabling Technologies
See more of this Group/Topical: Fuels and Petrochemicals Division

Sterne Kessler Goldstein Fox
ATTORNEYS AT LAW

Leveraging Intellectual Property for Biofuel Innovators

American Institute of Chemical Engineers Annual Meeting

Gaylord Opryland Hotel and Convention Center
November 10, 2009

Peter A. Jackman, Esq.
Director
Sterne, Kessler, Goldstein & Fox PLLC
Washington, DC
pjackman@skgf.com
202.772.8582
www.skgf.com

SKGF.COM

Sterne Kessler Goldstein Fox
ATTORNEYS AT LAW

Leverage Your IP

- 75% of all deal values are made up of intangible assets (Price Waterhouse Coopers).
- In 2005 80% of the market value of the S&P 500 resided in its intangible assets compared with only 17% in 1975 (CPA).
- "In the current economic climate, technology, innovation and creativity are critical in creating opportunities for economic renewal and addressing pressing global issues such as climate change." (Australian Director General of WIPO)

SKGF.COM

iManage No. 1030397v1

Sterne Kessler Goldstein Fox
ATTORNEYS AT LAW

Leverage Your IP

- With concerns about oil prices and energy independence, investment in clean technologies such as biofuels has risen dramatically (*e.g.* ARRA).
- Innovator companies have two major obstacles to their ultimate success:
  - Create technologies that are both functionally and economically feasible.
  - Navigating the vast body of relevant technology patents that have issued or are pending.

SKGF.COM

iManage No. 1030397v1

Sterne Kessler Goldstein Fox
ATTORNEYS AT LAW

Leverage Your IP

- Goals are to maximize the creation of new valuable intellectual assets and improve market competitiveness.
- Develop IP strategy consistent with business goals:
  - Internal Focus:
    - Identify sources of existing/potential IP
    - Protect intellectual assets offensively and defensively (*e.g.*, copyrights, trademarks, patents, trade secrets)
  - External Focus:
    - Freedom-to-operate issues

SKGF.COM

iManage No. 1030397v1

Sterne Kessler Goldstein Fox
ATTORNEYS AT LAW

Internal Focus: Conduct an IP Audit

- Identify existing IP and potential sources of future IP:
  - – Assists in developing/refining IP strategy for company's business goals
    - Strategy depends on biofuel company's products, services, current and future revenue goals
  - – Confirm core technology is protected
  - – Identify IP gaps
  - – Re-asses IP periodically
  - – Determine which IP is no longer worth pursuing/maintaining (*e.g.* foreign markets)

SKGF.COM

iManage No. 1030397v1

Sterne Kessler Goldstein Fox
ATTORNEYS AT LAW

Potential Sources of IP

- Business methods
- Machines (*e.g.* reactors, equipment)
- Test data (even negative results)
- Customer lists
- Industrial processes (*e.g.* methods of hydrolyzing cellulosic material, methods of fermenting, optimized process conditions)
- Modified biological organisms (*e.g.* bacteria, yeast, algae)
- New use of known process or composition
- Improvements to any of the above (need not be revolutionary, can be evolutionary)
  - But see ***KSR v. Teleflex***, 550 U.S. (2007).

SKGF.COM

iManage No. 1030397v1

Sterne Kessler Goldstein Fox
ATTORNEYS AT LAW

Patent Fundamentals

- Right of patent owner to <u>exclude</u> others from:
  - Making
  - Using
  - Selling or offering for sale in the U.S. or
  - Importing into the U.S.
  - a patented invention (determined by claims)
- For a period of 20 years from the filing date
- U.S. Patent Office accords "special" status (expedited prosecution) to patent applications for inventions which materially contribute to (a) the discovery or development of energy resources, or (b) the more efficient utilization and conservation of energy resources.
  - Consider if need quick issuance (*e.g.* funding or enforcement)

SKGF.COM

iManage No. 1030397v1

Sterne Kessler Goldstein Fox
ATTORNEYS AT LAW

Business Advantages of Patent Protection

- Commercial advantage over competitors
  - Establish proprietary leadership position in biofuel market
  - Force competitors to spend money to design around or license your technology
  - Blocking patents (broad coverage or non-core technology)
- Financial rewards for bringing innovative and useful products into the marketplace:
  - Directly by making or selling the invention
  - Indirectly by licensing to others to make/sell the invention
  - Creates shareholder value
  - Recoup dollars spent on R&D
- Validates the technology
- Stops unfair competition by those who compete by imitation
- Provides valuable trading assets to help assure freedom to operate (*e.g.* cross-license)

SKGF.COM

iManage No. 1030397v1

Sterne Kessler Goldstein Fox
ATTORNEYS AT LAW

Requirements for Patentability

- A patent is granted to the first inventor (U.S. is still first-to-invent) of:
  - Novel
  - Useful; and
  - Nonobvious
- Process (method), machine, article of manufacture or composition of matter (*e.g.* chemical compound) and new and useful improvements thereof.

SKGF.COM

iManage No. 1030397v1

Sterne Kessler Goldstein Fox
ATTORNEYS AT LAW

What is Patentable?

- "Anything under the sun that is made by man."
  - *Diamond v. Chakrabarty*, 47 U.S. 303 (1980)
- Chemical compositions, mixtures of ingredients, formulations
- New uses of processes, compositions (*e.g.* use of known microbial enzyme in industrial ethanol fermentation process)
- Modified biological organisms (*e.g.* bacteria, algae and yeast)
- Modified plants (*e.g.* sugarcane, corn, switchgrass)
- Industrial or laboratory processes (*e.g.* methods of pre-treatment)
- "Isolated" or "purified" DNA, proteins and microorganisms

SKGF.COM

iManage No. 1030397v1

Sterne Kessler Goldstein Fox
ATTORNEYS AT LAW

Utility Patents - When to Patent?

- In U.S., there is a one-year grace period to file which begins from the earliest:
  - Printed publication (*e.g.* publishing a paper or disseminating over the internet)
  - Sale (selling a product of that embodies the invention)
  - Offer for Sale or
  - Public Use (putting the invention into public use)
  - Which discusses or embodies the invention
- Most other countries require absolute novelty and offer no grace period (*e.g.* Japan).
- If world-wide patent protection is sought, file *before* any publication, sale, offer for sale or public use (*e.g.* poster abstracts, seminar presentations).
- Keep confidential (educate employees) until ready to file.
- Obtain CDA before disclosing to potential partner (*e.g.* joint venture, license agreement).

SKGF.COM

iManage No. 1030397v1

Sterne Kessler Goldstein Fox
ATTORNEYS AT LAW

Trade Secrets

- A "Trade Secret" is something that is not generally known to the public that gives the possessor of the information a competitive edge or market advantage.
  - Value – economic advantage
  - Secret – capable of being kept secret (reverse engineer?)
- Perpetual protection (as long as secrecy is maintained)
  - Have a consistent policy in place (e.g. mark "confidential", limit access)
  - Educate employees about policy
- Unlike patents, trademarks and copyrights (federal-based IP), trade secret protection is governed by state law.
  - Every state recognizes some form of trade secret protection.

SKGF.COM

iManage No. 1030397v1

Sterne Kessler Goldstein Fox
ATTORNEYS AT LAW

Trade Secrets

- Nearly all companies have secrets; many of these secrets are trade secrets.
- Trade secrets may include:
  - Product Designs
  - Test Results (including unsuccessful tests)
  - Material Formulas
  - Manufacturing Processes
  - Customer Lists
  - Supplier Lists
  - Merchandising Techniques
  - Purchasing Information
- Biofuel company needs to decide to seek either patent or trade secret protection early while still confidential (patents and trade-secrets are mutually exclusive).
  - Example – cell line/optimized culture media used as internal tool which would not be used commercially.

SKGF.COM

iManage No. 1030397v1

Sterne Kessler Goldstein Fox
ATTORNEYS AT LAW

Business Advantages of Trade Secrets

- Protect valuable technical information that cannot be protected under other forms of IP (*e.g.* patents).
- Protect ideas that offer a competitive advantage, thereby enabling your company to get a head start on the competition
- Keep competitors from learning that a product or process is under development and discovering its functional or technical attributes
- Protect valuable business information such as marketing plans, cost and price information and customer lists
- Protect "negative know-how" - information learned during the course of R&D on what not to do or what does not work optimally
- Protect any other information that has some value and is not generally known by your competitors

SKGF.COM

iManage No. 1030397v1

Sterne Kessler Goldstein Fox
ATTORNEYS AT LAW

Trade Secrets vs. Patents

- **Trade Secret**
- Term: as long as it remains secret
- Cost: cost of keeping the information secret
- Registration: no registration or application required; automatic
- Protection: protects against stealing and copying of proprietary, trade secret information
- Enforcement: must prove existence of trade secret
- Requirements: secrecy; commercial value, not generally known or easily ascertained

- **Patent**
- Term: 20 years from filing
- Cost: drafting costs; prosecution costs; maintenance costs
- Registration: application must be filed, examined and granted
- Protection: protects functionality, operation, structure, idea - even if independently created
- Enforcement: presumption of validity
- Requirements: utility; novelty; nonobvious; written description; enablement; best mode;

SKGF.COM

iManage No. 1030397v1

Sterne Kessler Goldstein Fox
ATTORNEYS AT LAW

External Focus: Freedom-to-Operate (FTO) Issues

- **Patentability ≠ freedom to operate**
  - Patent is only a grant of the right to <u>exclude</u> others from making, using, or selling the patented invention.
  - Patent does not give patent owner the <u>affirmative</u> right to do anything (even to make, use, sell own invention)
  - KEY: A company can receive a patent for its technology and still be blocked from practicing that technology by a broader (dominant) patent.
    - Later patent to expressing novel termite cellulases in yeast would be blocked by earlier patent directed to expressing any exogenous protein in yeast
- BUT – subordinate patent can still have value
  - If subordinate patent is directed to improvement in the technology that is considered essential or particularly advantageous (e.g. specific enzyme with increased cellulolytic activity)
- Valuable subordinate/improvement patent can be a tool to clear FTO (e.g., through cross-license)

SKGF.COM

iManage No. 1030397v1

Sterne Kessler Goldstein Fox
ATTORNEYS AT LAW

Freedom-to-Operate Issues

- The commercial market for the biofuel industry has increased dramatically, as has the number of patents to production technologies.

- A thorough understanding of the patent landscape is essential:
  - FTO search should be done early and focus on what is claimed (claims are often much broader than disclosure)
  - Do not expend resources developing technology only to find out your company is blocked from commercializing core technology by a third party patent.
  - Potential investors and partners generally consider FTO as important as, or more important than, a dominant IP portfolio.
  - Know risks at outset to determine strategies to remove impediments (*e.g.* license, re-exam, design around).

SKGF.COM

iManage No. 1030397v1

Sterne Kessler Goldstein Fox
ATTORNEYS AT LAW

Resources to Help with FTO

- Expired U.S. patents
  - Technology in expired patents is in the public domain.
- Foreign patents
  - Patent laws are territorial.

- Each of these help the biofuel innovator understand clearly what can be practiced without permission of others.

SKGF.COM

iManage No. 1030397v1

Sterne Kessler Goldstein Fox
ATTORNEYS AT LAW

Strategies for Clearing FTO

- **Noninfringement/Design Around Analysis**
  - Determine whether claims actually cover contemplated commercialization activity
  - Consider alternatives that will avoid claims (i.e., "design around")
  - Review prosecution history at USPTO (road map to design-around strategies)
- **Invalidity Analysis**
  - Consider whether claims are anticipated by or obvious over prior art
  - Consider whether claimed subject matter is adequately described and enabled
  - Consider whether there was any fraud in procuring the patent at the USPTO which would render patent unenforceable

SKGF.COM

iManage No. 1030397v1

Sterne Kessler Goldstein Fox
ATTORNEYS AT LAW

Strategies for Clearing FTO

- **Reexamination Process**
  - Process by which a third party can request that the USPTO reexamine an issued patent
  - Faster and less costly than litigation (similar to oppositions in Europe)
  - The requestor must show that there is a substantial new question of patentability
    - Easier in view of *KSR*
    - >90% of re-exam requests are granted
  - If the request is granted, the patent is looked at without a "presumption of validity"
  - Used to cancel or amend claims in the patent that are not patentable in view of the submitted prior art
    - In >70% of the re-exams, the claims are canceled or amended

SKGF.COM

iManage No. 1030397v1

Sterne Kessler Goldstein Fox
ATTORNEYS AT LAW

Strategies for Clearing FTO

- **Licensing**
  - Consider whether taking a license is desirable/possible
    - Particularly if noninfringement/design around/invalidity positions are uncertain
    - Removes uncertainty of litigation
  - License more likely to be granted where patent holder is not a competitor and does not plan to exploit technology in your intended field of use
  - BUT – must keep in mind royalty stacking problem (particularly in biofuel production space)
    - Allow for need of possible licenses in future

SKGF.COM

iManage No. 1030397v1

Sterne Kessler Goldstein Fox
ATTORNEYS AT LAW

Types of Licensing Agreements

- Non-exclusive
- Exclusive
- Field of Use
- Hybrid (know-how and patent rights)
- Cross-license
- Equity

SKGF.COM

iManage No. 1030397v1

Sterne Kessler Goldstein Fox
ATTORNEYS AT LAW

Pros and Cons of In-Licensing

- Advantages:
- Allows for making/using/selling of own technology
- Extends pool of resources
- Expansion into new markets
- Leverage others' manufacturing/production capabilities
- Provides rights to improvements
- Controls competition
- Security from litigation

- Disadvantages:
- May create competition (e.g. cross-license)
- Reduces level of control
- Adds potentially unsupportable expenses (royalties)
- Undermines independence
- Expenditure of resources to design around licensed patents

SKGF.COM

iManage No. 1030397v1

Sterne Kessler Goldstein Fox
ATTORNEYS AT LAW

Financial Considerations

- Unlike the pharma industry (profit margin >20%) the biofuel industry has a low profit margin.
- Develop licensing strategies that will minimize out-of-pocket expenses:
  - Royalty caps / royalty stacking provisions
  - Single payment/paid-up
  - Sliding scale royalties
  - Milestone payments
  - Equity arrangements

SKGF.COM

iManage No. 1030397v1

Sterne Kessler Goldstein Fox
ATTORNEYS AT LAW

Be IP Savvy

- Know what IP your company has and protect future IP offensively and defensively
- Know critical dates for protecting IP in the U.S. and abroad
- Understand the lost opportunity costs if IP is not protected
- Identify FTO issues early and develop clearing strategies

SKGF.COM

iManage No. 1030397v1

Sterne Kessler Goldstein Fox
ATTORNEYS AT LAW

Thank you!

Questions?

pjackman@skgf.com

202.772.8582

www.skgf.com

SKGF.COM

iManage No. 1030397v1

Consecutive Very-High-Gravity Batch Ethanol Fermentations with the Self-Flocculating Yeast SPSC01

Xumeng Ge Sr., Arkansas Biosciences Institute, Arkansas State University, Jonsboro, AR

Consecutive very-high-gravity batch ethanol fermentation was developed with the self-flocculating yeast SPSC01. Deflocculating of the yeast flocs that resulted in the degeneration of their sedimentation performance was observed with the prolongation of the fermentation system operation, which extended the settling time required for the yeast flocs to separete from the fermentation broth, and exaggerated the toxity of the high level ethanol in the yeast cells, making their viability decreased and the reliability of the fermentation system operation compromised. Experimental results indicated that among the factors affecting the sedimentation performance including the density, concentration and dimension size of the yeast flocs and the apparent viscosity and density of the fermentation broth, the yeast floc size that decreased continuously within the duration of the fermentation and governed by flocculating of the yeast cells dominated the process.

See more of this Session: Alternative Fuels and Enabling Technologies
See more of this Group/Topical: Fuels and Petrochemicals Division

Influence of Antioxidants Type and Concentration On the Oxidative Stability of Castor Oil FAME

Susana V. Araujo, Universidade Federal do Ceará, Fortaleza, Brazil

Francisco Murilo T. Luna, Chemical Engineering, Universidade Federal do Ceará, Fortaleza, Brazil

Estelio M. Rola Jr., Universidade Federal do Ceará, Fortaleza, Brazil

Diana C. S. Azevedo, Chemical Engineering, Universidade Federal do Ceará, Fortaleza, Brazil

Célio L. Cavalcante Jr., Chemical Engineering, Universidade Federal do Ceará, Fortaleza, Brazil

The oxidation stability of castor oil fatty methyl ester (FAME), doped with four different phenolic antioxidants, was evaluated using a rapid method of thermal and air-contact degradation. The methodology is based on the induction times observed when the samples are contacted with pure oxygen at elevated pressures and temperatures. The pressure drop in a sample exposed to pure oxygen at 700 kPA and 140ºC is used to evaluate the oxidation stability of castor oil FAME. The use of antioxidants increased from 6-15 times the stability of castor oil FAME, with the best results indicating the use of BHA (butylated hidroxyanisol) at 2000 ppm for this type of oil. Comparison with previously published reports indicate that, in general, the use of this rapid test does indeed represent the behavior normally observed for tests using available standard methods (Rancimat) and may save time in evaluating biodiesel samples and antioxidants performance.

See more of this Session: Alternative Fuels and Enabling Technologies
See more of this Group/Topical: Fuels and Petrochemicals Division

Hydroprocessing of Vegetable Oil

Elvan Sari, Chemical Engineering and Materials Science, Wayne State University, Detroit, MI

Craig DiMaggio, National Biofuel Energy Laboratory, Detroit, MI

Manhoe Kim, National Biofuel Energy Laboratory, Detroit, MI

Shuli Yan, Wayne State University, Detroit, MI

Steven O. Salley, Chemical Engineering and Materials Science, Wayne State University, Detroit, MI

K.Y. Simon Ng, Chemical Engineering and Materials Science, Wayne State University, Detroit, MI

The increase in energy demand due to the growth of world's population results in more researches for alternative energy. One of the ways to solve this energy problem is improvement of catalytic processes in production of biofuels. Generally there are two ways in diesel fuel like biofuel production from vegetable oils and animal fats. First one is transesterification which gives biodiesel as a product from processing of vegetable oil with methanol or ethanol using either an acid or base catalyst. Second one is hydrogenation which gives green diesel prepared via the catalytic processing of hydrogen with vegetable oils or animal fats. Researches in the production of biofuels are commonly focused on transesterification reaction rather than hydrogenation. The goal of this study is to design a stable catalyst that converts vegetable oil selectively to biofuel product with low hydrogen consumption. With this purpose, the production of green diesel through hydrogenation of soybean oil with Al2O3 supported Ni-Mo catalyst has been studied in a high pressure flow reactor under the reaction conditions 1000 psi pressure and various temperatures. While the gaseous products were analyzed by a gas chromatography, the liquid products were analyzed by GC-MS. As products C-17, C18, water and hydrogen rich gas were seen. Both hydrogenation and hydrodecarboxylation of fatty acids play role in formation of hydrocarbons and the extents of these reactions depend on the catalyst, reaction conditions.

See more of this Session: Alternative Fuels and Enabling Technologies
See more of this Group/Topical: Fuels and Petrochemicals Division

Carbon Dioxide Sequestration From Air Using Microalgae

Lihai Fan, College of Materials Science and Chemical Engineering, Zhejiang University, Hangzhou, China

Lihua Cheng, College of Materials Science and Chemical Engineering, Zhejiang University, Hangzhou, China

Lin Zhang, College of Materials Science and Chemical Engineering, Zhejiang University, Hangzhou, China

Huanlin Chen, College of Materials Science and Chemical Engineering, Zhejiang University, Hangzhou, China

Carbon dioxide (CO2) level in atmosphere is considered as the main cause of climate change, and it is becoming a worldwide concern. Using microalgal photosynthesis for CO2 sequestration has recently gained renewed interest because of its high fixation efficiency and the potential of biofuel production as by-product. In our research group, we have designed three different types of membrane-photobioreactor (MPBR) for microalgae cultivation. One is a 10L airlift photobioreactor integrated with a hollow fiber membrane module outside. The effects of the inlet CO2 concentration and the introduction of the membrane module on microalgae CO2 fixation were investigated. Compared with an ordinary photobioreactor, the retention time of the smaller and more uniform gas bubbles in the new membrane-photobioreactor increased from 2 s to more than 20 s, resulting in the enhancement of the CO2 fixation rate from 80 to 260 mg•l-1•h-1. Another is a 5.6 L airlift photobioreactor integrated with hollow fiber membrane and optical fibers inside the reactor. The effects of gas flow rate, light intensity, quality of inner light source, and characteristics of membrane module were studied. The results showed that the rate of CO2 fixation in MPBR was 0.95-5.40 times higher than other conventional reactors. In order to establish if the limitation of CO2 fixation using microalgae was improved in MPBR, we also constructed a membrane-sparged helical tubular photobioreactor, and it indicated that the overall mass transfer coefficient could be obviously increased by introducing a membrane sparger.

See more of this Session: Alternative Fuels and Enabling Technologies
See more of this Group/Topical: Fuels and Petrochemicals Division

Investigation of Esterification of Biomass Pyrolysis Acids Over Sulfonic Acid-Functionalized Mesoporous Silicas

Shaojun Miao, Department of Chemical and Biological Engineering, Iowa State University, Ames, IA

Brent H. Shanks, Chemical and Biological Engineering Department, Iowa State University, Ames, IA

Bio-oil produced from fast pyrolysis of biomass contains significant amount of carboxylic acids which are corrosive and also can catalyze reactions such as acetalization and polymerization resulting in the instability of bio-oil during storage. Esterification, which is an acid-catalyzed reaction, provides a potential route to stabilize bio-oil. Ideally, heterogeneous acid catalysts would be utilized for this reaction rather than homogeneous catalysts such as sulfuric acid. A possible heterogeneous acid catalyst systemis organosulfonic acid-functionalized mesoporous silicas. In this presentation, detailed studies are presented on the kinetics and mechanism of acetic acid esterification with methanol and ethanol over organosulfonic acid-functionalized mesoporous silicas, since acetic acid is a prevalent carboxylic acid in bio-oil. The esterification of acetic acid with methanol and ethanol were examined. In the kinetic studies, the reaction orders for acetic acid and alcohol, as well as the activation energy were determined. Varying concentrations of acidic sites were employed to investigate the mechanism for acetic acid esterification with methanol and ethanol, i.e., a single-site or dual-site mechanism. In addition, since water is a significant component in pyrolysis bio-oil, an investigation of the effect of water on the esterification was also performed.

See more of this Session: Catalytic Conversion of Oxygenates
See more of this Group/Topical: Fuels and Petrochemicals Division

Combined One-Step Hydrogenation/Esterification Over Bifunctional Mesoporous Organic-Inorganic Hybrid Silica: Model Reaction for Bio-Oil Upgrading

Yang Tang, Institute of Catalysis, Zhejiang University, Hangzhou, China

Shaojun Miao, Department of Chemical and Biological Engineering, Iowa State University, Ames, IA

Brent H. Shanks, Chemical and Biological Engineering Department, Iowa State University, Ames, IA

Bio-oil is a promising energy alternative for fossil fuels, but its corrosive(high content of organic acid) and reactive nature (many unsaturated components such as aldehydes) necessitates upgrading for its subsequent use. Combined one-step hydrogenation/esterification (OHE) has been demonstrated to be a possible approach for bio-oil upgrading, by stabilizing the two main reactive components, aldehydes and organic acids simultaneously. Supported metallic platinum catalyst and organosulfonic acid functionalized mesoporous silicas are reported to exhibit combined activity for hydrogenation and esterification. In this work, the supported metallic platinum and organosulfonic acid groups were combined to catalyze the OHE reaction of acetic acid and acetaldehyde, which are model compounds for bio-oil.

Different catalyst synthesis procedures were investigated and the textural properties, platinum loading, reducing agent were considered as the factors affecting catalytic performance. A strong thiol-platinum bonding effect was determined to inhibit both the oxidation of thiol groups and the reduction of platinum ions, leading to a reduction in the number of sulfonic acid sites and metallic platinum sites on the catalysts. The characterization and catalytic test results showed that the propyl-thiol groups should be tethered and oxidized to propylsulfonic acid groups before platinum loading to avoid the apparent strong thiol-bonding effect. An aqueous reduction and deposition of platinum with sodium borohydride and ethylene glycol as the reducing agents was the suitable method for bifunctionalization. It was found that the presence of propylsulfonic acid groups did not affect the hydrogenation ability of the platinum sites, while platinum appeared to promote the acidic strength of the propylsulfonic acid groups thereby improving esterification ability. The catalytic test results showed that the conversion of acetaldehyde and acetic acid through the OHE can reach high levels with negligible amount of byproducts, suggesting that this bifunctional hybrid catalyst has potential for improving OHE activity.

See more of this Session: Catalytic Conversion of Oxygenates
See more of this Group/Topical: Fuels and Petrochemicals Division

De-Oxygenation Catalysis for Gasoline Production On Titania

Prashant Reuben Daggolu, Dave C. Swalm School of Chemical Engineering, Mississippi State University, Mississippi State, MS

Mark G. White, Dave C. Swalm School of Chemical Engineering, Mississippi State University, Mississippi State, MS

Short carbon chain oxygenates are formed as by products when syngas is converted to mixed alcohols on Molybdenum (1) or Rhodium (2) based catalytic systems. These oxygenates are usually acetone, acetic acid, ethyl acetate and acetaldehyde apart from alcohols. While mixed alcohols can be easily converted to gasoline on H+/ZSM-5, considerable coking is formed when other oxygenates are used (particularly acetic acid). Hence a titania based catalytic system is being examined to convert the non-alcohol oxygenates to gasoline range hydrocarbon on a single multifunctional catalyst. The reactions are performed in similar conditions where alcohol to gasoline reaction is most productive.

Aldol condensation of acetone to mesitylene was examined at high pressure conditions. It was shown that a physical mixture of Titania and H+/ZSM-5 can produce good yield of gasoline range hydrocarbon. Acetic acid can be converted to acetone on titania. Hence titania is being studied to directly convert acetic acid to mesitylene on a single bed. Esters can be converted to acids and alcohols. Aldehydes can be converted to hydrocarbon on zeolite. Hence a mixture of all these oxygenates is being reacted on titania, Ti/ZSM-5 system. Initially binary pairs of oxygenates were reacted in a batch reactor. The results confirmed the feasibility of using titania for their conversion to hydrocarbon. Then a sample oxygenate mixture was made based on the product mixture formed when syngas is converted to mixed alcohols on a Rhodium catalyst. The oxygenate mixture was tested on titania and considerable hydrocarbon formation was observed. More work is being performed to optimize conversions and yields.

1. Liu, Shetian, Gujar, Amit C., Thomas, Peter, et al., Synthesis of gasoline-range hydrocarbons over Mo/HZSM-5 catalysts. Applied Catalysis A: General 357 (1), 18 (2009).

2. Hu, Jianli, Wang, Yong, Cao, Chunshe, et al., Conversion of biomass-derived syngas to alcohols and C2 oxygenates using supported Rh catalysts in a microchannel reactor. Catalysis Today 120 (1), 90 (2007).

See more of this Session: Catalytic Conversion of Oxygenates
See more of this Group/Topical: Fuels and Petrochemicals Division

Pd/TOMPP-Catalyzed Telomerization of 1,3-Butadiene with Polyols and Carbohydrates: New Opportunities for Catalytic Biomass Valorization

Pieter C. A. Bruijnincx, Inorganic Chemistry & Catalysis, Debye Institute for Nanomaterials Science, Utrecht University, Utrecht, Netherlands

Andrei N. Parvulescu, Inorganic Chemistry & Catalysis, Debye Institute for Nanomaterials Science, Utrecht University, Utrecht, Netherlands

Peter J. C. Hausoul, Inorganic Chemistry & Catalysis, Debye Institute for Nanomaterials Science, Utrecht University, Utrecht, Netherlands

Regina Palkovits, Inorganic Chemistry & Catalysis, Debye Institute for Nanomaterials Science, Utrecht University, Utrecht, Netherlands

Robertus J. M. Klein Gebbink, Chemical Biology & Organic Chemistry, Debye Institute for Nanomaterials Science, Utrecht University, Utrecht, Netherlands

Bert M. Weckhuysen, Inorganic Chemistry & Catalysis, Debye Institute for Nanomaterials Science, Utrecht University, Utrecht, Netherlands

The sustainable production of valuable bulk chemicals from biomass as an alternative for petrochemicals is an important and pressing scientific challenge. We therefore aim to develop clean and efficient catalytic processes for the conversion of renewable feedstocks. A particularly interesting technology is the conversion of various biomass-derived polyhydric compounds by palladium-catalyzed telomerization of 1,3-butadiene [1]. This 100% atom-efficient reaction converts alcohols in one step to the corresponding C8-ethers which, after hydrogenation, have many potential applications as surfactants and cosmetics, but also as chemical building blocks for the pharmaceutical industry.

Since its initial discovery the Pd-catalyzed telomerization reaction has matured and is currently commercially applied for the production of 1-octanol. Phosphine-based ligand systems are most commonly used, e.g. the water-soluble tris(m-sulfonato-phenyl)phosphine ligand (TPPTS), but more recently also imidazolium-based carbenes have been reported [1]. In this context, we wish to report here the results of broad substrate screening obtained using our newly developed homogeneous catalyst system: Pd/TOMPP (tris(o-methoxyphenyl)phosphine) [2-5]. We initially communicated that Pd/TOMPP is a highly active and selective catalyst for the conversion of glycerol, an important biomass-derived platform molecule [2,3]. It is important to note here that these telomerization reactions were achieved without the use of solvent and without the need for an added base. Encouraged by these results, we have now greatly expanded the scope of the Pd/TOMPP catalyst system to include many of the major renewable oxygenates, such as various polyols, sugar alcohols and carbohydrates [4,5]. Catalytic results show that the Pd/TOMPP combination rivals the activity and selectivity of any of the known catalysts over the whole range of substrates.

The telomerization activity of the Pd/TOMPP catalyst was screened using around 20 different substrates which include various (sugar) alcohols, polyols and the representative carbohydrate classes. In case of liquid substrates, reactions were carried out both solvent- and base-free. For solid substrates the use of a solvent was necessary to allow for their participation in the telomerization reaction. After optimization of reaction conditions, high conversions were obtained in very short reaction times; depending on the substrate 12-70 min were needed for near full conversion of the alcohol substrate. In addition, the reactions could be performed with low butadiene-to-substrate ratios. High selectivities towards the desired monotelomer (well over 80% for most substrates) were obtained for the polyol substrates and relatively low degrees of substitution were observed for the carbohydrates and sugar alcohols. For example, selectivities for the mono-telomer for 1,2-butanediol and 1,2-propanediol were close to 90 % with TONs of 7200 and 7800, resp. Also, exceptionally high TOFs of up to 350000 h-1 were obtained, illustrating the high activity of Pd/TOMPP system. Catalyst activity was found to depend on the type and properties of the alcohol (e.g. coordination properties, acidity, hydrophobicity) and activity followed the order 1,2-butanediol > 1,4-butanediol > 1,2-propanediol > 1,3-propanediol > ethanediol [4]. A structure/activity relationship could also be established for the telomerization activity with the different classes of carbohydrate substrates. In addition, a strong dependence of the reaction rate on the length of the substrate chain was observed for the sugar alcohols [5].

In conclusion, we have successfully applied the Pd/TOMPP system in the telomerization of 1,3-butadiene with various biomass-based oxygenates. Low catalyst loadings were used and high TONs and TOFs were obtained together with high selectivities to the desired products. A relation between the structure and physicochemical properties of the substrate and catalytic activity was identified.

We would like to thank the ASPECT-ACTS Program for financial support.

[1] Behr, A.; Becker, M.; Beckmann, T.; Johnen, L; Leschinski, J.; Reyer, S. Angew. Chem. Int. Ed. 2009, 48, 3598-3614.

[2] Palkovits, R.; Nieddu, I.; Klein Gebbink, R. J. M.; Weckhuysen, B. M. ChemSusChem 2008, 1, 193-196.

[3] Palkovits, R.; Nieddu, I.; Kruithof, C.A.; Klein Gebbink, R. J. M.; Weckhuysen, B. M. Chem. Eur. J. 2008, 14, 8995-9005.

[4] Palkovits, R.; Parvulescu, A. N.; Hausoul, P. J. C.; Kruithof, C.A.; Klein Gebbink, R. J. M.; Weckhuysen, B. M. Green Chem. 2009, in press

[5] Hausoul, P. J. C.; Bruijnincx, P. C. A.; Klein Gebbink, R. J. M.; Weckhuysen, B. M., to be submitted

See more of this Session: Catalytic Conversion of Oxygenates
See more of this Group/Topical: Fuels and Petrochemicals Division

CONVERSION OF PROPYLENE AND PROPANAL ON HZSM-5

Trung Q. Hoang[a], Xinli Zhu[a], Tawan Sooknoi[b], Daniel E. Resasco[a], Lance L. Lobban[a] and Richard G. Mallinson[a]

[a]Center for Biomass Refining, School of Chemical, Biological, and Materials Engineering
The University of Oklahoma, Norman, Oklahoma, 73019 USA
[b]King Mongkut Institute of Technology Ladkrabang, Bangkok, Thailand

Introduction

Substantial production of oxygenates from biomass conversion and biodiesel production has created the need for upgrading these compounds to fuels and chemicals to maximize yield and value. Bio-oil production from fast pyrolysis of biomass has been estimated to have lower operating and capital costs compared to other biofuel processes [1-2]. Bio-oil contains a large amount of oxygenated compounds with functional groups such as aldehyde, ketone, acid, and polyol. They are thermally unstable and need to be upgraded before their properties are suitable for fuels [3-4]. In addition, converting light oxygenates to larger, more stable molecules can further improve liquid yield, in contrast to hydrotreating [3-5], that consumes hydrogen while lowering liquid yields. The complexity of bio-oil has created a great challenge in understanding reaction pathways and tailoring catalysts toward desirable fuel products. Thus, model compounds have been studied in substantial research on different type of catalysts: metal[6-7], metal oxide[8], and zeolites[9].

Differing from noble metal and metal oxide catalysts, acidic zeolites can directly convert oxygenates to primarily isoalkanes and aromatics in the gasoline boiling range using the well-known acid catalyzed dehydration, oligomerization and aromatization. HZSM-5 has been extensively studied and employed in conversion of methanol in industry, known as the MTG process (Methanol-to-Gasoline)[10]. It has been found that substantial amounts of aromatics were observed in the final products. In general, the major aromatization pathway reported is that oxygenates first convert to olefins and these then oligomerize and dehydrocyclize to form aromatics. However, other studies have found that greater aromatization was observed with propanal compared to other C_3 oxygenates (ketone, alcohol, acid, ester)[11-13] as well as oxygenates compared with alkanes [14-15]. Thus, in this study, we investigate the conversion of propanal and study the pathways for aromatics formation in comparison with propylene on an HZSM-5 catalyst.

EXPERIMENTAL

Commercial HZSM-5 (Si/Al 45) zeolite was supplied by SUD-CHEMIE. HZSM-5 (Si/Al 25) was synthesized by the sol-gel hydrothermal method described elsewhere [20]. The steady flow reactor was a 1/4 inch inner diameter, quartz tube fixed bed continuous flow reactor. Propanal, obtained from Sigma-Aldrich, was fed into the reactor by using a syringe pump (KD Scientific) at a constant flow rate of 0.12ml/h. Propylene was obtained from Airgas with ultra high purity.

In the pulse experiments, the same diameter quartz tube was used as in the steady flow configuration. Liquid propanal was introduced into the reactor as 5 μl (liquid) pulses using a GC syringe (Hamilton 10μl). Propylene was introduced into the system using a 6-port valve with 5 ml sample loop. For temperature programmed reaction-desorption experiments, propanal was adsorbed with 20 pulses at 250 ^{o}C. Then the reactor was heated up to 600 ^{o}C with ramp rate 10 ^{o}C/min.

For each half hour time on stream, products were analyzed online by GC/FID (HP6890). The collected liquid products in solvent (Methanol) were then identified by GC/MS (Shimadzu Q-2010). Both GCs are equipped with an HP-INNOWAX column. In the pulse experiments, the products were analyzed by mass spectrometer (MS) detector (MKS Cirrus 200).

RESULTS

Characterization

The synthesized and commercialHZSM-5 were characterized by X-ray using a powder diffractometer (Bruker D8 Discover). The XRD pattern is similar to those in the literature [10]. Scanning electron microscopy was used to observe the morphology of the synthesized zeolite. Though all the synthesized samples have a similar crystallite size (1-2μm), the commercial HZSM-5 has much smaller crystallites (100-150nm). The acidity of the HZSM-5 was determined by temperature programmed desorption (TPD) of iso-propyl amine (IPA) using an online MS detector (MKS Cirrus 200)[16]. The TPD was performed in a quarter inch quartz tube reactor with 50 mg of catalyst. The catalyst sample was first pretreated in He at 400 °C for 1 hr and then cooled down to 100 °C. The SEM image of HZSM-5 (25) and acidity of both catalysts are shown in Figure 1.

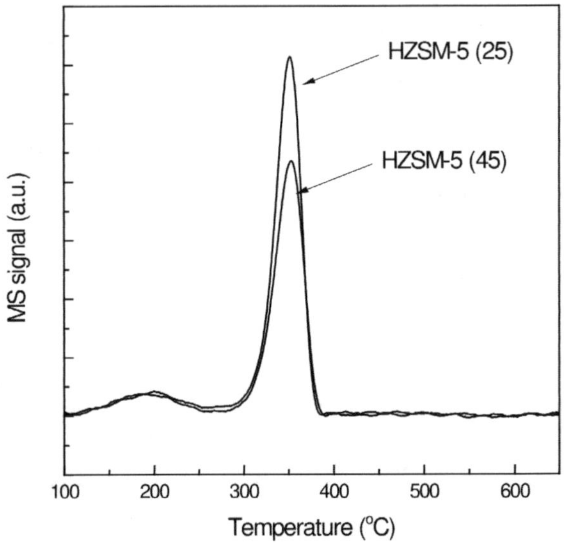

Figure 1. SEM image of HZSM-5 (25) and IPA-TPD acidity profile for both catalysts

Catalyst Activity in a Steady Flow reactor

Propanal was converted to aromatics (C_6-C_9), gas (C_1-C_3), and isoalkenes (C_4-C_9) on HZSM-5 (45) at 400 °C. The product yields versus space time from 0.07 h to 0.25 h with the data taken after one hour time on stream are shown in Figure 2. The conversion of propanal increased from 59.5% to 99.8% with increasing W/F (g_{CAT}/g_{Feed}.h). Aromatics and gas (C_1-C_3) yields increased more rapidly compared to the isoalkenes (C_4-C_9) yield as W/F increased. In fact, aromatics were always observed as major products even at the shortest W/F. In this range of W/F, the highest yield of aromatics was 51.9% obtained at 0.25 h W/F. The second major products were gases (C_1-C_3) in which the majority was propylene. The isoalkenes (C_4-C_9) were always relatively minor products.

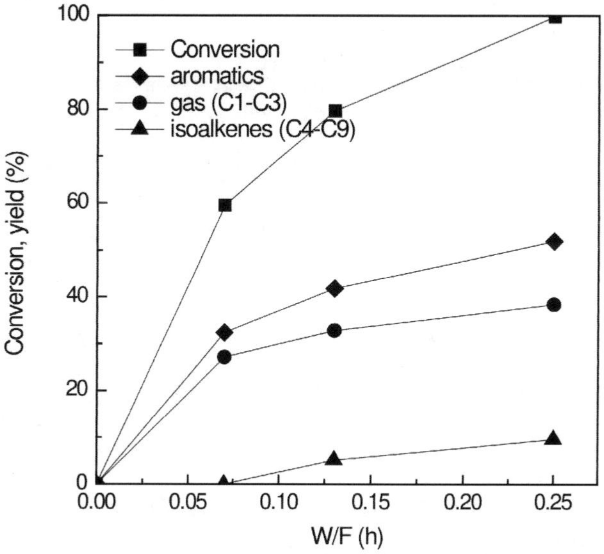

Figure 2. W/F series of propanal conversion on HZSM-5

Conversion of propylene was performed at the same reaction conditions as the propanal experiments. However, conversion was much lower and negligible amounts of aromatics were found (<1%). In order to observe significant yields of aromatics, more severe operating conditions were achieved by using a synthesized HZSM-5 with a higher acid density (Si/Al=25) and a higher temperature of 500 °C. The W/F results are shown in Figure 3. In contrast to the propanal experiments, at shorter W/F (but still much longer than with propanal on the other catalyst at lower temperature), the major products are alkenes and light gases C_1-C_2. At longer W/F, yields of both aromatics and light gas continued to increase while the yield of alkenes reached a maximum and then declined. This suggests that the alkenes were consumed to produce more aromatics and cracking products, consistent with the pathway in the literature.

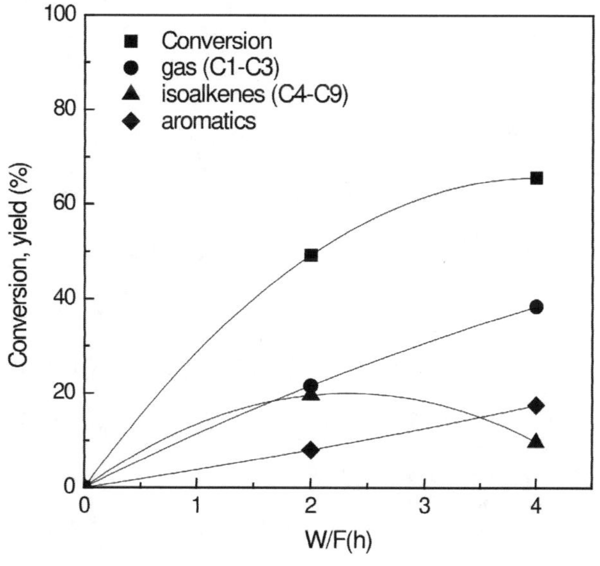

Figure 3. W/F series of propylene conversion on HZSM-5

Catalyst Activity in non-Steady State Experiments

The pulse reactor was loaded with 10 mg of HZSM-5 (45) and pretreated at 400 °C in He at the flow rate 30 sccm. Propanal was introduced into the reactor, at 400 °C (or other temperature), with consecutive 5μl pulses until the catalyst was deactivated. The products obtained were consistent with the flow reactor results mentioned above. As shown in Figure 4, the largest signals were primarily C_9 (TMB&MEB) and C_8 (X), less of C_7 (T), and negligible C_6 (B). As the number of propanal pulses increased, catalyst deactivation was observed by the decrease in the size of product peaks and increasing propanal peak. The water signal was reduced but did not completely disappear. In addition, the dimer (D) product was detected only in the last few pulses as deactivation proceeded, although the signals were small. At lower temperature, 350 °C and 300 °C, faster deactivation occurred and signal of C_9 aromatics was more pronounce compared to C_8 and C_7 aromatics.

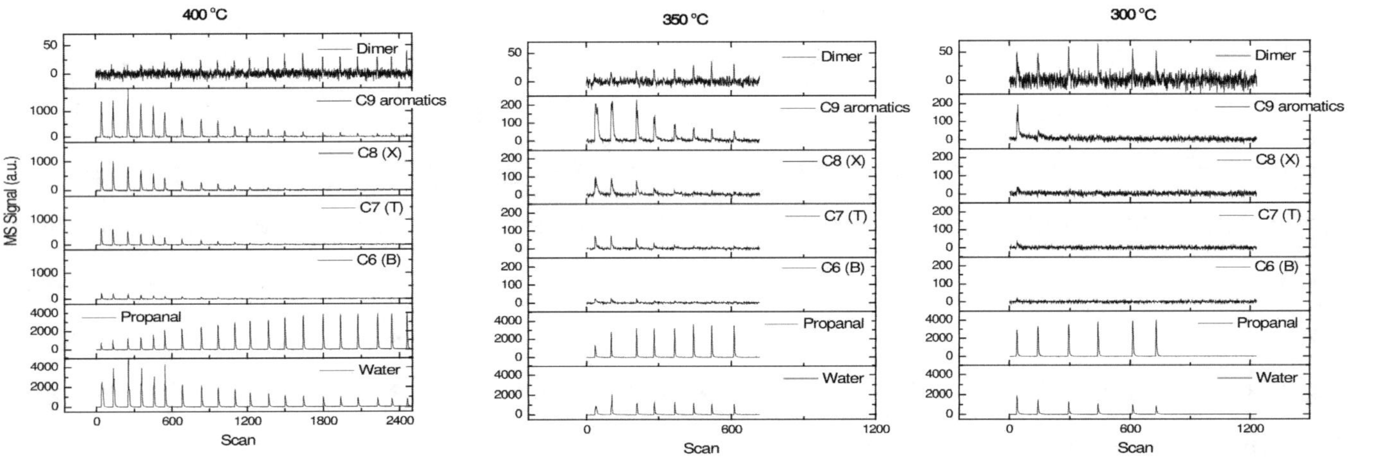

Figure 4. Products from pulse experiment of propanal on commercial HZSM-5 (45)

A temperature programmed reaction-desorption experiment was conducted under the pulse reactor conditions by pulsing propanal at 250 °C, with 20 pulses of 5 μl, until the catalyst was deactivated and then the temperature was increased at a ramp rate of 10 °C per minute. During the isothermal period shown on the left in Figure 5, the dimer and C_9 aromatic were the only products that were observed initially. While the dimer continued to appear, aromatization was rapidly deactivated as shown by the disappearance of the C_9 aromatic peak. This suggests that while the weaker acid sites like the Lewis acids may catalyze the aldol condensation of propanal to form the dimer, 2-methyl-pentenal, as well as dehydration to form propylene, stronger acids site like the Bronsted acid sites are required for cyclization and dehydration to produce the C9 aromatics and are the first to deactivate. This suggests that aldol dimerization and then aromatization of an aldol trimer may be major initial pathway for aromatics formation, as has been shown on other acidic systems [17-18].

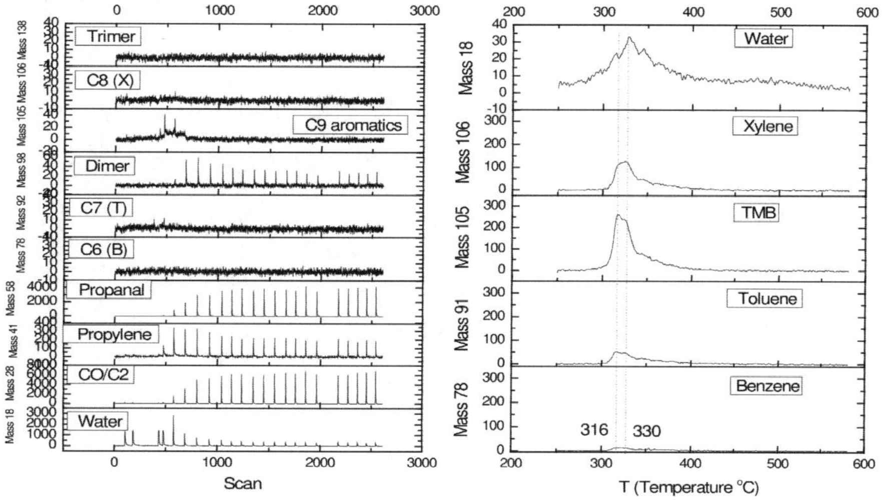

Figure 5. TPD of deactivated catalyst at 250 °C of propanal conversion on commercial HZSM-5 (45)

After the isothermal reaction phase, Helium was allowed to flow for 10 minutes before the temperature ramp was begun. As shown on the right in Figure 5, water along with significant amounts of C_9 aromatic appear concurrently from 316-330 °C. Smaller peaks of C_8 (X) and C_7 (T) were also observed. This suggests that some adsorbed intermediate, that caused deactivation of aromatic production, converts into the C_9 aromatic at higher temperature. The fact that water was produced along with formation of the C_9 aromatic indicates the dehydration of an oxygenated intermediate. This is consistent with an aldol pathway forming a cyclic trimer oxygenate, as a precursor to dehydration and aromatization.

In the case of propylene pulse experiments, propylene was converted to C_6 (B) and C_7 (T) aromatics as the major products and only a small amount of C_8 (X) and C_9 aromatics were observed. This result is consistent with the flow reactor results as mentioned above. This also further confirms that the initial aromatic formed from propylene conversion is C_6 (B), but that propylene does not convert to aromatics at 400 °C on HZSM-5 with a Si/Al of 45.

Conclusion

Conversion of propanal has been investigated on HZSM-5 and compared to the behavior of propylene at 400 °C. Aromatics are the major products from propanal in which C_9 aromatics has been identified as the primary aromatic. Under these conditions, propylene does not produce significant aromatics. It is proposed that C_9 aromatics are produced by direct cyclization and aromatization of an aldol condensation trimer catalyzed by the acid sites. C_9 aromatics serve as the source of other products produced by subsequent dealkylation, disproportionation and cracking. This major additional pathway for aromatics formation is different from the conventional acid catalyzed carbenium ion pathway for alkane conversion on zeolite, occurring at milder conditions. The overall pathway of propanal conversion shows in following figure.

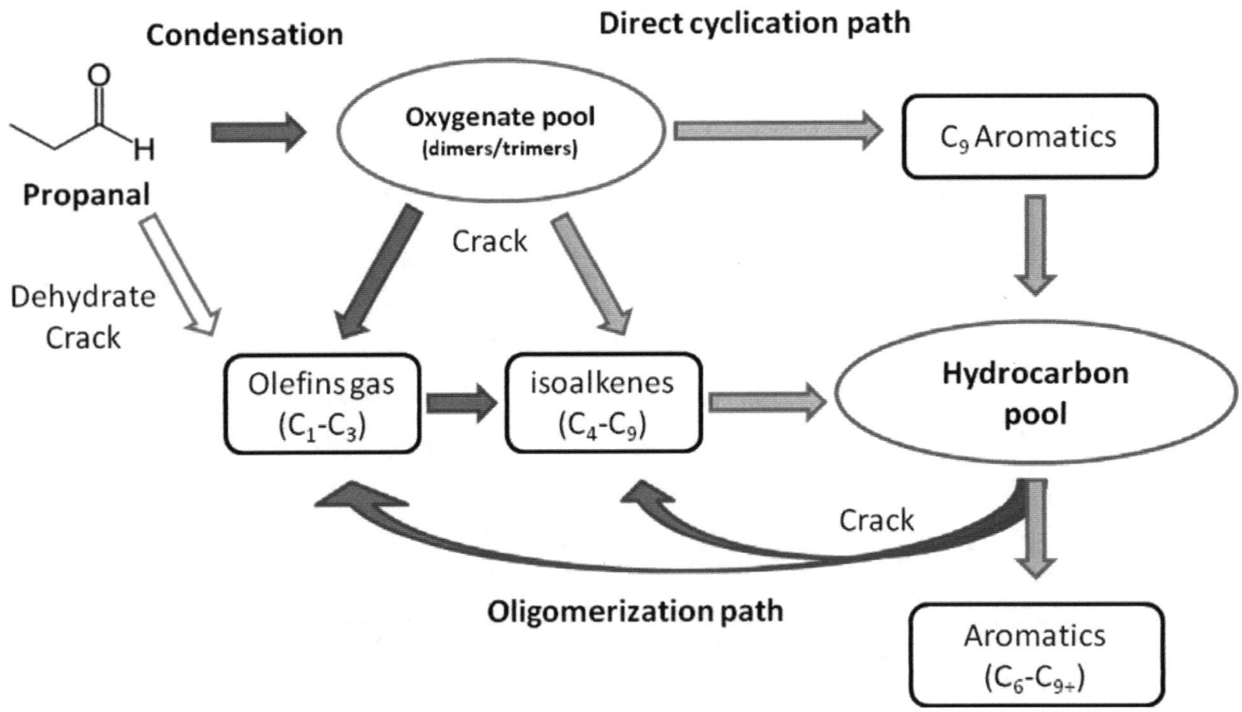

Figure 6. The proposed pathways of propanal conversion on HZSM-5

Acknowledgements

The supports from the Oklahoma Secretary of Energy and the Oklahoma Bioenergy Center are greatly appreciated.

References

(1) Simonetti, D. A.; Dumesic, J. A. *Chemsuschem* **2008**, *1*, 725.
(2) Chheda, J. N.; Huber, G. W.; Dumesic, J. A. *Angew Chem Int Edit* **2007**, *46*, 7164.
(3) Mohan, D.; Pittman, C. U.; Steele, P. H. *Energy & Fuels* **2006**, *20*, 848.
(4) Czernik, S.; Bridgwater, A. V. *Energy & Fuels* **2004**, *18*, 590.
(5) Maggi, R.; Delmon, B. *Hydrotreatment and Hydrocracking of Oil Fractions* **1997**, *106*, 99.
(6) Huber, G. W.; Dumesic, J. A. *Catal Today* **2006**, *111*, 119.
(7) Kunkes, E. L.; Gürbüz, E. I.; Dumesic, J. A. *J Catal* **2009**, *266*, 236.
(8) Dooley, K. M.; Bhat, A. K.; Plaisance, C. P.; Roy, A. D. *Appl Catal a-Gen* **2007**, *320*, 122.
(9) Adjaye, J. D.; Bakhshi, N. N. *Biomass Bioenerg* **1995**, *8*, 131.
(10) Meisel, S. L. In *Studies in Surface Science and Catalysis*; D.M. Bibby, C. D. C. R. F. H., Yurchak, S., Eds.; Elsevier: 1988; Vol. Volume 36, p 17.
(11) Bandermann, F.; Fuhse, J. *Chemie Ingenieur Technik* **1987**, *59*, 607.
(12) Chang, C. D.; Silvestri, A. J. *J Catal* **1977**, *47*, 249.
(13) Hutchings, G. J.; Johnston, P.; Lee, D. F.; Warwick, A.; Williams, C. D.; Wilkinson, M. *J Catal* **1994**, *147*, 177.
(14) Danuthai, T.; Jongpatiwut, S.; Rirksomboon, T.; Osuwan, S.; Resasco, D. E. *Appl Catal a-Gen* **2009**, *361*, 99.
(15) Sooknoi, T.; Danuthai, T.; Lobban, L. L.; Mallinson, R. G.; Resasco, D. E. *J Catal* **2008**, *258*, 199.
(16) Farneth, W. E.; Gorte, R. J. *Chem Rev* **1995**, *95*, 615.
(17) Dejaifve, P.; Vedrine, J. C.; Bolis, V.; Derouane, E. G. *J Catal* **1980**, *63*, 331.
(18) Salvapati, G. S.; Ramanamurty, K. V.; Janardanarao, M. *J Mol Catal* **1989**, *54*, 9.

Deoxygenation of Furfural On Pd and Pd-Cu Catalysts

Surapas Sitthisa, School of Chemical, Biological and Materials Engineering, University of Oklahoma, Norman, OK

Tawan Sooknoi, School of Chemical, Biological and Materials Engineering, University of Oklahoma, Norman, OK

Lance L. Lobban, School of Chemical, Biological and Materials Engineering, University of Oklahoma, Norman, OK

Richard G. Mallinson, School of Chemical, Biological and Materials Engineering, University of Oklahoma, Norman, OK

Daniel E. Resasco, School of Chemical, Biological and Materials Engineering, University of Oklahoma, Norman, OK

Aldehydes are commonly found in bio-oils, but they exhibit low stability which is undesirable for fuel uses. Deoxygenation is needed to make bio-oil more fungible with standard gasolines and diesels. It can be achieved by hydrotreating over standard hydrogenation catalysts at high temperatures and pressures. However, using noble metal catalysts, the reaction conditions can be much less severe. In the present work, supported Pd-Cu bimetallic catalysts have been used for converting fufural. TPR suggests that the addition of Cu to supported Pd catalyst results in formation of Pd-Cu bimetallic alloys. The activity and selectivity of the catalysts are largely influenced by the addition of Cu. The yield of furan from decarboylation was significantly reduced when Cu was incoporated. However, no loss of hydrogenation activity to furfuryl alcohol was observed. This is because the addition of Cu readily reduces the exposed Pd ensembles. Hence, the decabonylation activity that requires a large ensemble of Pd is much more affected. In contrast, hydrogenation remains unchanged as the dilution of ensembles has little effect on structure insensitive reactions. The deoxygenation activity also depends on the molecular structure of the feed. When 5-methyl furfural was used, significantly higher decarbonylation rates to methyl furan were obtained. No hydrogenated products were found, but only small amounts of dimethyl furan, obtained by hydrogenolysis.

See more of this Session: Catalytic Conversion of Oxygenates
See more of this Group/Topical: Fuels and Petrochemicals Division

CONVERSION OF OXYGENATED AROMATICS

Teerawit Prasomsri, Roberto E. Galiasso Tailleur, Walter E. Alvarez [a], Tawan Sooknoi, and Daniel E. Resasco

Center for Biomass Refining and, School of Chemical, Biological and Materials Engineering, University of Oklahoma, Norman, OK, United States
[a] Conoco Phillips Research Center, Bartlesville, OK 74004, United States

Abstract

This work focused on the conversion of oxygenated aromatics over HY zeolites (Si/Al = 15 and 40). 1- and 2-tetralone are used as representative compounds. The reactions were conducted in a fixed-bed reactor at 300 - 500 °C under atmospheric pressure. The feeds were introduced to the reactor by using a bubbler in order to minimize condensation or solvent problems. It was found that tetralones are thermally unstable and 2-tetralone has a higher reactivity than 1-tetralone. Dehydrogenation (i.e. naphthol product) and coupling (i.e. heavies) were the main reaction paths. Deoxygenation (i.e. to tetralin and naphthalene) products was a slower path. In the presence of co-fed compounds (e.g. decalin, tetralin and dimethyltetralin), naphthalene production is promoted, suggesting the hydride transfer facilitates deoxygenation of naphthol. In addition, isomerization of naphthol can be expected under the acid-catalyzed reaction.

Introduction

The projected demand for transportation fuels increases dramatically in the next decade [Couch at al., 2009]. Security and sustainability concerns enhance the need for access to alternative domestic sources known as "unconventional crude oils" (heavy oil, bitumen, tar sands, shale oils, and coals) that are abundant in North America and may reduce the consumption of imported feedstocks. The transition is undoubtedly underway, but it will not be easy or inexpensive. The use of heavy oil as a crude produces high aromaticity in the internal streams in the refinery as well as a net increase in the volume of the heavier fractions. One possible solution would be to recycle some of these heavier fractions through the catalytic cracking unit (FCC). However, it is well established that benzene, naphthalene, phenanthrene, and anthracene do not crack under typical FCC conditions, but instead they undergo condensation and form coke. Therefore, reducing the molecular weight of heavy molecules is just one of the problems. In contrast to aromatics, naphthenic rings are much more easily cracked as their aromaticity stability is broken. The functionalization of polycyclic (also termed polynuclear aromatics) with heteroatoms, such as oxygen may greatly affect the reactivity of the aromatic rings, thus opening opportunities to unzip the stable structure of the heavy molecules under milder conditions than those used in hydrotreating. Hence, this work focused on reaction of tetralones, as the representative molecules of oxygenated aromatics and co-feeding of representative hydrocarbon families such as decane, decalin, tetralin and methyltetralin.

Experimental

Feed rate calibration

The molecules used in this study are typically heavy and varied in boiling points. To avoid solvent usage and minimize condensation in the system, the feeds were introduced to the reactor by using a glass bubbler saturator. Vapor concentration of individual compounds was modified by adjusting the heating bath temperature. For co-feed reaction experiment, the co-fed compound was controlled separately by another bubbler. Additional balanced helium line was added to keep the total gas flow rate constant.

Catalytic activity testing

The reactions were conducted in a ¼" stainless steel reactor. Commercial HY zeolite from Zeolyst International (Si/Al = 15 and 40) was used as the catalyst and packed in the center of the reactor. The top and bottom parts of the bed were filled with 3 mm-diameter glass beads. Operating conditions were atmospheric pressure in hydrogen, temperature range 300-500 °C, with various space velocities by changing the amount of catalyst. Prior to each run, the feed was directed through the bypass until the concentration reached steady-state, and then switched to the reactor. Products were analyzed periodically on-line with an HP5890 gas chromatograph equipped with HP-5 column and an FID detector. The temperature program started at 40 °C, linearly increased to 280 °C with a 15 °C/min heating ramp, and finally held for 8 min.

Results

Thermal reactions of 1- and 2-tetralone

The conversion and product yields obtained from thermal reactions of 1- and 2-tetralon (tested in a blank reactor, filled with inert glass beads) are compared in Table 1. It is clear that the both compounds can be thermally decomposed, and the conversion readily increases with temperature. Conversion of 2-tetralone is significantly higher than that of 1-tetralone in all temperature range. This is because 2-tetralone possesses a benzylic carbon, which required relatively less C-H dissociation energy [Bruckner, 2002], and hence can be rapidly activated thermally [Kjærsbo *et al.*, 2003].

Table 1 Thermal reactions of 1- and 2-tetralone

Feed	1-Tetralone			2-Tetralone		
Temperature	300 °C	400 °C	500 °C	300 °C	400 °C	500 °C
Conversion	5.6%	11.0%	42.4%	15.0%	22.6%	56.0%
Product Yields (wt%)						
RO	0.0%	0.3%	1.7%	0.0%	0.0%	0.0%
RC	0.0%	0.0%	0.3%	0.0%	0.0%	0.4%
Tetralin	0.1%	0.1%	0.2%	0.0%	0.0%	0.0%
Dihydronaphthalene	0.6%	0.8%	1.8%	0.0%	0.0%	0.0%
Naphthalene	0.5%	0.8%	5.6%	0.0%	0.2%	0.2%
Tetralone	94.4%	89.0%	57.6%	85.0%	77.4%	44.0%
Tetralol	0.0%	0.0%	0.0%	0.0%	0.0%	0.0%
Naphthol	4.4%	8.3%	31.8%	5.8%	7.1%	26.0%
Heavies	0.0%	0.7%	1.0%	9.2%	15.4%	29.3%

Reaction conditions: 1 atm H_2, TOS 0.5 h

For 1-tetralone conversion, naphthol is the main product. At 500 °C, the yield of naphthalene was observed. This suggests that a hydrogen transfer is promoted at high temperature [Behar and Pelet, 1988]. Unlike 1-tetralone, the heavy products are the major product from the conversion of 2-tetralone. The selectivity toward 2-naphthol is in a lower extent, and no naphthalene is present in all range of temperature studied. Traced amounts of dihydronaphthalene, tetralin, ring-contraction (RO) and ring-opening (RO) products can be observed from 1-tetralone thermal reaction, while none of them appeared from 2-tetralone.

Effect of temperature on the 1- and 2-tetralone reactions over HY

The presence of HY catalyst dramatically improves the conversion of the both 1- and 2-tetralone (Fig. 1). Similarly to the thermal conversion, 2-tetralone manifests a higher reactivity than that of 1-tetralone. It is clear that adsorption of these aromatic ketones would be feasibly promoted over the acid sites [Bezouhanova, 2002; and Lei and Turro, 1992]. 2-tetralone is not only more reactive than 1-tetralone thermally, but also more sensitive to acid-catalysis.

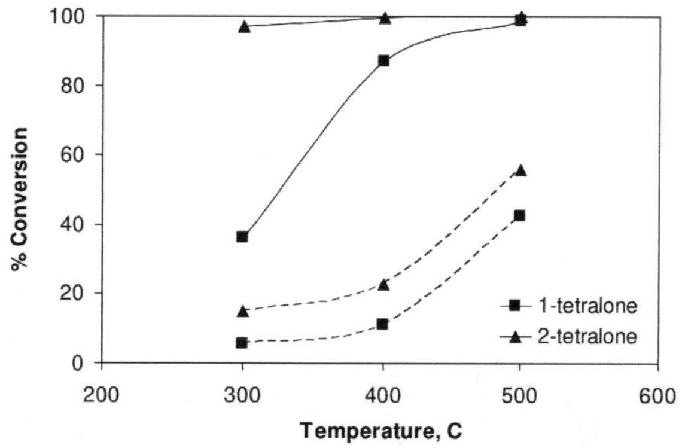

Fig. 1 Conversions of 1- and 2-tetralone as a function of temperature. Reaction conditions: 1 atm H_2, TOS 0.5 h, W/F = 1.723 h. Solid lines and broken lines represent catalytic and thermal conversion, respectively.

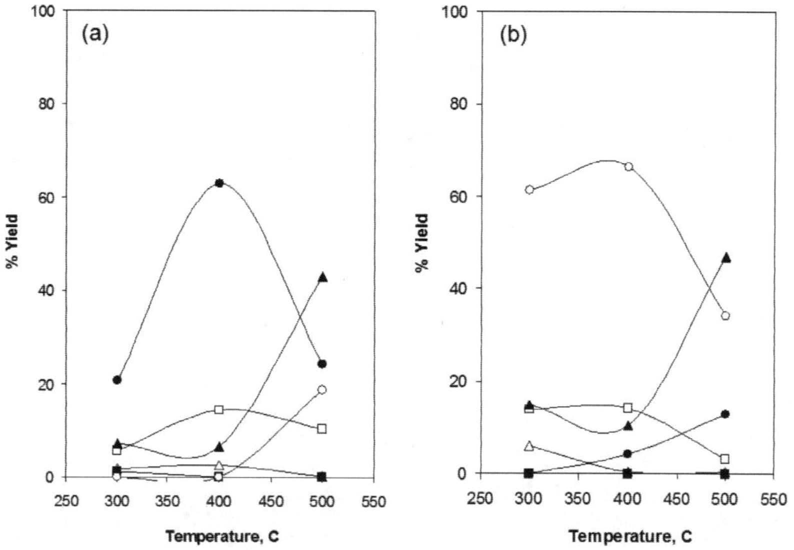

Fig. 2 Product yields from the reaction of (a) 1-tetralone and (b) 2-tetralone as a function of temperature. Reaction conditions: 1 atm H_2, TOS = 0.5 h, W/F = 1.723 h. Tetralin (△), dihydronaphthalene (■), naphthalene (□), 1-naphthol (●), 2-naphthol (○) and heavies (▲)

Fig. 2 exhibits the product distribution from 1- and 2-tetralone reactions as a function of temperature over the HY zeolite. The acid-catalyzed reaction promotes the formation of their corresponding naphthols, suggesting an enhanced hydrogen transfer over acid catalyst. Interestingly, 2-naphthol was observed from 1-tetralone and vice versa for 2-tetralone. This is particularly observed at 500°C where the yield of the corresponding naphthol was also decreased. This suggests that the isomerization of the corresponding naphthol could well take place at high temperature over the acid HY zeolite. Heavy products, ascribed as 4-ring compounds, from both 1- and 2-tetralone were largely produced at high temperature, suggesting extensive hydrogen transfer / coupling over acid catalyst [Kjærsbo *et al.,* 2003]. Naphthalene and tetralin yields are noticeably enhanced over HY, as compared to the thermal reaction. This again suggests that hydrogen transfer is readily facilitated over acid site [Puente and Sedran, 2000; Pajak *et al.*, 1996; Suarez *et al.*, 1990] and deoxygenated products can be obtained. Traced amounts of ring-contraction and ring-opening products were found [Santikunaporn *et al.*, 2007] in this acid catalyzed reaction.

Co-feed reactions of 1-tetralone over HY

Various co-fed compounds (e.g. decane, decalin, tetralin, and methyltetralin) were also studied in this project. Form the reaction of co-fed compounds alone, 13% of naphthalene can be contributed from tetralin. Table 2 shows the product distribution of co-feed reaction. Crack yield fraction contains small alkenes, cyclopentene/hexane, and mono/dimethylbenzene. In the presence of co-fed compounds, naphthalene yield with respect to 1-tetralone conversion increased (i.e. compared to 4% from pure feed 1-tetralone reaction). Naphthalene production is in the order of co-fed m-tetralin > tetralin > decalin > decane.

Table 2 Catalytic co-feed reaction

Feed	1-tetralone			
Co-fed	Decane	Decalin	Tetralin	m-Tetralin
Product Yields				
Cracks	2.3%	4.6%	4.2%	5.4%
1-Ring aroms.	0.0%	0.0%	0.0%	5.1%
n-Decane	49.4%	0.0%	0.0%	0.0%
trans-Decalin	0.0%	38.3%	0.0%	0.0%
cis-Decalin	0.0%	9.3%	0.0%	0.0%
Tetralin	0.0%	2.9%	13.4%	0.0%
m-Tetralin	0.0%	0.0%	0.0%	0.9%
Naphthalene	5.1%	25.5%	44.0%	34.4%
2-Ring aroms.	0.0%	0.0%	0.0%	44.3%
1-Tetralone	0.9%	2.2%	2.7%	2.3%
1-Naphthol	30.9%	11.7%	16.4%	5.6%
2-Naphthol	8.9%	1.6%	5.4%	1.5%
Heavies	2.4%	3.8%	3.9%	0.9%
1-tetralone conv.	98.1%	95.0%	96.0%	94.7%
Product Yields				
Naphthalene	10.6%	56.8%	57.9%	77.6%
1-Tetralone	1.9%	5.0%	4.0%	5.3%
1-Naphthol	64.1%	26.1%	24.3%	12.7%
2-Naphthol	18.5%	3.7%	8.0%	1.5%
Heavies	0.3%	8.5%	5.8%	2.8%

Reaction conditions: 450 °C, 1 atm He, TOS 0.5 h, Co-fed concentration ~50%

Next studies were to investigate the effect of acidity (see Fig. 3) and co-fed (see Fig. 4) concentration. It is noted that naphthalene formation and isomerization of naphthol (i.e. 2-/1-naphthol ratio) are significantly depend on acidity of HY zeolites, suggesting that there are more available protonated naphthols on the surface which can undergo deoxygenation (i.e. to form naphthalene) and isomerization. In addition, co-fed concentration of decalin, as a hydride donor, remarkably affect to naphthalene formation. This behavior is more pronounced for the higher acidity HY zeolite (i.e. naphthalene yield increased more than that over lower acidity HY zeolite). Interestingly, isomerization of naphthol decreased as co-fed concentration increased, suggesting that the hydride transfer promotes naphthol deoxygenation rather than isomerization.

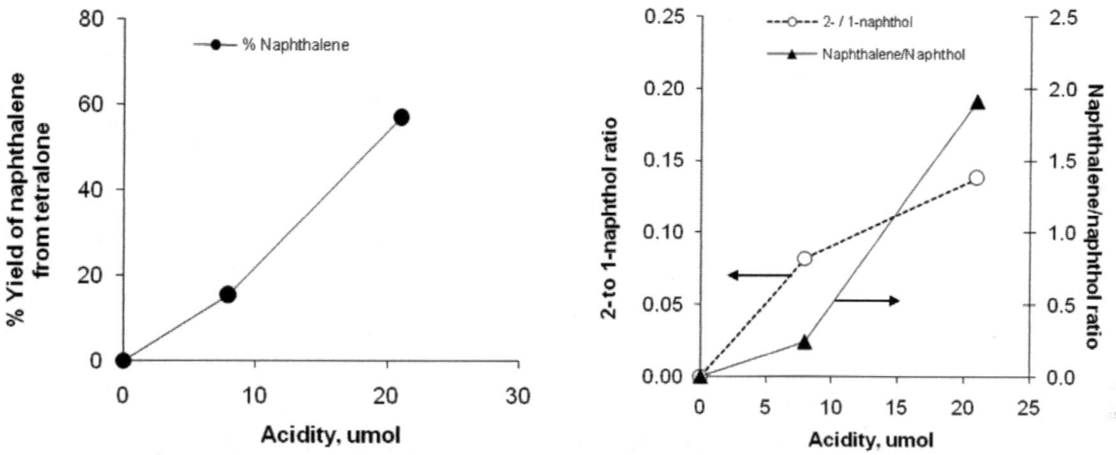

Fig. 3 Yield of naphthalene (left); 2- to 1-naphthol ratio and naphthalene to naphthols ratio (right) as functions of acidity. Reaction conditions: Feed = 1-tetralone/decalin (50/50), 450 °C, 1 atm He, TOS = 0.5 h.

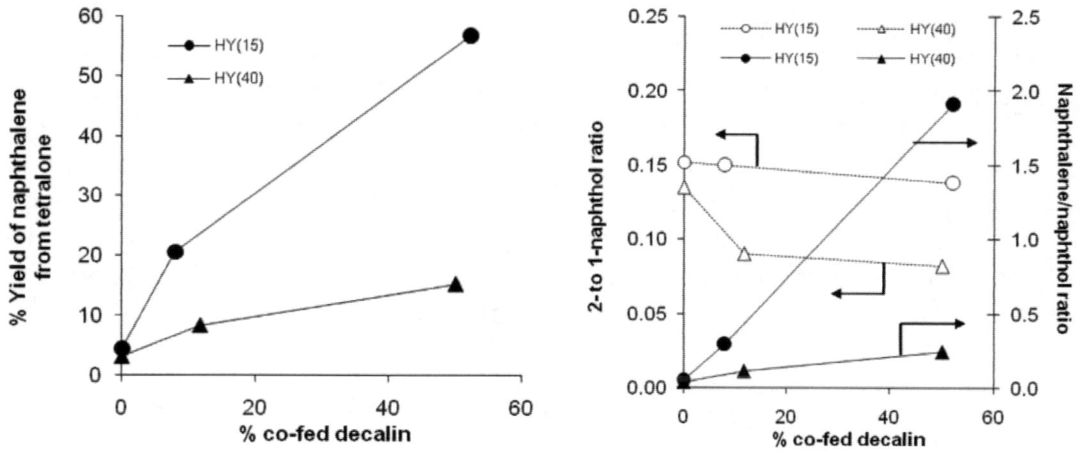

Fig. 4 Yield of naphthalene (left); 2- to 1-naphthol ratio and naphthalene to naphthols ratio (rigth) as functions of %co-fed decalin. Reaction conditions: 450 °C, 1 atm He, TOS = 0.5 h.

Discussion

Comparison of the activity between 1- and 2-tetralone shows that 1-tetralone is relatively more stable due to its conjugated double bonds between carbonyl group and the aromatic ring, which stabilize the molecule structure. In addition, the presence of benzylic-carbon (a carbon adjacent to C=O and aromatic ring) in 2-tetralone leads to a better stabilized free radical intermediates [Bruckner, 2002], and hence higher activity in thermal reaction. 1-tetralone is expected to form the less stable intermediate which has cross-conjugated double bonds. The possibility to form the free radical intermediate from 2-tetralone is dramatically higher than that from 1-tetralone, and this explains the formation of heavies that mainly observed in the thermal conversion of 2-tetralone. The heavies are ascribed as the derivatives from the couplings of 2-ring compounds via benzylic radical intermediate [Kjærsbo *et al.*, 2003]. Unlike 2-tetralone, as 1-tetralone contains no benzylic-carbon, coupling product is not readily observed.

In parallel, dehydrogenation (i.e. C-H thermal cracking) [Xn *et al.*, 2006] can be facilitated at high temperature, particularly for a precursor of an aromatic. This would lead to the formation of naphthols from both 1- and 2-tetralone as shown;

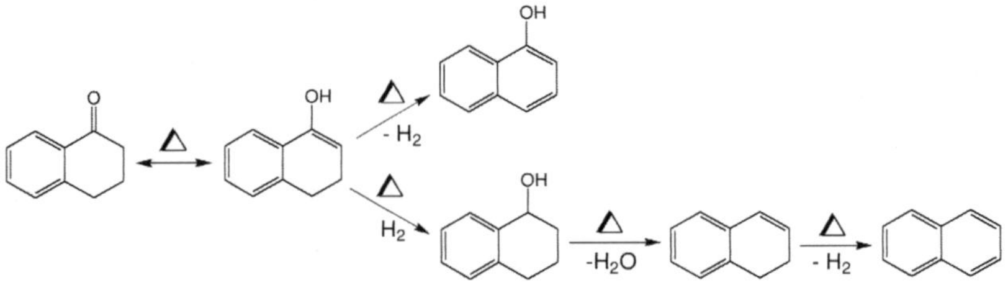

In a similar manner, reactions of 1- and 2-tetralone over HY zeolite are subjected to the combination of thermal and acid catalyzed reactions. The presence of acid HY zeolite generally increases the rate of the hydrogen transfer [Puente and Sedran, 2000; Pajak *et al.*, 1996; Suarez *et al.*, 1990]. Hence, the conversion is increased, as well as the formation of naphthol as major product. This also enhances the parallel reaction (i.e. hydrogen transfer to form tetralol) that produces dihydro-naphthalene and naphthalene, as discussed previously. Moreover, dihydronaphthalene can readily disproportionate (i.e. intermolecular hydride transfer) to produce naphthalene and tetralin [Corma *et al.*, 200; Santikunaporn *et al.*, 2004].

Thermal transformation of 1-naphthol to naphthalene can be also expected under high hydrogen pressure [Schlosberg and Kurs, 1984] or at high temperature [Cypres, 1981]. However, the condition studied in this work does not show any evidence of thermal conversion of 1-naphthol to naphthalene. Interestingly, the presence of acid HY zeolite at ~500 °C promotes 1-naphthol isomerization, as well as deoxygenation to naphthalene. It is plausible that ketoform of 1-naphthol can be hydrogenated, most likely by hydrogen transfer from carbon deposits or heavies, to form 1-hydroxy-1,2-dihydronaphthalene intermediate [Brandt *et al.*, 2002; Pirinccioglu and Thibblin, 1998]. This species would be readily dehydrated over acid sites to form naphthalene.

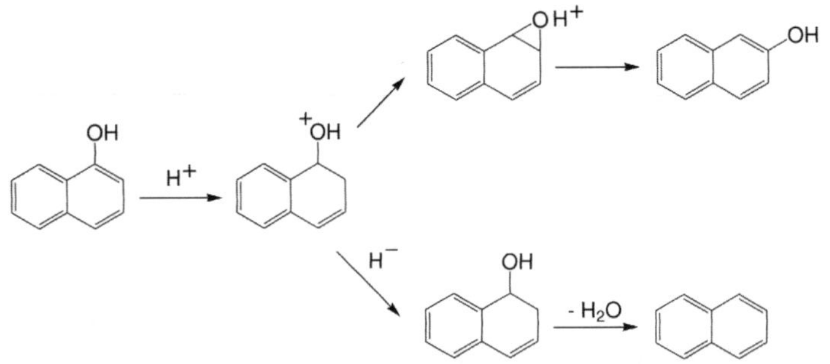

Alternatively, protonation of 1-naphthol can be expected in the presence strong acid and naphthalene-1,2-oxide intermediate [Brandt *et al.*, 2002; Rao *et al.*, 1993; Kasperek and Bruice, 1972] can be formed. A rapid ring opening of this intermediate species would result in the formation of 2-naphthol. As assisted by hydrogen transfer over the acid site, the isomerization and the deoxygenation of 1-naphthol can be promoted at high W/F and acidity. Accordingly, yield of 1-naphthol is significantly decreased, while yield of naphthalene and 2-naphthol is largely pronounced. In the presence of co-fed molecules, the hydride transfer facilitates deoxygenation of naphthol to produce naphthalene, and somehow suppress isomerization of naphthol resulting from reduction in carbenium ion on the surface. The ring structure of donor shows the better hydride transfer ability, as compare to the linear molecule. The presence of methyl group creating tertiary carbon also improves the hydride transfer ability.

Since the heavy products derived mainly form thermally coupling of oxygenated aromatic intermediates (i.e. free radical and carbocation) [Kjærsbo *et al.,* 2003], the formation of heavies would be suppressed under rapid hydrogen transfer. Accordingly, a similar product distribution was obtained from both 1- and 2-tetralone reactions over acid catalyst despite higher conversion is observed over 2-tetralone.

Conclusion

As compared to 1-tetralone in which the double bond of the carbonyl group being conjugated with the aromatic rings, 2-tetralone, possessing benzylic α-carbon is more active and sensitive to acid HY zeolite. The conversion increases as the reaction temperature increased, and is enhanced in the presence of the acid catalyst HY. This reveal a marked role of the hydrogen transfer on the reaction pathway and product selectivity, which are also affected by temperature, and space time (W/F). Dehydrogenation of tetralone to form its corresponding naphthol is the major pathway. The hydrogen transfer promoted by acid catalyst, leads to the formation of deoxygenated products, namely naphthalene, traced amounts of dihydronaphthalene. Heavy products, formed by coupling of the free radical intermediates, can be readily suppressed over acid-catalyzed reaction at relatively lower temperature. Interestingly in the presence of co-fed compounds, ascribed as hydride donors, the acidity of catalyst HY plays a marked role in naphthol transformation. The hydride transfer facilitates deoxygenation of naphthol to form naphthalene, as well as isomerization of naphthol. In addition, increasing the co-fed concentration promotes naphthalene formation, rather than isomerization of naphthol.

Acknowledgements

The authors are grateful for financial support provided by the State of Oklahoma OCAST, from the Secretary of Energy of the State of Oklahoma.

References

Behar, F., and Pelet, R. (1988) Hydrogen-Transfer Reactions in the Thermal Cracking of Asphaltenes?, *Energy & Fuels*, 2, 259-264.

Bezouhanova, C. P. (2002), Synthesis of aromatic ketones in the presence of zeolite catalysts, *Applied Catalysis A: General*, 229(1-2), 127-133.

Brandt, P., Jia, Z. S., and Thibblin, A., (2002), Kinetic and thermodynamic stability of naphthalene oxide and related compounds: A comparative microcalorimetric and computational (DFT) study, *Journal of Organic Chemistry*, 67(22), 7676-7682.

Bruckner, B. (2002), Radical substitution reactions at the saturated C atom, *Advanced Organic Chemistry*, 1-42.

Corma, A., Gonzalez-Alfaro, V., and Orchilles, D. V. (2001), Decalin and tetralin as probe molecules for cracking and hydrotreating the light cycle oil, *Journal of Catalysis*, 200, 34-44.

Couch, K. A., Glavin, J. P., Johnson, A. O. (2009), The feed impact of bitumen-derived feeds on FCC unit, UOP LLC, a Honeywell Company, Des Plaines, Illinois, USA

Cypres, R. (1981) Hydrogenopyrolysis of coal and coal derivatives, Prepr. Pap. - Am. Chem. Soc., Div. Fuel Chem.; (United States); Journal Volume: 26:3; Conference: National American Chemical Society summer meeting, New York, NY, USA, 23 Aug 1981.

Kasperek, G. J. and Bruice, T. C. (1972) *Journal of the American Chemical Society*, 94, 198.

Kjærsbo, T., Daasbjerg, K., and Pedersen, S. U. (2003), Study of the coupling reactions between electrochemically generated aromatic radical anions and methyl, alkyl and benzyl radicals. *Electrochimica Acta*, 48(13), 1807-1816.

Lei, X., and Turro, N. J. (1992) Photochemical hydrogen abstraction by aromatic carbonyl compounds in zeolite slurries, *Journal of Photochemistry and Photobiology A: Chemistry*, 69(1), 53-56.

Pajak, J., Krebs, V., Marêché, J. F., and Furdin, G. (1996), Hydrogen transfer from tetralin and decalin to high-temperature coal tars. Relation with carbon deposit formation. *Fuel Processing Technology*, 48(1) 73-81.

Pirinccioglu, N., and Thibblin, A. (1998), Acid-catalyzed solvolysis of allylic ethers and alcohols. competing elimination and substitution via a thermodynamically "stable" carbocation, *Journal of the American Chemical Society*, 120(26), 6512-6517.

Puente, G. de la, and Sedran U. (2000), Evaluation of hydrogen transfer in FCC catalysts. A new approach for cyclohexene as a test reactant, *Chemical Engineering Science*, 55(4), 759-765.

Rao, S. N., More O'Ferrall, R. A., Kelly, S. C., Boyd, D. R., and Agarwal, R. (1993), *Journal of the American Chemical Society,*115, 5458.

Santikunaporn, M., Herrera, J. E., Jongpatiwut, S., Resasco, D. E., Alvarez, W. E., and Sughrue, E. L. (2004), Ring opening of decalin and tetralin on HY and Pt/HY zeolite catalysts, *Journal of Catalysis*, 228(1), 100-113.

Santikunaporn, M., Alvarez, W. E., and Resasco, D. E. (2007), Ring contraction and selective ring opening of naphthenic molecules for octane number improvement, *Applied Catalysis A: General*, 325(1), 175-187.

Schlosberg, R. H., and Kurs, A. (1984), Pyrolysis studies of organic oxygenates. 5. Thermal chemistry pathways of 1-naphthol under hydrogen, *Journal of Organic Chemistry*, 49(16), 3032–3033.

Suarez, W., Cheng, W-C., Rajagopalan, K., and Peters, A. W. (1990), Estimation of hydrogen transfer rates over zeolite catalysts, *Chemical Engineering Science*, 45(8), 2581-2588.

Xn, B., Sievers, C., Hong, S. B., Prins, R., and van Bokhoven J. A. (2006), Catalytic activity of Brønsted acid sites in zeolites: Intrinsic activity, rate-limiting step, and influence of the local structure of the acid sites. *Journal of Catalysis,* 244, 163–168.

Nanoporous Carbon for Energy Applications

Galen J. Suppes, University of Missouri-Columbia, Columbia, MO

Michael Gordon, University of Missouri-Columbia, Columbia, MO

Eric Leimkuehler, University of Missouri-Columbia, Columbia, MO

Ali Tekeei, University of Missouri-Columbia, Columbia, MO

Bryan Sawyer, University of Missouri-Columbia, Columbia, MO

Carbon materials (e.g. activated carbon, carbon black) are used in most batteries, many supercapacitors, and gas storage. This presentation summarizes recent advances in record-breaking storage of methane and hydrogen in ALLCRAFT carbon. New battery applications are covered.

See more of this Session: Alternative Fuels and Enabling Technology II
See more of this Group/Topical: Fuels and Petrochemicals Division

Modeling and Optimization of Fischer-Tropsch Synthesis Processes

Wei Yuan, Department of Chemical Engineering, Auburn University, Auburn University, AL

Norman E. Sammons Jr., Department of Chemical Engineering, Auburn University, Auburn University, AL

Mario R. Eden, Department of Chemical Engineering, Auburn University, Auburn University, AL

A generic process simulation model of a gas-phase Fischer-Tropsch Synthesis (FTS) process has been developed based on previous work at Auburn University. The model consists of the following main sections: 1. Synthesis gas production through reforming, 2. Gas clean-up, 3. Fischer-Tropsch reaction, 4. Separation and recycle of unreacted synthesis gas, 5. Fractionation of fuel range products from heavy waxes, 6. Cracking of heavies and fractionation of fuel range products. The model has been developed using data available in the open literature and its performance is currently being validated against the latest experimental data generated by researchers in the Consortium of Fossil Fuel Science (CFFS). The FTS reactor is modeled using data for the ARGE reactor (Ruhrchemie and Lurgi) and the downstream processing results in two products, i.e. a gasoline range fraction and a jet fuel similar to JP-8. The product distribution is obtained using the Anderson-Schulz-Flory (ASF) distribution for a Cobalt catalyst. A hydrocracking reactor is added to convert the heavier compounds back to fuel range products. In addition, the light fraction is converted to synthesis gas for recycle to the FTS reactor by steam reforming.

The gas-phase model is used as the foundation for a model of the supercritical phase Fischer-Tropsch Synthesis (SCF-FTS) process being developed at Auburn University. By operating the FTS reaction in a dense phase, the direct benefits are two-fold, i.e. vapor-like transport properties and liquid-like thermal properties. This means that the product distribution is much more narrow than traditional gas-phase FTS. Using experimental data generated within CFFS, models have also been developed for analysis of the supercritical water reforming (SCWR) for hydrogen production from biomass. The high-pressure hydrogen produced from supercritical water reforming is intended for injection in the SCF-FTS process during the cracking/isomerization step.

Economic and environmental analysis was performed through quantifying the level of environmental impact using EPA's WAR algorithm. In this way, experimental and theoretical efforts can supplement each other, providing direction for further research.

See more of this Session: Alternative Fuels and Enabling Technology II
See more of this Group/Topical: Fuels and Petrochemicals Division

Transforming Chemistry from Postdictive to Predictive: Extensible Automated Reaction Mechanism Generation for Butanol Pyrolysis

Michael R. Harper[1], Kevin M. Van Geem[2], Guy B. Marin[2], William H. Green[1]

[1]Department of Chemical Engineering, Massachusetts Institute of Technology, Cambridge, MA 02139
[2]Laboratory for Chemical Technology, Universiteit Gent, Krijgslaan 281, S5, 9000 Gent, Belgium

Chemical kinetics' paradigm is primarily postdictive: the ignition time of a potential alternative fuel source as a function of temperature is obtained by running a series of experiments and fitting the data to a global kinetics model. These set of experiments consume substantial financial and personnel resources, including the fuel. Unfortunately, many obstacles arise that may cause the series of experiments, and thus the alternative fuel source, to be deemed unsuccessful, *e.g.* slow ignition time or increased particulate formation. Sometimes, a successful set of experiments can be cause for even greater concern, because there is usually very little understanding of why the fuel performed in the manner that it did: Will all C_4 alcohols behave similarly; what is the effect of adding an additional carbon atom to the structure's backbone; and how will impurities affect the results, are questions that may arise and, until recently, could only be answered by running additional experiments.

Shifting the chemical kinetics' paradigm from postdictive to predictive involves obtaining validated, detailed chemical mechanisms, or a system of coupled ordinary differential equations (ODEs), that accurately predict the fuel's performance[1, 2]. Answering the previously posed questions then only requires changing the initial conditions or parameters passed to the ODE solver, without having to run a single experiment. Constructing these detailed chemical mechanisms by hand is possible, however this method is difficult to repeat, extremely time costly, and susceptible to human error. Over the past decade, a handful of groups have developed automated tools to facilitate a more robust means of constructing reaction mechanisms[3]. Herein we discuss the algorithm and applications of a free, open-source, extensible software package, Reaction Mechanism Generator[4, 5].

The Reaction Mechanism Generator (RMG) software package is an automated, rate-based kinetic model generator that constructs pressure-dependent reaction mechanisms for isothermal, isobaric batch systems[6, 7]. The user must supply: the system temperature and pressure; the initial species concentrations; and a termination goal, either time of reaction or species conversion. The user may also input a list of preferred species' thermochemistries and a list of preferred reaction rate coefficients. The output files of a RMG simulation include species concentration profiles and a reaction mechanism (in the form of a Chemkin file); pressure-dependent reaction rates are expressed using Chebyshev polynomials and high-pressure limit reaction rates are expressed using modified Arrhenius parameters.

The RMG software enlarges the reaction mechanism one species at a time, starting with the initial species, see

Figure 1. For a given RMG iteration, the software has two species classifications: (1) "core species," or the species that RMG predicts will be significant in the overall reaction mechanism, and (2) "edge species," or all other species generated in the mechanism. Utilizing a list of extensible reaction family templates, the RMG software generates all possible reactions for the "core" species present in the mechanism. For each species generated, its rate of formation, as defined by the change in concentration with respect to time, is computed. The largest of these rates is then compared to the "minimum rate of formation," R_{char}, derived using a user-defined tolerance. If the rate of formation is greater than R_{char}, the species is labeled a "core" species and the mechanism enlarging algorithm is continued; if not, it remains an "edge" species. The software continues this process until the user-specified time of reaction or species conversion is reached.

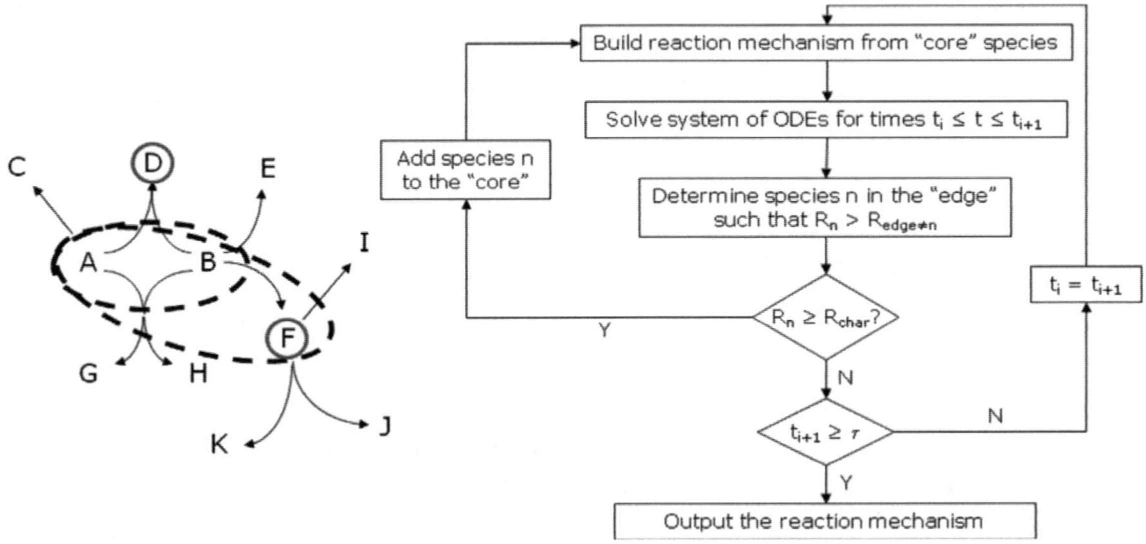

Figure 1: Automated, rate-based, reaction mechanism generation with RMG. The cartoon on the left-hand side depicts the reaction network. The species within the dashed oval are the "core" species; all others present are "edge" species. In the cartoon, species A and B were supplied by the user. The initial "edge" species formed by RMG were species C, D, E, F, G, and H. Species F had the largest flux; this flux was greater than R_{char} so species F was added to the core (the larger dashed oval). During the next iteration, RMG added the species I, J, and K to the "edge" species. Species D now has the largest flux of all "edge" species.

To perform the rate analysis and to predict species concentrations as a function of time, the software requires reaction rate coefficients and thermodynamic parameters for all reactions and species generated in the model, respectively. When a new species is generated, RMG initially checks for any user supplied thermochemical parameters; this allows the user to override any thermochemical value RMG would implement in a mechanism with their personal preference. If the user has not supplied thermochemical data for the given species, RMG next checks against the community's recommended data set, as defined by the Process Informatics Model (PrIMe) project (discussed in the next paragraph). If this search is also unsuccessful, RMG will estimate the species' thermochemistry using Benson's group-additivity approach. A similar hierarchy is employed for estimating high-pressure limit reaction rate coefficients.

The PrIMe project [8, 9] is a community-collaborative initiative that collects, stores, and validates data with the ultimate goal of processing and assembling the data into kinetic models. Two components of the PrIMe project are the data depository, which houses the community's raw data, and the data library, which stores evaluated data. The data library is the data set alluded to previously; RMG mines this data set for thermochemical data after any user-specified inputs but before estimation routines. RMG and PrIMe communicate using a unique species identifier, the IUPAC International Chemical Identifier (InChI). RMG's description of a species is converted to an InChI, which is used as the query when mining the PrIMe data library. The PrIMe data depository has approximately 11,000 species, over 7,000 of which are labeled with an InChI, including all C/H/N/O species. Mining the PrIMe data library is performed offline and all evaluated data is converted into appropriate RMG format and stored within the RMG database.

Case studies on 1- and 2-butanol have been performed. While there is serious commercial interest in developing 1-butanol into a high-volume, bio-based alternative transportation fuel[10, 11], 2-butanol is manufactured primarily as a precursor to the industrial solvent methyl ethyl ketone. A detailed pressure-dependent reaction mechanism was constructed for both species using the RMG software and PrIMe data library. The mechanisms were validated against experimental data, including pyrolysis experiments performed in a bench-scale set-up at the Laboratory of Chemical Technology of Ghent University[12] and butanol-doped methane diffusion flames previously reported by McEnally and Pfefferle[13].

In addition to requiring species thermochemistry and reaction rate coefficients, some of the reactor models used in simulating the experimental apparatus require transport properties for every species in the mechanism. The Chemkin-II libraries [14] were used to estimate species transport properties. The transport properties for species present both in our mechanism and the PrIMe data library were taken from the PrIMe data library. For all remaining stable species, we estimated the Lennard-Jones parameters using an empirical relationship that depends on a species' critical temperature and pressure [15], Equations (1) and (2). In these equations, k_B is the Boltzmann constant.

$$\sigma = 2.44\left(\frac{T_c}{P_c}\right)^{1/3} \quad \textbf{(1)}$$

$$\varepsilon_A / k_B = 0.77T_c \quad \textbf{(2)}$$

The critical temperature and pressure for each molecule were estimated using a group-additivity approach devised by Joback[16], Equations (3) – (5). $v_i^{(j)}$ is the number of groups of type j in compound i, n_a is the total number of atoms in the molecule, and T_b is the normal boiling point.

$$T_c = \frac{T_b}{0.584 + 0.965\sum_j \nu_j^{(i)}\Delta_{T_{cj}} - \left(\sum_j \nu_j^{(i)}\Delta_{T_{cj}}\right)^2} \quad \text{(in K)} \tag{3}$$

$$T_b = 198 + \sum_j \nu_j^{(i)}\Delta_{T_{bj}} \quad \text{(in K)} \tag{4}$$

$$P_c = \frac{1}{\left[0.113 + 0.0032 n_a - \sum_j \nu_j^{(i)}\Delta_{P_{cj}}\right]^2} \quad \text{(in bar)} \tag{5}$$

Note that this estimate of transport properties does not account for intermolecular interactions due to the charge distributions on the molecules. It was assumed that rotational relaxation to the bath gas temperature was fast (single collision). The indicator regarding the geometrical configuration of the molecule was determined from the molecule's structure. For radical species, the transport properties of the stable parent molecule were used.

Modified Arrhenius parameters were computed for reactions of interest using conventional Transition State Theory (TST). The species' geometries were optimized at the CBS-QB3 level and frequency calculations were performed at the B3LYP/CBSB7 level. The formula used to compute the reaction rate coefficient as a function of temperature is given in (6):

$$k = \frac{k_B T}{hC^0} \exp\frac{-\Delta G^\dagger}{RT} \tag{6}$$

k is the reaction rate coefficient, k_B is the Boltzmann constant, T is the temperature (in Kelvin), h is the Planck constant, C^0 is the standard concentration, R is the universal gas constant, and $\Delta G^\dagger$ is the free energy of activation, defined as the difference between the free energy of the transition state and the free energy of the reactants.

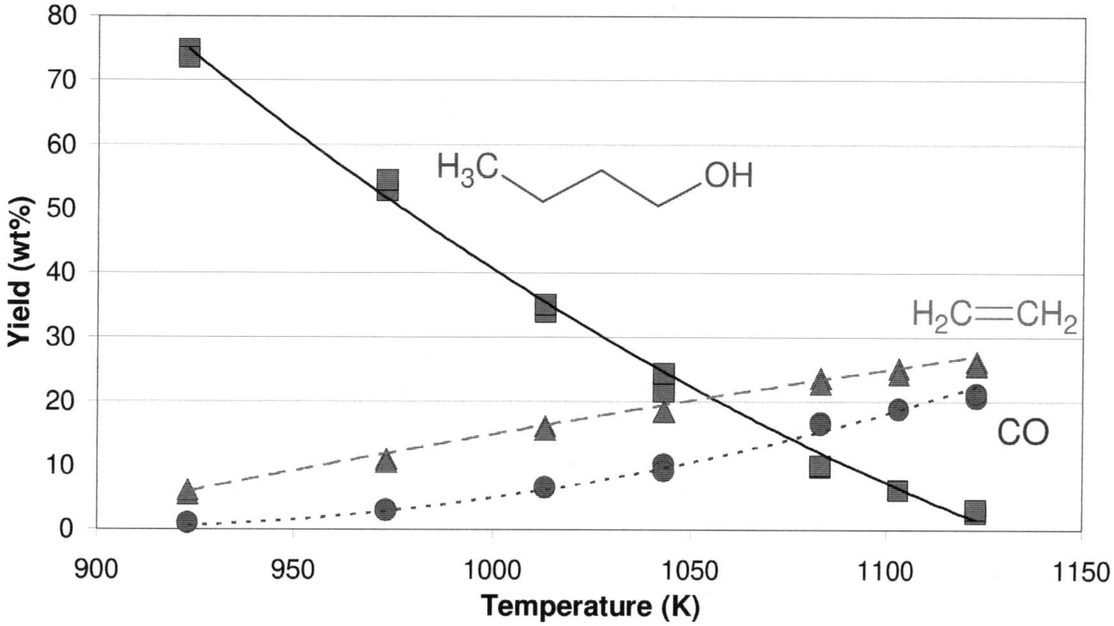

Figure 2: Comparison between simulated and experimentally determined product yields as a function of the maximally measured temperature in the pyrolysis reactor [— simulated, ▲●■: experimental]

For simulating the pyrolysis experiments in the bench scale set-up the reaction network was implemented in Chemkin[17]. The laminar flow reactor was simulated with Chemkin using the cylindrical shear flow option. Excellent agreement was observed between simulated and measured products, *e.g.* the results in Figure 2 obtained for 1-butanol, ethylene, and CO. Small radial concentration gradients were simulated, indicating that the assumption of using plug flow for this reactor is indeed valid as assumed by Chen and Froment[12], see Figure 3.

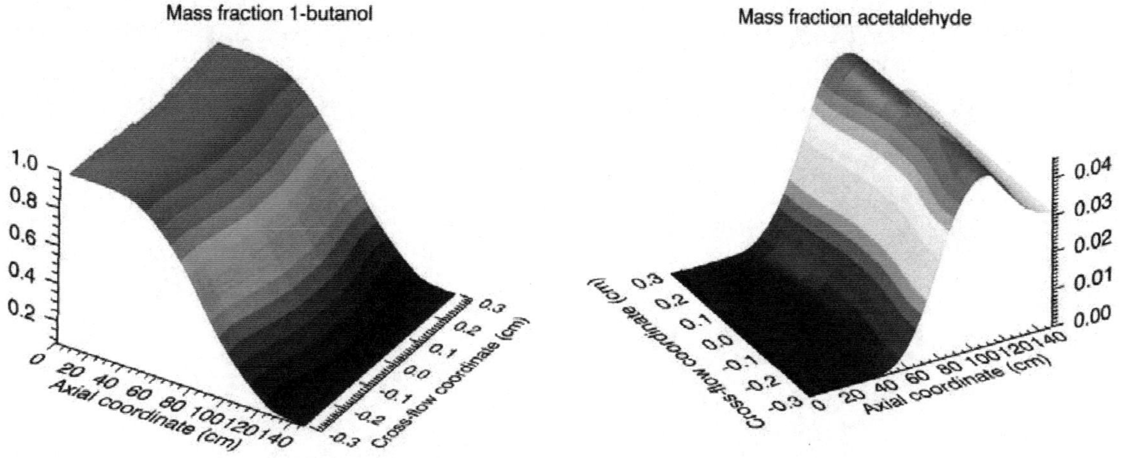

Figure 3: Verification of negligible radial concentration gradients for a major (1-butanol) and a minor (acetaldehyde) component of the reactor effluent

For the methane flame doped with 1-butanol, accurate simulation results were also obtained as can be seen in Figure 4. The reactor model used to emulate the doped methane flame is the same reactor model constructed and utilized by Sharma *et al.* [7] in their work on constructing a reaction mechanism for three of the hexadiene isomers.

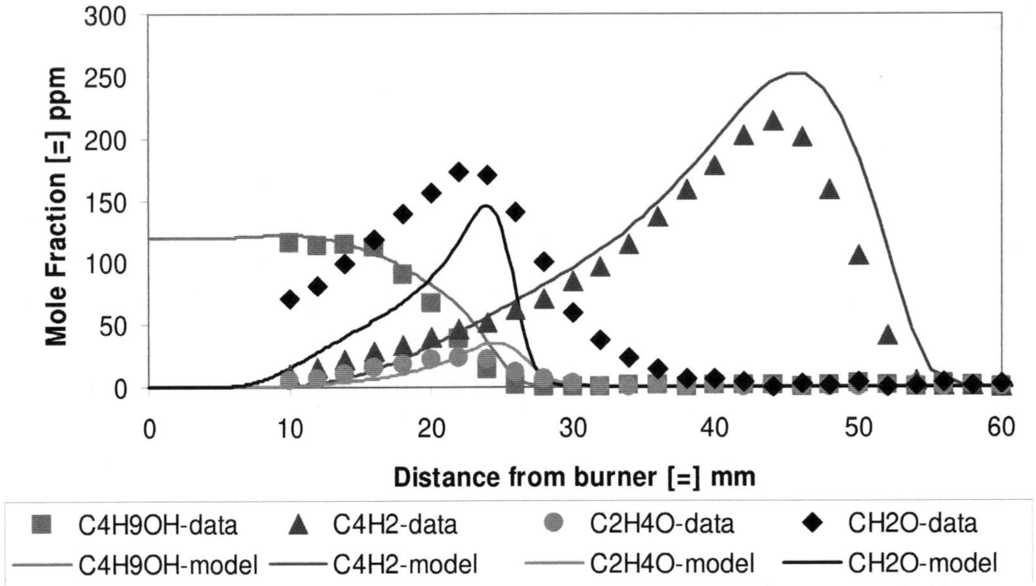

Figure 4: Mole fraction profiles of 1-butanol (squares), diacetylene (triangles), acetaldehyde + vinyl alcohol (circles), and formaldehyde (diamonds) along the centerline of the doped 1-butanol methane diffusion flame as modeled by the solver referenced in Ref. [7]. Experimental data is from Ref. [13]. Note: the 1-butanol profile has been reduced by a factor of 10 for visualization purposes.

The 1-butanol mechanism was also validated against the butanol combustion data, performed in a jet-stirred reactor, previously reported by Sarathy *et al.*[18]. The reactor consists of a 4 cm diameter fused silica sphere equipped with four nozzles of 1 mm internal diameter that promote rapid mixing of the gases as they enter the reactor. The Chemkin[17] perfectly-stirred reactor module was used to simulate the jet-stirred reactor. Again a good agreement is observed between experimental and simulated data as can be seen from Figure 5.

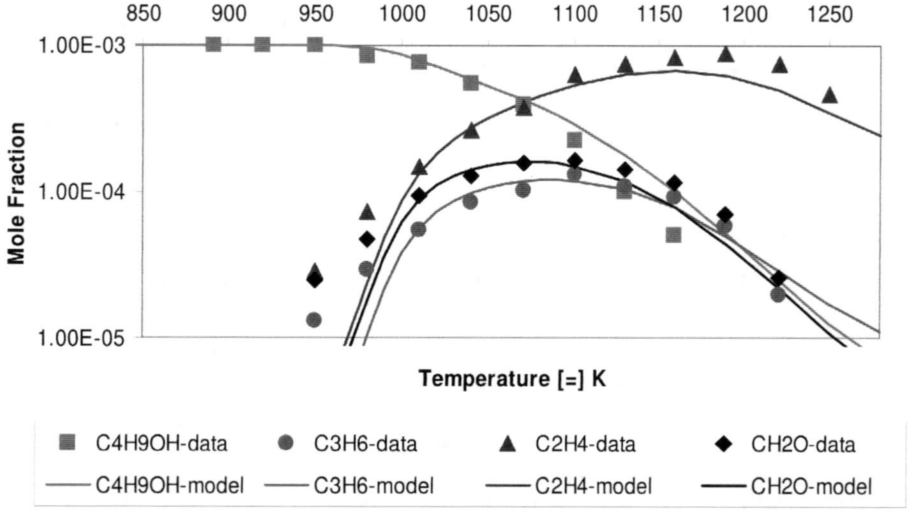

Figure 5: Mole fraction profiles of 1-butanol (square), propene (circle), ethylene (triangle), and formaldehyde (diamond) vs. temperature as modeled in the Chemkin Perfectly-stirred reactor module: Pressure=1atm, Equivalence ratio = 2.0, Residence time = 0.07 seconds. Experimental data is from Ref. [18].

Conclusion

In this work a detailed pressure-dependent reaction mechanism was automatically generated for 1-butanol and 2-butanol using the RMG software and PrIMe data library. The mechanisms were validated against experimental data, including pyrolysis experiments performed in a bench-scale set-up, and butanol-doped methane diffusion flames. Good agreement was observed between simulated and experimental data in all cases. This example illustrates RMG's capabilities of generating predictive chemical kinetics for both combustion and pyrolysis systems.

Acknowledgement

Kevin M. Van Geem holds a Postdoctoral Fellowship of the Fund for Scientific Research, Flanders, Belgium and a Fulbright fellowship for performing postdoctoral research at M.I.T. MRH and WHG would like to acknowledge the National Science Foundation and the Department of Energy for funding.

References:

1. Green, W.H., Jr., *Predictive kinetics: a new approach for the 21st century.* Adv. Chem. Eng., 2007. **32**(Chemical Engineering Kinetics): p. 1-50.
2. Green, W.H., *Building and solving accurate combustion chemistry simulations.* Nihon Nensho Gakkaishi, 2008. **50**(151): p. 19-28.
3. Pierucci, S. and E. Ranzi, *A review of features in current automatic generation software for hydrocarbon oxidation mechanisms.* Computers & Chemical Engineering, 2008. **32**(4-5): p. 805-826.

4. Song, J., *Building robust chemical reaction mechanisms: Next generation of automatic model construction software*, in *Chemical Engineering, Ph.D. thesis*. 2004, Massachusetts Institute of Technology: Cambridge, MA.
5. Green, W.H., *RMG - Reaction Mechanism Generator http://rmg.sourceforge.net/*. 2009.
6. Van Geem, K.M., et al., *Automatic reaction network generation using RMG for steam cracking of n-hexane.* AIChE Journal, 2006. **52**(2): p. 718-730.
7. Sharma, S., M.R. Harper, and W.H. Green, Submitted to Combustion and Flame.
8. Frenklach, M., *PrIMe: Process Informatics Model http://www.primekinetics.org/*. 2006-2009.
9. Frenklach, M., et al., *Process Informatics Model (PrIMe): A customer for ThermoML.* Abstracts of Papers, 231st ACS National Meeting, Atlanta, GA, United States, March 26-30, 2006, 2006: p. CINF-036.
10. *Technology Review: BP's Bet on Butanol http://www.technologyreview.com/energy/18443/.*
11. *ButylFuel, LLC http://www.butanol.com/.*
12. Chen, Q. and G.F. Froment, *Thermal-Cracking of Substituted Aromatic-Hydrocarbons .1. Kinetic-Study of the Thermal-Cracking of I-Propylbenzene.* Journal of Analytical and Applied Pyrolysis, 1991. **21**(1-2): p. 27-50.
13. McEnally, C.S. and L.D. Pfefferle, *Fuel decomposition and hydrocarbon growth processes for oxygenated hydrocarbons: butyl alcohols.* Proceedings of the Combustion Institute, 2005. **30**(Pt. 1): p. 1363-1370.
14. Kee, R.J., F.M. Rupley, and J.A. Miller, *Chemkin-II: A FORTRAN Chemcal Kinetics Package for the Analysis of Gas-Phase Chemical Kinetics, SAND89-8009*. 1989, Sandia National Labs.
15. Welty, J.R., et al., *Fundamentals of Momentum, Heat, and Mass Transfer*. 4th ed. 2001: John Wiley & Sons, Inc.
16. Joback, K.G., *A unified approach to physical property estimation using multivariate statistical techniques*, in *Chemical Engineering Department*. 1984, Massachusetts Institute of Technology: Cambridge, MA.
17. Kee, R.J., et al., *CHEMKIN Release 4.1.1*. 2007, Reaction Design, Inc.: San Diego, CA, USA.
18. Sarathy, S.M., et al., *An experimental and kinetic modeling study of n-butanol combustion.* Combustion and Flame, 2009. **156**(4): p. 852-864.

Limitations and Challenges Associated with the Disposal of Mercaptan-Rich Acid Gas Streams by Injection – A Case Study

Felise Man, Gas Liquids Engineering Ltd., Calgary, AB, Canada
John J. Carroll, Gas Liquids Engineering, Ltd., Calgary, AB, Canada

1.0 Introduction

As oil and gas reserves become depleted, producers are increasingly driven to process difficult or unconventional petrochemical fuel sources to sell to the market in order to meet rising demand. One such source is the production of sour gas wells that contain high levels of mercaptan (thiol) species.

In sour gas processing applications, a waste acid gas stream (ie. a stream containing high concentrations of hydrogen sulphide and carbon dioxide) is produced as an effluent stream in the gas sweetening unit. Generally speaking, when these volumes are relatively low (less than 5 MMSCFD or less than 20 t/d sulphur equivalent), a suitable method of waste handling of the stream is by injection to a wellbore formation. Acid gas injection is a mature technology with over 50 applications in Western Canada and another 20 around the world.

In the application where mercaptan-rich sour gas wells are produced, the resultant effluent stream from a gas sweetening process will be one that contains high levels of H_2S, CO_2, mercaptans and hydrocarbons. While investigating the ability to dispose this stream in a similar manner to that of a typical acid gas injection scheme, a number of challenges have been identified. These challenges stem from three main sources. First, little experimental data is available with regards to phase equilibrium, water content, hydrate formation conditions, and transport properties of both pure mercaptan species and mercaptan mixtures. Second, current simulation programs used for process design can employ poor assumptions or contradictory data in generating the models. And third, even when these programs are used to generate results around the injection scheme, several additional issues are revealed which requires attention. These issues are investigated in a hypothetical case study, comparing the injection of a mercaptan-rich acid gas stream to a conventional acid gas stream.

2.0 Properties of Mercaptans

Mercaptans are sulfur-containing organic compounds with the general R-SH formula. They are relatively common in natural gas and petroleum systems as naturally occurring species in low concentrations. Mercaptans are also routinely added to gas products as an odourant for leak detection. As these species are generally encountered only in small quantities, much of the experimental research has been conducted in concentrations of the hundreds of parts per million ranges. In order the predict the behaviour and processing capabilities of streams containing mercaptans, studies including those that investigate the pure component properties, phase equilibrium with other natural gas components (hydrocarbons and non-hydrocarbons), water properties (solubility and water content), and hydrate formation are of particular importance.

2.1 *Pure Component Properties*

A cursory review of the literature available on pure mercaptan components yields some data on boiling points, critical temperature and pressures, acentric factors and molar mass. Sources include studies by Poling et al. (2001), Tsonopoulos and Amborse (2001), Linstrom (2005), or estimation using equation 1. Of importance to note is that reported critical constant values were measured at or before the turn of the 20th century, and have not been verified by modern methods.

$$\omega = \frac{3T_{boil,R}}{7(1 - T_{boil,R})} log(P_C) - 1 \qquad \text{(Eqn. 1)}$$

2.2 *Phase Equilibrium*

Experimental data for vapour liquid equilibrium results are necessary in order to develop accurate and reliable process designs. In the case of mercaptan systems, little data exists for binary systems composed of two mercaptans, mercaptans and carbon dioxide, mercaptans and hydrogen sulphide, or mercaptan and hydrocarbons.

Of significant importance is the research conducted is that of Lee et al. (1978). This study is the only one found in the literature survey conducted that explores VLE data for H_2S and light mercaptan binary systems, at relatively high H_2S concentrations. Another interesting study was that of Denyer et al. (1949) where azeotropes in mercaptan + hydrocarbon mixtures were studied. No data on systems containing two mercaptans, nor mercaptan and CO_2 mixtures were found.

2.3 *Water Content and Solubility*

There are only a few data points available for the equilibrium between light mercaptans and water. A thorough review of the literature revealed only the six studies investigating experimental measurements for phase equilibria involving water and mercaptans. From the point of view of this study, the most important findings would be those of Gillespie and Wilson (1984) findings, where water content data for methyl and ethyl mercaptan only were reported. No data are available for the water content of n-propyl and n-butyl mercaptan and none for mixtures of acid gas with mercaptans nor for hydrocarbons with mercaptans.

Figure 1 shows the water content of methyl mercaptan at 311K as determined by Gillespie and Wilson (1984). The curves were fit through the data points determined in the study and do not represent a rigorous thermodynamic model. For comparison, also shown on this plot is the water content for H_2S. These data points shown are from Gillespie and Wilson (1982), however the curves are calculations from *AQUAlibrium*. *AQUAlibrium* is a thermodynamic model and thus there is some error between the experimental data and the calculations from the model.

2.4 *Hydrates*

A review of the literature found that there is some discussion of the formation of gas hydrates from methyl mercaptan, but very few specific details (such as a temperature-pressure locus). Having said that, Stackelberg and Müller (1954) report that the hydrate of methyl mercaptan dissociates at the relatively high temperatures of 273K and 34 kPa and 283K at

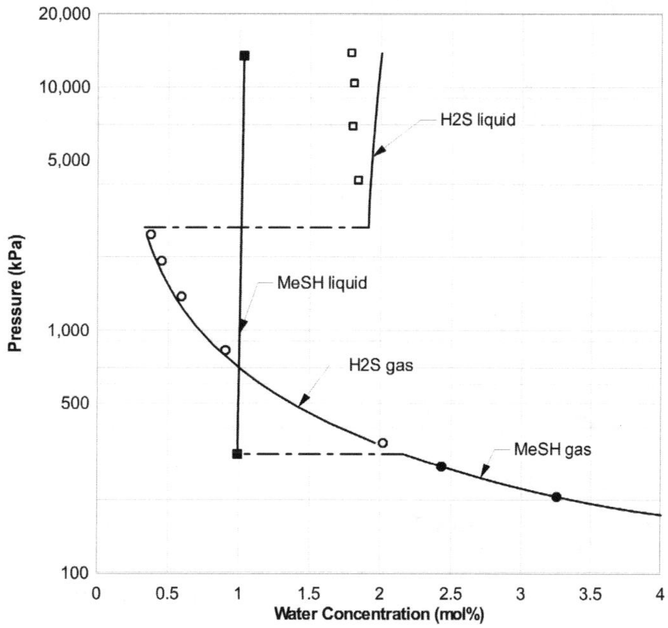

Fig. 1 Water Content of Hydrogen Sulphide (open points) and Methyl Mercaptan [MeSH] (solid points) at 311K

atmospheric pressure (101 kPa). In comparison, the hydrate of H_2S dissociates at 273K, 98.6 kPa and at 283K, 275 kPa (Carroll and Mather, 1991).

There is no specific mention of the hydrates of other mercaptans in the literature, but based upon their size and chemical similarity to methyl mercaptan, ethyl, n-propyl, and i-propyl mercaptans are surely all hydrate formers. No experimental data were found to verify this.

3.0 Limitations of Process Simulation Tools and Process Design

Almost all of the simulation work presented in this paper was completed using *VMGSim* with the Advanced Peng-Robinson (APR) property set, as recommended by *Virtual Materials Group* (*VMG*). Some phase equilibria were calculated using *AQUAlibrium*, but these were for systems that did not contain any mercaptans. The injection pressure for the acid gas case was calculated using *GLEWPro*. Since *GLEWPro* uses the *AQUAlibrium* calculation engine it also could not be used for systems that contain mercaptans.

An interesting paper by Twu et al. (2004) demonstrates some of the difficulties in modeling the vapor-liquid equilibrium in systems containing mercaptans and light hydrocarbons. Although their model is not used in this work, the reader should review the paper in order to understand some of the complexities involved in the systems investigated by the study.

3.1 Vapour-Liquid Equilibria and Hydrate Formation

As described earlier, the lack of experimental data available around VLE of binary component mercaptan systems greatly impacts the accuracy of simulation tools used. In situations where data do not exist, simulation models may rely on assuming ideal behaviour of the component. As azeotropy is known to occur in many mercaptan-hydrocarbon systems, this is clearly a poor assumption and may lead to erroneous process models that do not reflect plant behaviour.

The lack of even basic data makes it impossible to build models to predict both the hydrate formation in the pure mercaptan systems and in multicomponent mixtures that contain mercaptans. In fact, many programs either do not include mercaptans in their database or treat them like non-formers. *VMGSim* treats mercaptans as non-formers. Thus, the predictions of hydrate formation in mixtures containing mercaptans are approximate at best. In cases where the mercaptan concentration is low, its effect may be assumed to be negligible. In cases where the concentrations are sufficiently high, such as the one described in the case study that follows, ignoring the effect of mercaptans may result in highly flawed simulation results.

3.2 Water Content and Dehydration Processes

Dehydration in acid gas injection design is critical to ensure corrosion protection of the injection pipeline. In situations where the design model is questionable, a thorough review of the methods available to mitigate against potential design or equipment failures is required. As witnessed in Table 1, the values used by *VMGSim* are sufficiently variable compared to the interpolated value from data determined by Gillespie and Wilson (1984).

Furthermore, when Figure 1 is reviewed it can be noted that there is a clear difference between the water holding capacity of methyl mercaptan to that of H_2S. H_2S systems hold more water in its liquid phase than in its gas, while the opposite is true for mercaptan systems. Conventional H_2S-rich injection schemes generally take advantage of this ability to dehydrate the acid gas prior to injection through compression and cooling alone. Given that the mercaptan systems do not follow the same behaviour, along with the lack of experimental data and poor model predictions, the process decision is made to add an additional dehydration operation beyond the typical compression and cooling schemes normally encountered. The three options available include triethylene glycol (TEG) absorption, refrigeration, and molecular sieves. These options are investigated in the following case study.

Table 1. Water Content of Methyl and Ethyl Mercaptan at 311K

	Water Content (mol%)	
Methyl Mercaptan	Gillespie and Wilson (1984)	From *VMGSim*
689 kPa	0.99	21.1
6890 kPa	1.01	21.4
10 340 kPa	1.02	21.5
13 780 kPa	1.03	21.6
Ethyl Mercaptan	Gillespie and Wilson (1984)	From *VMGSim*
689 kPa	0.360	23.8
6890 kPa	0.365	23.8
10 340 kPa	0.368	24.0
13 780 kPa	0.371	24.1

3.0 Case Study

With these modelling inaccuracies already described in mind, a case study has been prepared to provide an idea of the mercaptan rich stream's potential behaviour when processed for acid gas injection. While the information may have severe errors built-in from using a simulation program, the exercise does serve to reveal further issues that may arise while preparing the waste stream for disposal, including difficulties during compression and dehydration.

The input data are based on a current injection project in northwestern Alberta. The process conditions and compositions of the two streams used in this study are given in Table 2. The acid gas stream is hypothetical but is typical for an acid gas mixture. The mercaptan-rich stream is similar to the first but contains mercaptans and a higher concentration of hydrocarbons. The high hydrocarbon content was included as it is anticipated that the sweetening of a mercaptan-rich stream would be completed using a mixed solvent or a physical solvent, both of which are notorious for absorbing more hydrocarbons than the more common aqueous alkanolamine solvents.

Table 2. Process Conditions and Compositions of Two Cases

	Acid Gas	Plus Mercaptan
Flowrate (Sm^3/d)	14.16×10^3	14.16×10^3
Pressure (kPa)	152	152
Temperature (K)	322	322
Composition (mol%)		
Hydrogen Sulfide	62.55	52.92
Carbon Dioxide	34.48	26.18
Methane	2.39	3.80
Ethane	0.47	1.81
Propane	0.11	0.40
Methyl Mercaptan	–	8.03
Ethyl Mercaptan	–	4.14
n-Propyl Mercaptan	–	1.93
n-Butyl Mercaptan	–	0.79

3.1 Injection Pressures

The injection pressure for the acid gas stream was calculated using *GLEWPro*, a software package designed specifically for acid gas wells (Wang and Carroll, 2006). For the acid gas mixture given in Table 2, the fluid remained in a single phase down the length of the reservoir, where a wellhead injection pressure of 9.1 MPa was calculated.

Since the mercaptans are not library components in *GLEWPro*, a different approach was required to estimate the injection pressure. The calculation involved the integration of the original flow equations using Euler's method. The physical properties of the stream were calculated using *VMGSim* and the pressure drop due to friction was estimated using their pipe model. The frictional pressure drop is quite small relatively to the hydrostatic contribution, common to this type of injection. The resultant injection pressure was estimated to be 9.17 MPa.

In spite of the differences in the calculation methods, the injection profiles for the two mixtures are essentially the same. Based on these calculations, the acid gas compressor is designed to a discharge of 12.4 MPa to meet injection pressure following frictional and head losses through the discharge and transport piping.

3.2 Phase Envelopes and Compression

VMGSim was used to generate the phase envelopes for the two mixtures given in Table 3. The acid gas mixture has the typical banana-shaped phase envelope, whereas the plus mercaptans mixture has a much broader phase envelope that extends to higher temperatures and pressures. This broader shape is not due to the addition of hydrocarbons as the increase in methane in the mercaptan-rich stream should produce a more volatile mixture (ie. lower dew point temperatures), while propane and ethane should have a negligible effect. Thus the differences observed between the two phase envelopes are largely due to the addition of the mercaptans.

Figure 2 shows the side-by-side plots of compression of the conventional acid gas stream versus the mercaptan-rich gas stream, plotted along with their respective phase envelopes. Both streams were saturated with water. For both cases, the same fixed discharge cooler temperatures and pressure drops, compressor ratios and compressor efficiencies were used. In addition, condensation of non-aqueous fluids was avoided for both streams, despite the fact that the accuracy of mercaptan-rich data used in the model is suspect. The conventional acid gas stream reaches the desired discharge pressure following five stage of compression and cooling, at a total power requirement of 101 kW.

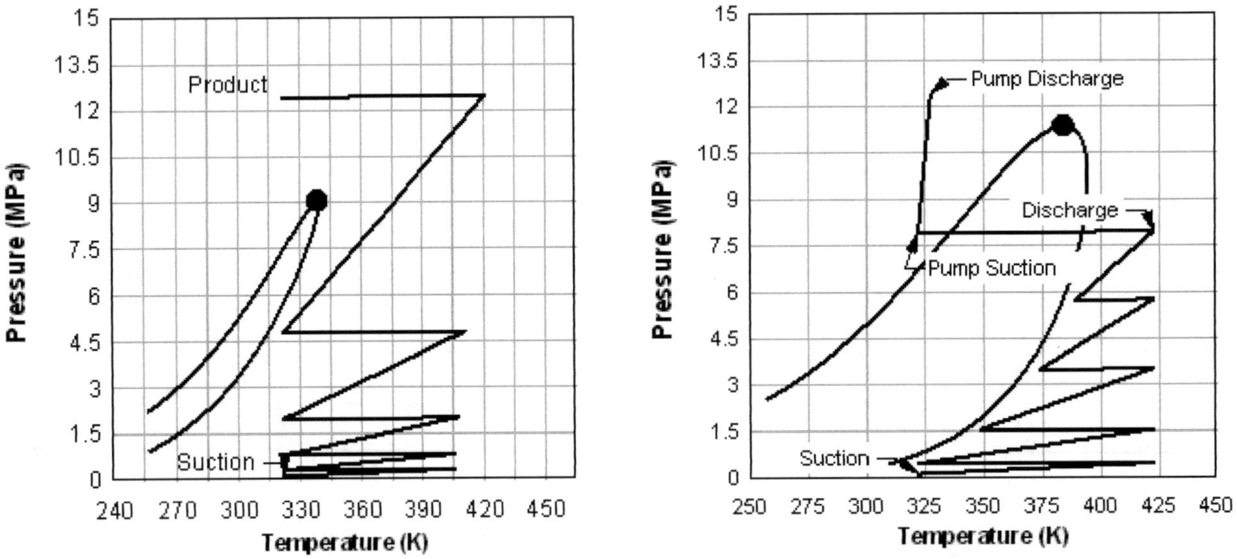

Figure 2. Compression Curves with Phase Envelopes for a Conventional Acid Gas Stream (L) and for a Mercaptan-rich Acid Gas Stream (R)

The increase in interstage cooling temperatures used to avoid condensation of non-aqueous fluids had a notable effect on the mercaptan-rich stream, wherein 5-stages of compression is not sufficient to attain injection pressure. Instead, an acid gas pump had to be installed to boost the liquefied product to pressure. Total power requirements for the mercaptan-rich system is 97 kW (compression) with a 2.5 kW pump. Other options for compression of the mercaptan-rich system were investigated however found to be not as suitable as the 5-stage compressor + pump option, based on cost or operability.

3.3 Dehydration

Dehydration for the conventional acid gas stream can be completed through compression and cooling alone, by strategically fixing an interstage pressure where cooling can occur to knockout an optimal volume of water from the vapour phase. This results in the compressed and cooled acid gas to be undersaturated in its liquefied form, which in turn prevents corrosion and hydrate issues in the stream for disposal.

Dehydration of the mercaptan-rich blended stream cannot be completed in a similar manner due to the lack of understanding of the water holding capacity of this stream as described earlier, and because of the relatively high temperatures required on the interstage. Therefore even if it were possible to achieve near-optimal water knock out, the water content may still be too high for pipeline transport. Three types of processes are available for dehydration, including glycol, mole sieve, and refrigeration units.

All three processes have specific issues associated with dehydrating mercaptan-rich streams. First, refrigeration units cannot be used as the process would result in the formation of a non-aqueous liquid. Second, the suitability of mole sieve units for this application is poor due to the low pressure, high water content nature of the stream. Furthermore mercaptan species are readily adsorbed on most mole sieve materials. Finally, the operation of glycol dehydration units is also limited due to the high interstage pressures, along with the glycol's affinity to both acid gases and mercaptans.

Despite the limitations, a TEG dehydration unit was modelled to gain an understanding of TEG dehydration capabilities, TEG circulation rate requirements, potential impurity loadings, and rough equipment sizing. The model was designed to include a contactor column, lean-rich heat exchanger, stripper column with reboiler and condenser, and a gas-glycol heat exchanger. Two cases were generated, one to process the stream exiting the first stage of compression and another case to process the output from the second stage of compression. Based on the findings, the system would have better operability with the first processing case.

An outlet water content of 0.8 g/Sm^3 (50 lb/MMSCF) gas was specified for the dehydrated stream. This relatively lenient specification allowed for a low TEG circulation rate, approximately 13.4 L/kg (1.6 gal/lb) water removed, and a low lean glycol concentration. This circulation rate, in fact, represented the lowest volume that could be simulated within the capabilities of *VMGSim*. The overhead stream exiting the regenerator column was quite sour (11 mol% H_2S and 1 mol% CO_2), and contained a total of 32 mol% of mercaptans, indicating the high absorption levels of these compounds within the TEG solution. However, as with other aspects of this project, there was no experimental data for the solubility of mercaptans in glycol that could be used to verify this observation. The model predicted that the overhead stream contained only 57% of water.

4.0 Conclusions

An approach for handling naturally occurring mercaptan species in natural gas production is the disposal of the stream via injection, similar to acid gas injection schemes common to

Western Canada. Under further investigation, several design issues and processing challenges have been identified under this approach.

Through a literature review, it was determined that very little data exist on mercaptans that were generated by experimental studies, either as a pure compound or in binary mixtures. These include phase behaviour, water content, and hydrate formation conditions. Information that does exist is either dated, or represents only a few data points over the range of conditions of interest. This lack of experimental data can be carried through to the process design of these systems, as many of the deficiencies have not been captured in the process simulation programs; rather assumptions are made or conflicting data have been designed into the program.

As a result, those designing processes around mercaptan-rich systems must be skeptical and discerning when reviewing the data and models simulated. A case study wherein a compression scheme was generated, however phase behaviour, water content and hydrate data were questioned highly. While some information was gathered using the model involving compression and cooling of the mercaptan-rich stream, the decision was made to ignore data generated by the model on dehydration of the stream. To err on the conservative side, additional dehydration was determined to be necessary to prevent corrosion and hydrates from occurring.

5.0 REFERENCES

1. Carroll, J.J. and A.E. Mather (1991), "Phase Equilibrium in the System Water-Hydrogen Sulphide: Hydrate-Forming Conditions", *Can. J. Chem. Eng.*, **69**, pp. 1206-1212.
2. Denyer, R.L., F.A. Fidler, and R.A. Lowry (1949), "Azeotrope Formation between Thiols and Hydrocarbons", *Ind. Eng. Chem.*, **41**, pp. 2727-2737.
3. Gillespie, P.C. and G.M. Wilson (1982), "Vapor-Liquid and Liquid-Liquid Equilibria: Water-Methane, Water-Carbon Dioxide, Water-Hydrogen Sulfide, Water-nPentane, Water-Methane-nPentane", *GPA Research Report RR-48*, Gas Processors Assoc., Tulsa, OK.
4. Gillespie, P.C. and G.M. Wilson (1984), "Sulfur Compounds and Water V-L-E and Mutual Solubility MESH-H_2O; ETSH-H_2O;CS_2-H_2O; and COS-H_2O", *GPA Research Report RR-78*, Gas Processors Assoc., Tulsa, OK.
5. Lee, J.I., A.E. Mather, F.D. Otto (1978), "Vapor-Liquid Equilibria in the System Hydrogen Sulfide-Methanethiol", *J. Chem. Eng. Data*, **23**, pp. 78-79.
6. Linstrom, P.J. (2005), "A Guide to the NIST Chemistry Web Book", http://webbook.nist.gov/chemistry/guide/.
7. Poling, B.E., Prausnitz, J.M., and O'Connell (2001), *The Properties of Gases and Liquids*, 5th ed., McGraw-Hill, New York, NY.
8. Stackelberg, M. and H.R. Müller (1954), "Feste Gashydrate II Struktur and Raumchemie", *Z. Elektrochemie*, **58**, pp. 25-39. – in German.
9. Tsonopoulos, C. and Ambrose, D. (2001), "Vapor-Liquid Critical Properties of Elements and Compounds. 8. Organic Sulfur, Silicon, and Tin Compounds (C + H + S, Si, and Sn)", *J. Chem. Eng. Data*, **46**, pp. 480-485.
10. Twu, C.H., W. Tassone, and W.D. Sim (2004), "Accurately Predict the VLE of Thiol-Hydrocarbon Mixtures", *Chem Eng, Prog.*, **100 (9)**, pp. 39-45.
11. Wang, S. and J.J. Carroll (2006), "Model Calculates Acid-Gas Injection Profiles", *Oil & Gas J.*, **104 (33)**.

Renewable Carbohydrates Are a High-Energy Density Storage Carrier for Generation of Hydrogen or Electricity

Y-H. Percival Zhang, Biological Systems Engineering, Virginia Polytechnic Institute and State University, Blacksburg, VA

Cellulosic material is the most abundant renewable carbon-containing polysaccharide. Each year more than 100 billion tons of cellulosic material is produced by photosynthesis. We have demonstrated a novel cell-free synthetic pathway biotransformation (SyPaB) reaction for generation of hydrogen from starch or cellulosic materials and water as C6H10O5 (aq) + 7 H2O (l) --> 12 H2 (g) + 6 CO2 (g) [1,2]. These endothermic entropy-driven chemical reaction achieve the chemical energy output (hydrogen)/input (cellulosic materials) > 1 by absorbing ambient-temperature thermal energy for the first time [2,3]. Here we proposed to use cellulosic polysaccharide as a high-energy density hydrogen carrier (e.g., 14.8 mass H2%) through cell-free synthetic pathway biotransformation. With integration of PEM fuel cells for water recycling and heat exchange, the super-energy density sugar battery with at least 10-fold energy density of current lithium batteries is under development.

[1] Y.-H. P. Zhang, B. R. Evans, J. R. Mielenz, R. C. Hopkins, M. W. W. Adams, PLoS One 2007, 2, e456.

[2] X. Ye, Y. Wang, R. C. Hopkins, M. W. W. Adams, B. R. Evans, J. R. Mielenz and Y.-H. P. Zhang, ChemSusChem, 2009, 2, 149-152.

See more of this Session: Alternative Fuels and Enabling Technology II
See more of this Group/Topical: Fuels and Petrochemicals Division

Use of Alternative Fuels to Replace Coal in the Manufacture of Cement

Steve R. Duke, Chemical Engineering, Auburn University, Auburn University, AL

Scott C. Morris, Chemical Engineering, Auburn University, Auburn University, AL

Aaron Folta, Civil Engineering, Auburn University, Auburn University, AL

Srikanth Akkapeddi, Civil Engineering, Auburn University, Auburn University, AL

Anton Schindler, Civil Engineering, Auburn University, Auburn University, AL

Don Stafford, US East Business Unit, Lafarge North America, Alpharetta, GA

A multiyear DOE-supported study is underway to identify, determine, and understand fundamental burn characteristics and properties of alternative fuel sources to replace coal for energy generation, with emphasis on impacts in cement processing. An operating cement plant (Lafarge North America's Roberta Plant in Calera, Alabama) serves as model to reduce dependence on traditional fuels through the use of bio-based and waste fuels. Portland cement manufacturing involves the combustion of solid fuels along with raw materials to produce clinker. Temperatures exceeding 1,400°C are needed to carry out reactions for cement chemistry. Energy needs for a manufacturing plant are 12,000 MMBTU/day. Potential alternatives to fossil fuels must provide adequate and economical energy but they must also result in products of combustion that have negligible impact on cement quality.

Results will be presented from a comprehensive evaluation of the raw materials, fuels, clinker, cement, and emissions from test burns at Lafarge's Roberta Cement Plant when using scrap tires, waste plastics, woodchips, switch grass, poultry litter, forests trimmings, and scrap railroad ties as fuel alternative to coal. The evaluation includes analyses and long-term testing of the concrete product, impacts on process operations, emissions effects, and materials handling. Sixteen three-day trial burns were conducted in 2007, 2008, and 2009. Each trial burn contained one or more of the alternative fuels. An elaborate sampling and analysis plan was carried out for each trial burn with close cooperation between the Lafarge plant, the Lafarge testing laboratory, Auburn University, and outside laboratories.

All of the alternative solid fuels were deemed technically viable as supplements for coal for direct burn in a manufacturing plant. Replacement ratios (on a BTU basis) were up to 8% for the bio-based fuels and up to 18% for the waste plastics and the synergistic burns where two or more of the alternative materials were used together. No negative impacts were seen on product quality or emissions. Alternative Solid Fuel (ASF) handling and process operations were established for the use of each of these fuels.

See more of this Session: Alternative Fuels and Enabling Technology II
See more of this Group/Topical: Fuels and Petrochemicals Division

Enabling Crude Recovery through Inhibition of Asphaltenes

Sara M. Hashmi, Chemical Engineering, Yale University, New Haven, CT

Abbas Firoozabadi, Chemical Engineering, Yale University, New Haven, CT

As light, sweet crude oils become depleted and heavy crude oils become more prevalent in the fuel landscape, practical solutions to the problems of asphaltene precipitation and deposition become increasingly important. We test a series of dispersing chemicals on crude mixtures in heptane, using index of refraction and dynamic light scattering measurements to quantify the effectiveness of the dispersants. We furthermore present insights into the mechanism of action of the dispersants as enabling an increase in the surface area of colloidal scale asphaltene particles. These insights may guide future selection of dispersants with appropriate chemical and material properties.

See more of this Session: Alternative Fuels and Enabling Technology II
See more of this Group/Topical: Fuels and Petrochemicals Division

Synthesis of Biodiesel with Fat Based Feed of High Content of Free Fatty Acids

M. Siebenhofer, University of Technology Graz, Department of Chemical Engineering and Environmental Technology, Graz, Austria

Irmela Kofler, VTU Engineering, Grambach/Graz, Austria

Wolfgang Glasl, VTU Engineering, Grambach/Graz, Austria

The conversion of triglycerides with alkaline catalysts and methanol is a state of the art process in biodiesel production in industrial scale, preferably applied for processing raw material with low content of free fatty acids. Several feedstock for methylester-production may contain up to 100% of free fatty acids. Alkali catalysis would lead to saponification instead of esterification. Therefore acidic catalysts are used instead. In this project the kinetics of esterification of feedstock with a high content of free fatty acids with methanol is investigated. To avoid demand of excessive catalyst, reaction the temperature of reaction was raised. To study the influence of different reaction conditions on the rate and yield runs were made in a lab scale batch vessel. The amount of FFA was reduced from 90% to less than 10% after one hour of reaction time. The experimentally obtained data indicate kinetics of second order reversible reaction. The rate of esterification increases with temperature. Even high excess of alcohol can not decrease FFA to less than 5% in single step operation. After removing the byproduct water, the amount of FFA can be reduced to 2% in a second esterification step. Conversion was also investigated with heterogeneous catalysts.

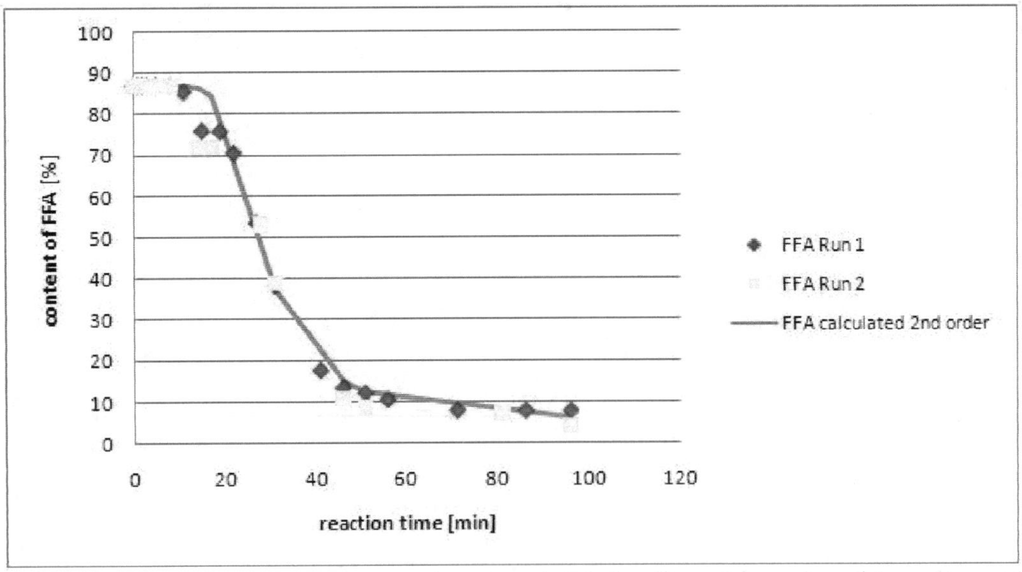

Figure 1: Comparison of measured and modelled data for the esterification of free fatty acids with methanol

See more of this Session: Alternative Fuels and Enabling Technologies III
See more of this Group/Topical: Fuels and Petrochemicals Division

Deterioration of B20 Under Compression Ignition Engine Operation Conditions

Kapila Wadumesthrige, Department of Chemical Engineering and Materials Science, Wayne State University, Detroit, MI

Mark Winston-galant, Alternative Energy Technology, Wayne State University, Detroit, MI

Haiying Tang, Department of Chemical Engineering and Materials Science, Wayne State University, Detroit, MI

Steven O. Salley, Department of Chemical Engineering and Materials Science, Wayne State University, Detroit, MI

K. Y. Simon Ng, Department of Chemical Engineering and Materials Science, Wayne State University, Detroit, MI

Biodiesel has been widely accepted as an alternative for fossil-derived diesel fuel for use in compression ignition (CI) engines. Several properties, such as oxidative stability, cloud point, and pour point restrict the use of biodiesel beyond current B20 blend levels (20% biodiesel in 80% ULSD) for on-road vehicle applications. Maintaining the properties of B20 as specified by ASTM D 7476-08 is important because once out of spec, B20 will result injector coke formation, fuel filter plugging, increased exhaust emissions, and overall loss of engine performance. Although the properties of initial B20 are within the specifications, under engine operating and longer storage conditions, B20 could deteriorate. In a diesel engine, the fuel that goes to the injector and does not enter the cylinder is then recycled back to the fuel tank. The recycled fuel returns to the fuel tank at an elevate temperature, which can cause thermal oxidation. In this study the oxidative stability, cetane number, viscosity, cold flow properties, and acid number of a B20 fuel on a heavy duty CI engine with a 200 gallon fuel tank were measured as a function of engine run time. The oxidative stability decreased by 50% within 80 hours of engine run time. Also the viscosity, cetane number, and acid number were increased with time. The thermal degradation of B20 is due to several factors, such as high temperature, depletion of reactivity of antioxidants, and dissolution of metals from engine parts at high temperatures. These property changes are explained with the compositional changes. These results will help understand the methods of minimizing degradation of B20.

See more of this Session: Alternative Fuels and Enabling Technologies III
See more of this Group/Topical: Fuels and Petrochemicals Division

Bench-Top Engine System for Fast Screening of Alternative Fuels and Fuel Additives

Hongmei An, Department of Chemical and Biomolecular Engineering, University of Houston, Houston, TX

Rachel L. Muncrief, Department of Chemical and Biomolecular Engineering, University of Houston, Houston, TX

Miguel Cruz, Department of Chemical and Biomolecular Engineering, University of Houston, Houston, TX

Charles Rooks, Department of Chemical and Biomolecular Engineering, University of Houston, Houston, TX

Michael Harold, Department of Chemical and Biomolecular Engineering, University of Houston, Houston, TX

Huzeifa Ismail, Industry Technology, Baker Hughes Inc., Sugar Land, TX

There is a renewed interest and use in diesel-powered vehicles because of better fuel economy and the emergence of alternative fuels such as biodiesel. At that same time, more stringent diesel emissions requirements are in place for both new and existing diesel vehicles and equipment. This has motivated the development of emission after-treatment technologies as well as alternative diesel fuels and diesel fuel additives. Fuel properties have an important effect on the generation of the two key diesel pollutants, which are NOx and particulate soot. However, existing testing methods involve the use of cumbersome dynamometers. At the University of Houston, a high throughput fuel-testing system was developed to evaluate the combustion and emission characteristics of a variety of diesel fuels and additives. The system enables rapid and accurate information, which has great utility in the evaluation of alternative diesel fuels, such as biodiesel and other oxygenated hydrocarbons and diesel fuel additives.

In this presentation, we will describe the system and the testing of several fuels and additives. The main components of the apparatus are a 5kW engine generator, an additive/fuel injector, a load controller, a NOx analyzer, and a remote monitoring and controlling system as shown in Figure 1. This testing system has the capability to generate fuel/additive blends of different ratios and feed them into a single cylinder engine. The blend ratio or additive concentration can be changed in situ. The engine can be operated at different loads using the programmable load control. NOx concentration, exhaust flow rate, and other parameters are monitored and recorded to evaluate the impact of the additive/alternative fuel being tested.

An injection testing protocol was developed for screening the effect of additive dose or fuel blend ratio on NOx emission. One example is shown in Figure 2. A commercial additive NOxtrol™ emissions control additive (Baker Hughes Inc.) was tested in this experiment, and its concentration in the baseline fuel (Valero #2 diesel) is changed stepwise. Each step is carried out for 20 min. Figure 2 shows how the NOx changes with the additive concentration at a fixed engine load. In this experiment, the baseline fuel was used first and the engine was run at near full load. After 20 min of baseline testing (pre-injection baseline), the NOxtrol™ additive was continuously injected into the baseline fuel line using the ISCO pump. Initially, the additive was injected at a rate of 0.033ml/min for 20min, which gives a concentration of about 1000ppm. The additive injection rate was then changed to 0.067ml/min (2000ppm) for an additional 20min. These steps were repeated for 3000ppm and 4000ppm. The final step was to stop the additive injection and run the baseline fuel for another 20 min (post-injection baseline). NOx concentration ([NOx]) for each step was calculated by averaging the measured NOx concentration in the last 10 min of each 20 min step. The NOx concentration for baseline fuel

($[NOx]_{base}$) is calculated by averaging [NOx] in pre-injection and post-injection steps. The NOx reduction in each injection step is marked in Figure 2, which is calculated by the equation:

$$\text{NOx reduction} = ([NOx]_{base} - [NOx]) / [NOx]_{base} * 100\%$$

The experiment results are repeatable and the experiment error for the NOx reduction is ±0.9%.

The results generated from this system are comparable with the data from a 500HP AC chassis dynamometer. Typical data will be presented for several diesel fuels and blends with selected results compared to vehicle emissions data from a heavy-duty chassis dynamometer.

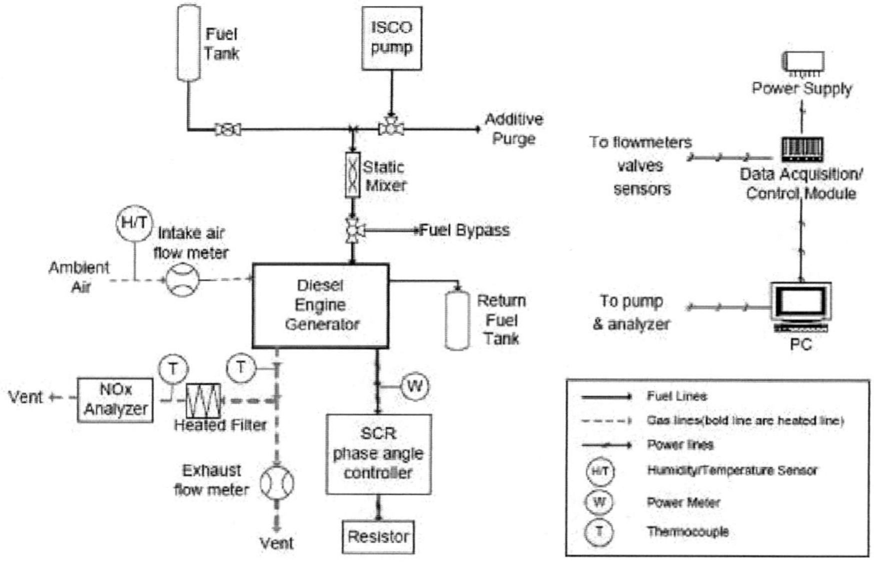

Figure 1- schematic of bench-top engine generator testing system

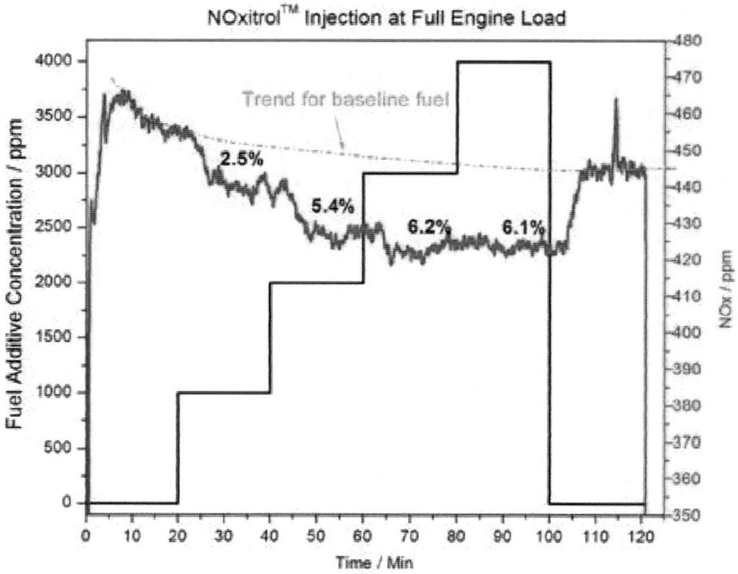

Figure 2-NOx concentration in the diesel exhaust changes with the additive (NOxitrol™, Baker Hughes Inc.) concentration during the 2-hour testing at full engine load (5kW).

NOxitrol is a trademark of Baker Hughes Incorporated.

See more of this Session: Alternative Fuels and Enabling Technologies III
See more of this Group/Topical: Fuels and Petrochemicals Division

Conversion of Crude Glycerol From Biodisel Production to Oil Via the Oleaginous Yeast *Rhodotorula Glutinis*

J. Mathew Thomas, Dave C. Swalm School of Chemical Engineering, Mississippi State University, Mississippi State, MS

Rafael Hernandez, Dave C. Swalm School of Chemical Engineering, Mississippi State University, Mississippi State, MS

Todd French, Dave C. Swalm School of Chemical Engineering, Mississippi State University, Mississippi State, MS

William Holmes, Mississippi State Chemical Laboratory, Mississippi State University, Mississippi State, MS

Earl Alley, Dave C. Swalm School of Chemical Engineering, Mississippi State University, Mississippi State, MS

Jacqueline Hall, Dave C. Swalm School of Chemical Engineering, Mississippi State University, Mississippi State, MS

Over the past decade the growing popularity of biodiesel as an alternative to petroleum-derived diesel fuel has created a dramatic increase in the production of crude glycerol. With the high cost of feedstock and low price of the crude glycerol byproduct many biodiesel producers have scaled back operations or shutdown completely. If biodiesel is to succeed as an alternative to petroleum-based fuels it is essential that the crude glycerol produced is processed/refined cost effectively into a value-added product. Oleaginous microorganisms have the ability to use many different types of carbon sources, such as glycerol, and convert them into oils that can then be used for the production of fuels. The purpose of this study is to investigate the feasibility of the oleaginous species *Rhodotorula glutinis* to utilize crude glycerol as a carbon source for the production of oils. The study will evaluate the kinetics of biomass growth, oil production, and glycerol consumption of *R. glutinis* grown on crude glycerol from biodiesel production. The crude glycerol kinetics will be compared to kinetics of *R. glutinis* grown on pure glycerol to determine any inhibitory effects of the crude glycerol contaminants.

See more of this Session: Alternative Fuels and Enabling Technologies III
See more of this Group/Topical: Fuels and Petrochemicals Division

Biodiesel Production by Using Methanol-Ethanol Mixture in a Two Stage Batch Reactors

Manhoe Kim, National Biofuel Energy Laboratory, NextEnergy, USA, Detroit, MI

Shuli Yan, Wayne State University, Detroit, MI

Steven O. Salley, Department of Chemical Engineering and Materials Science, Wayne State University, Detroit, MI

Simon K. Y. Ng, Chemical Engineering and Materials Science, Wayne State University, Detroit, MI

Almost all commercial biodiesel is currently produced by using a homogeneous base catalyzed process. The removal of the homogeneous catalyst after reaction may be a significant problem and a large amount of waste water was produced to separate and clean the catalyst and the products. If biodiesel can be produced by using smaller amount of catalyst, it can minimize the separation and purification cost. Ethoxide has stronger nucleophilicity than methoxide. Methanol is more acidic than ethanol. When methanol-ethanol mixture was used for transesterification of triglyceride, a synergetic effect of the mixed methanol-ethanol was reported. The synergy effect of mixed methanol/ethanol on the transesterification of triglycerides was explained by the increased reactive sites of the conjugated methoxide-ethanol structure. Methylester biodiesel has a drawback in the cold flow property. Biodiesel containing ethylester can give much better cold flow property then biodiesel composed of methylester only. In this study, the method which can reduce the amount of homogeneous catalyst and fulfill the specification for biodiesel (ASTM D6751-07a) was studied by using mixed methanol-ethanol. Two stages of batch reactors were set-up. The CH_3ONa catalyst was distributed to the first stage and the second stage in the ways of 3:7, 5:5, and 7:3. The amounts of alcohol were supplied to the 1st and 2nd stages in the ratio of 3:7, 5:5, and 7:3, respectively. Amounts of Alcohol to oil ratio and methanol to ethanol ratio were fixed to 6:1 and 10:1. The reaction temperature was fixed to 65°C. Three different amount of CH_3ONa (0.2, 0.5, and 0.7%) were applied. The concentrations of sodium content, biodiesel yield, free glycerin and total glycerin in oil phase were measured from the unwashed products. All the results obtained from using mixed methanol-ethanol were compared to those which were obtained with pure methanol. The results from this study will be discussed.

See more of this Session: Alternative Fuels and Enabling Technologies III
See more of this Group/Topical: Fuels and Petrochemicals Division

Optimization of the Pre-Treatment Process of Cottonseed Oil for Biodiesel Production

Regina C. R. Santos, Universidade Federal do Ceará, Fortaleza, CE, Brazil

Breno S. Rocha, Universidade Federal do Ceará, Fortaleza, CE, Brazil

Monica C. G. Albuquerque, Chemical Engineering Department, Universidade Federal do Ceará, Fortaleza, Brazil

Celio L. Cavalcante Jr., Chemical Engineering, Universidade Federal do Ceará, Fortaleza, CE, Brazil

Cotton seed has approximately 19 wt% of oil, which can be used for biodiesel production. However current refining processes used industrially for cotton oil production turn the biodiesel production from cotton oil not economically attractive. A batch-wise refining process using sodium hydroxide and clay washing steps is normally employed to remove gossypol and other acid contaminants of the cottonseed oil, under different conditions of temperature and agitation. The objective of the present study is to optimize the variables of this cotton oil pre-treatment process, and evaluate the influence of operations conditions on cottonseed oil quality. A factorial design and response surface methodology was utilized in order to determine the relationship between quality control of oil (color, acidity, amount of soap) and biodiesel molar yield. The results show that the initial and the final temperature operating conditions, and the sodium hydroxide concentration and excess affect significantly the oil quality in the pre-treatment process. Optimal values for these operating variables were used to achieve an oil quality that makes it possible to synthesize biodiesel with a maximum molar yield. Samples of this oil were evaluated according to the parameters of oil characterization (acidity, color, soap content).

See more of this Session: Alternative Fuels and Enabling Technologies III
See more of this Group/Topical: Fuels and Petrochemicals Division

Optimization and Reaction Kinetics of the Production of Biodiesel From Castor Oil

Scott D. Crymble, Dave C. Swalm School of Chemical Engineering, Mississippi State University, Mississippi State, MS

Rafael Hernandez, Dave C. Swalm School of Chemical Engineering, Mississippi State University, Mississippi State, MS

W. Todd French, Dave C. Swalm School of Chemical Engineering, Mississippi State University, Mississippi State, MS

Ricinus communis, commonly known as the castor plant, has been cultivated for millennia as a source of oil for lamps and other applications. The seeds of the castor plant contain 35% to 55% oil. The major fatty acid component of castor oil is ricinoleic acid, which comprises 90% of the fatty acids present in castor oil. Ricinoleic acid is a monounsaturated, 18-carbon fatty acid with a hydroxyl functional group at C-12. This hydroxyl group causes castor oil to be highly polar, simplifying the mixing of the oil with methanol for the biodiesel reaction. Castor oil is more viscous than typical oils, which poses problems in the resulting biodiesel. A study of the esterification reaction kinetics of converting the castor oil into biodiesel was performed to determine reaction kinetics. This information can be used to engineer a biodiesel production facility with castor oil as a feedstock. Factors considered in the kinetics study included reaction temperature, catalyst concentration, ratio of methanol to oil, and reaction time. The resulting biodiesel was tested for compliance with appropriate ASTM standards. Production of biodiesel from castor oil would provide two benefits: (1) a biodiesel feedstock oil that does not diminish the food supply, and (2) reduce the US dependency on foreign oil by providing a domestic alternative to petroleum diesel.

See more of this Session: Alternative Fuels and Enabling Technologies III
See more of this Group/Topical: Fuels and Petrochemicals Division

Analyses of Switchgrass Bio-Oils

Roberto Galiasso, University of Oklahoma, Norman, OK

Lance Lobban, University of Oklahoma, Norman, OK

Daniel Resasco, University of Oklahoma, Norman, OK

Richard Mallinson, University of Oklahoma, Norman, OK

Bio-oils from thermal processing of switchgrass have been analyzed by NMR and FTIR and other techniques to determine their chemical structural properties in order to design catalytic upgrading strategies as well as thermal processing conditions.

See more of this Session: Catalytic Biomass Pyrolysis and Gasification II
See more of this Group/Topical: Fuels and Petrochemicals Division

Influence of Pyrolysis Parameters On Individual Component of Bio-Oil

Suchithra T. Gopakumar, Chemical Engineering, Auburn University, Auburn, AL

Sushil Adhikari, Biosystems Engineering, Auburn University, Auburn, AL

Ram B. Gupta, Chemical Engineering, Auburn University, Auburn, AL

National energy security and rising environmental concerns associated with fossil fuels have urged us to find out alternative sources for energy production. Biomass is widely accepted as a potential source of energy and it is the only renewable energy source that can provide liquid fuels. Biomass can be converted into fuels thermally and biologically. Primarily, thermal conversion of biomass is done through combustion, gasification or pyrolysis. At present, fast pyrolysis of biomass is getting more attention because of its high oil yield and high energy density compared to virgin biomass. Fast pyrolysis is a high temperature process in which biomass is thermally cracked into smaller compounds in an inert atmosphere at high heating rate. The vapor from this process is quickly condensed into liquid and this viscous liquid is called pyrolysis oil or bio-oil. Bio-oil is a chemically complex mixture of water, aldehydes, acids, alcohols, carbohydrates, ketones, phenols and other lignin fragments. The fuel properties of bio-oil are comparable with conventional fossil fuels. In addition, bio-oil can be used as a feedstock for many commodity chemicals. Many of the applications of bio-oil are limited because of some undesirable properties such as high pH, low heating value, high viscosity, and unstable. High reactivity of oxygenated compounds in bio-oil makes it very unstable.

Current studies are being focused in optimizing pyrolysis conditions, upgrading of bio-oil and finding more applications of bio-oil. Biomass feedstock has a great influence on the properties of bio-oil. In the present study, change in physical properties of bio-oil while varying pyrolysis conditions has been noticed. The change in properties is a result of the change in chemical composition of bio-oil. However, only a handful studies have studied the effect of operating parameters on individual compound. In this study, careful control of pyrolysis parameters, such as temperature, pressure, heating rate and pyrolyzing environment (hydrogen, helium, air) will be carried out to study the effect on the yield of individual component in the bio-oil. Switchgrass, pine wood, peanut hulls and poultry litter are selected as feedstocks. The selected feedstocks are agricultural wastes and they are selected because of their relative abundance in Alabama and in the U.S. in general. This study will lead us to determine optimal pyrolysis conditions for high quality bio-oil, which would make downstream processes easier and economical for fuel production. Pyrolysis studies will be carried out in a pyroprobe which is interfaced with a GC/MS. The probe contains a computer controlled heating element which holds finely grounded biomass in a quartz tube. Reactant gas carries the pyrolysis vapors from the probe to a trap which is resting in normal temperature. The trap adsorbs the condensed bio-oil components and the non-condensed vapors are purged with reactant gas. The adsorbed bio-oil is desorbed through GC/MS carrier gas and is directed to GC for analysis. Pyroprobe has a large range and a high precision for changing all the pyrolysis parameters.

See more of this Session: Catalytic Biomass Pyrolysis and Gasification II
See more of this Group/Topical: Fuels and Petrochemicals Division

POTENTIAL USE OF SUGAR BEET (*BETA VULGARIS*) BAGASSE AS A RENEWABLE HYDROCARBON RESOURCE VIA FAST PYROLYSIS

Sedat H. Beis

Biological Sytems Engineering, Virginia Polytechnic and State University, Blacksburg VA

In today's competitive world, it is important to have both product flexibility and diversity of an industrial plant. Depending on market developments, new potential demands may arise for a commodity already in demand by another industry. The Sugar industry based on sugar beet (*Beta vulgaris*) possesses such an advantages feed stock which can be converted to end and side products table sugar, molasses and bagasse demanded by the food, fermentation and feed industries. As the need for alternative renewable fuels has reached to a maximum more than ever it is worthwhile to reconsider some biomass based side products as a feed stock with respect to production of biofuels and chemicals. Currently, sugar beet bagasse (SBB) is used as livestock feed.

In this study, the thermal conversion of SBB was studied to evaluate the conversion yields as well as chemical compositions of the pyrolysis products. A batch of SBB supplied from the Turkish Sugar Industries, Eskisehir Plant was air dried and used in the runs. Physical and chemical properties such as moisture content, ash content and elemental composition were determined after grinding the sample -500 microns. Thermal conversion experiments were carried out in a fluid bed fast pyrolysis unit. The unit consisted of a biomass feeder, a fluid bed reactor, a condenser and an electro static precipitator (ESP). Silica sand (+100-350 microns) was used as the heating medium. Fluidization is achieved by means of preheated nitrogen gas. 200 g of SBB batches were pyrolyzed at 450 ^{o}C to produce the pyrolysis products such as bio-oil, bio-char and syngas. The pyrolysis reactor was heated by a three zone electrical furnace. Runs were started after the reactor, the condenser and the ESP reached steady state operating conditions. Products were kept under nitrogen atmosphere until the system cooled down to room temperature. The chemical composition of the liquid product was determined by ^{13}C NMR spectroscopy. Gaseous components were separated and quantified by employing a gas chromatograph equipped with molsieve, Hayesep-D and MXT-1 columns and were detected by a thermal conductivity detector and a flame ionization detector with methanizer.

The results showed that fast pyrolysis of SBB produced 35 % organics and water, 21% char and 44% gas. The oil composition showed a different pattern owing to lower lignin content than woody biomass. The dominant gas components were CO, CO_2 and CH_4 with minor amounts of C_2-C_4 range hydrocarbons.

Product Distribution From Fast Pyrolysis of Hemicellulose

Pushkaraj R. Patwardhan, Department of Chemical and Biological Engineering, Iowa State University, Ames, IA

Justinus A. Satrio, Center for Sustainable and Environmental Technologies, Iowa State University, Ames, IA

Robert C. Brown, Center for Sustainable and Environmental Technologies, Iowa State University, Ames, IA

Brent H. Shanks, Department of Chemical and Biological Engineering, Iowa State University, Ames, IA

The goal of this study is to understand the product distribution arising from the fast pyrolysis of hemicellulose. Hemicellulose constitutes about 20-40% of the biomass. With the increasing interest in utilizing biomass as a renewable resource for fuels via the fast pyrolysis route, understanding the pyrolysis behavior of biomass constituents has particular significance. Although several studies have appeared in the literature regarding biomass pyrolysis, they mainly focus on the whole biomass and products classified into broad categories: bio-oil, char and gas. Understanding the distribution of chemical species produced from fast pyrolysis of individual constituents of biomass will potentially help gaining an insight on chemical reactions and possible pathways involved in fast pyrolysis process, which in turn can potentially be used as an information tool for controlling chemical/physical properties of the bio-oil. Previously we have examined cellulose pyrolysis, but relatively little is known regarding the chemical species that are produced from fast pyrolysis of the hemicellulose fraction of biomass. In the present work, hemicellulose fractions were isolated from different types of biomass feedstocks (hardwood, softwood, herbaceous and agricultural waste), pyrolyzed, and the resulting chemical species were quantified. Formic acid, glycolaldehyde, acetol and furfural were identified as the primary products from hemicellulose pyrolysis. The effect of feedstock chemistry, temperature and ash present in the biomass on the yields of these products will be discussed.

See more of this Session: Catalytic Biomass Pyrolysis and Gasification II
See more of this Group/Topical: Fuels and Petrochemicals Division

Reaction Kinetics of Coal Gasification with the N2/O2/CO2 Mixture

Muhammad Faisal Irfan, Living and Environmental Sciences, Fukuoka Women's University, Fukuoka, Japan

Professor Katsuki Kusakabe, Living and Environmental Sciences, Fukuoka Women's University, Fukuoka, Japan

The combustion characteristics of Datong coal in the different $N_2/O_2/CO_2$ environments have been extensively studied by using thermo-gravimetric technique. The effects of combustion environment, oxygen and CO_2 concentrations, particle size and different heating rates were considered. Results show that the coal pyrolysis in CO_2 environment can be divided into three stages: moisture release, devolatilization and char gasification by CO_2 at higher temperature zone. In the lower temperature zone, the mass loss rate of coal pyrolysis in CO_2 environment is higher than that in N_2 environment and the burning process of coal in O_2/CO_2 environment is almost same compared with that in O_2/N_2 environment for equivalent oxygen concentrations but slightly lesser than O_2/N_2 environment. With the increase of oxygen concentration, the burning rate of coal increases and burnout time is shortened. As the heating rate increases, coal particles are faster heated in a short period of time and burnt in a higher temperature region, but the increase in heating rate has almost no obvious effect on the combustion mechanism of coal. Applying different methods, values of the Arrhenius activation energy E and pre-exponentional factor A were estimated from the TGA data obtained via heating samples at different heating rates. A decrease of activation energy values were observed with the increase of different heating rates.

See more of this Session: Catalytic Biomass Pyrolysis and Gasification II
See more of this Group/Topical: Fuels and Petrochemicals Division

FAST PYROLYSIS OF WOOD LIGNIN: TOWARDS A CONTINUOUS PROCESS

Sedat H. Beis[1], Saikrishna Mukkamala[1], Nick Hill[1], Ta-Hsuan Ong[3], Adriaan Van Heiningen[1], Brian G. Frederick[2,4], M. Clayton Wheeler[1], Elizabeth Stemmler[3] and William J. DeSisto[1,4]

[1]Chemical and Biological Engineering, University of Maine, Orono, ME
[2]Department of Chemistry, University of Maine, Orono, ME
[3]Department of Chemistry, Bowdoin College, Brunswick, ME
[4]Laboratory for Surface Science and Technology, University of Maine, Orono, ME

Lignin is a major element of wood structure and has great potential as a renewable resource of biomass for the production of biofuels and valuable chemicals. Lignin has significant advantages over other biomass sources because it is not in the food supply chain and is an industrial side/by product in the paper and pulp industry as well as the biofuels fermentation industry. Although it exhibits challenges due to it is complex structure, conversion of lignin into biofuels and chemicals has received attention over the years. Most of the work done on lignin pyrolysis until now is limited to research at the analytical level. The nature of lignin molecule makes it difficult to be processed by using conventional fast pyrolysis methods. Therefore, there is a need for the development of a process for the efficient conversion of lignin into fuels and chemicals. In this respect, fast pyrolysis of lignin was studied on bench scale to assess the effects of process parameters such as moisture, feed rate, fluidization gas flow and temperature on the overall conversion and product quality. The viscosity and the average molecular weight of the lignin oil were determined. The liquid product was characterized for the chemical structure identification by employing ^{13}C NMR spectroscopy. Gas components were analyzed and quantified by using gas chromatography.

Catalytic Production and Upgrading of Biomass Derived Monofunctional Hydrocarbons

Elif I. Gurbuz, Chemical and Biological Engineering, University of Wisconsin - Madison, Madison, WI

Edward L. Kunkes, Chemical and Biological Engineering, University of Wisconsin - Madison, Madison, WI

Dante A. Simonetti, Department of Chemical, University of California at Berkeley, Berkeley, CA

Ryan M. West, Chemical and Biological Engineering, University of Wisconsin - Madison, Madison, WI

Juan Carlos Serrano-Ruiz, Chemical and Biological Engineering, University of Wisconsin - Madison, Madison, WI

Christian A. Gaertner, Chemical and Biological Engineering, University of Wisconsin - Madison, Madison, WI

James A. Dumesic, Chemical and Biological Engineering, University of Wisconsin - Madison, Madison, WI

We have studied the deoxygenation/ reforming of biomass derived carbohydrates to yield monofunctional hydrocarbons that can be utilized as fuels, commodity chemicals and solvents. Additionally, we have developed C-C coupling processes to upgrade these monofunctional species into fuel grade components that can be utilized in the current transportation infrastructure.

We have demonstrated the catalytic conversion of biomass-derived carbohydrates, including monosaccharides and sugar alcohols, into hydrophobic mixtures of monofunctional C4-C6 hydrocarbon species containing alcohols, ketones, heterocyclic compounds and carboxylic acids. The conversion occurs over carbon supported Pt-Re catalysts and removes more than 80% of the initial oxygen content of the sugars and polyols, yielding a spontaneously-separating organic phase. This process operates at moderate pressures (20-30 bar) and temperatures (283-523 K) and utilizes highly concentrated aqueous feeds (40-60%) of sorbitol or glucose. At 503 K and 18 bar, Pt-Re/C showed excellent stability for longer than one month time-on-stream and yielded an organic stream containing ~ 50% of the carbon found in the 60 wt% sorbitol feed. A yield of 70% of the maximum possible conversion of the carbon in sorbitol to monofunctional species was obtained, corresponding to the production of 1 kg of organic for every 4 kg of sorbitol.

Furthermore, we have studied catalytic C-C coupling processes to convert functional species (carboxylic acids, alcohols and ketones) derived from carbohydrate conversion into C7-C12 ketones, that can be converted into diesel grade alkanes via deoxygenation over solid acid supported metal catalysts such as Pt/NbOPO4. These processes include ketonization - in which two carboxylic acid molecules combine to form a linear ketone, CO2 and water, and aldol condensation/hydrogenation - in which two ketone or secondary alcohol molecules combine to form a singly-branched ketone. We have studied ketonization of the aforementioned carbohydrate derived organic species over a CeZrOx catalyst at 648-673K and found near 100% conversion of carboxylic acids into C7-C11 linear ketones. The aldol condensation/hydrogenation process occurs on bi-functional catalysts that contain acid/basic functionality as well as metal sites to dissociate hydrogen. Aldol condensation/hydrogenation was studied over a low loading (0.25 wt%) Pd/CeZrOx catalyst at 623K, and was found to convert ~60% of the condensable ketones and alcohols found in the aforementioned carbohydrate-derived mixtures into C8-C12 branched ketones.

Further investigation of the aldol condensation/hydrogenation reaction was performed by examining the reactivity of a representative ketone - 2-hexanone over Pd/CeZrOx and CeZrOx catalysts at temperatures between 573 and 673 K, and pressures of 5 to 26 bar. Reaction kinetics studies show that in addition to the expected C12 condensation product (7-methyl-5-undecaone), the CeZrOx-based catalysts produce C18 and C9 secondary species, along with light alkanes (<C7). Low loadings of Pd (e.g., 0.25 wt. %) lead to optimal activity and selectivity for the production of C12 species. The high activation energy of C9 formation

(140 kJ/mol) compared to the formation of C12 and C18 species (15 and 28 kJ/mol, respectively) indicate that these species may be formed as a result of the decomposition of heavier condensation products. The self-coupling of 2-hexanone was found to be positive order in both 2-hexanone and hydrogen. The addition of primary alcohols and carboxylic acids as well as water and CO_2 to the feed was found to reversibly inhibit the self-coupling activity of 2-hexanone.

See more of this Session: Production of Fungible Biofuels From Lignocellulose
See more of this Group/Topical: Forest and Plant Bioproducts Division

ENERKEM'S THERMOCHEMICAL BIOREFINERIES: Converting Heterogeneous Biomass Into Alcohols

David Lynch, Enerkem Inc., Montreal, QC, Canada

INTRODUCTION

Enerkem's proprietary thermo-chemical technology platform is designed to use non-homogeneous waste as feedstock to produce a clean syngas suitable for catalytic synthesis. This opens the door to the use of negative-cost feedstock such as municipal solid waste (MSW) and contaminated biomass such as de-commissioned electricity poles. Enerkem can produce many valuable transportation fuels and green chemicals using a wide variety of feedstock due its extensive gas conditioning process which produces a tailored syngas that can be converted into alcohols using established and readily available industrial catalysts at the proper temperatures and pressures. Enerkem's products include methanol and ethanol, other advanced fuels, such as synthetic diesel, dimethyl ether, and synthetic gasoline, as well as other green chemicals. Additionally, Enerkem's gasification technology allows it to produce renewable electricity. The end-product mix makes it possible for Enerkem to develop a wide range of projects based on market conditions from biorefineries to renewable electricity plants.

Enerkem’s proprietary technology platform has been developed completely in-house. Since 2000, the Company has followed a rigorous path to support commercial viability and proceeded through each developmental step: lab to pilot to demo commercial plant. The incremental steps in commercialization resulted in more than 3,600 hours of exhaustive piloting and optimizing of both design and feedstock processing.

With its multi-product and feedstock-flexible technology platform, Enerkem helps reduce greenhouse gas emissions, while contributing to a greener economy. By producing fuels locally rather than depending on foreign oil, Enerkem contributes to increasing energy independence and reducing imports.

OVERVIEW

The development of Enerkem's technology platform has been done completely in-house. R&D work started in 1999 with the bench scale syngas island. Over 1,000 testing hours were conducted. Twelve R&D staff was devoted to the work. At that time, the Company started to patent its technology.

In 2003, the Company started operating its pilot plant in Sherbrooke, Québec. The pilot plant produces syngas, methanol, and second-generation ethanol, and is equipped with various sampling ports for data collection. It is also well instrumented and is automated for testing and reporting. The system can feed solid materials, slurries, and liquids. To date, approximately 20 different feedstocks have been used to test and validate the technology, and for engineering design purposes. These feedstocks include:

- Municipal Solid Waste (4 cities);
- Forest biomass (as well as treated wood and C & D wood, both pelletized and chipped);
- Agriculture biomass (wheat straw, corn stovers);
- Plastics (various forms: pelletized, shredded, fluffed and mixed);
- Glycerine (rich fraction of biodiesel production after transesterification);
- Biosolids (dried, from Montreal and from pulp & paper operation);
- Petcoke (Alberta, with 8% humidity and 5-6% sulfur)

Unlike many project developers that license outside technology, Enerkem's technical team's intimate knowledge of the technology and its dedication to the technology development efforts give it the necessary internal capacity to engineer, build and operate its own plants.

Enerkem's first commercial-scale biofuels and biochemicals plant is currently producing its premium syngas. It is located in Westbury, Quebec. It will be the world's first ethanol plant to use negative-cost and unconventional materials - treated wood from used electricity poles. The plant will, at term, produce 1.3 million gallons of second-generation ethanol annually.

The Company will also build and operate a plant that will produce 10 million gallons of next generation biofuels annually, from sorted municipal solid waste. The City of Edmonton will supply a minimum of 100,000 tonnes of sorted municipal solid waste per year. Construction is expected to start in 2009. Another commercial project is under development. The Company will transform sorted municipal solid waste and woody biomass in Pontotoc, Mississippi. It plans to produce 20 million gallons of ethanol per year.

In Enerkem's bubbling fluidized-bed gasifier, a continuous fee of biomass is carried into a reactor where an inert heat carrier (i.e. sand) is "fluidized" in a zone known as "the bed" under relatively low temperatures and moderate pressures, hence called "low severity gasification". Enerkem's technology platform consists of three distinct process steps: gasification of biomass; cleaning and reforming of synthetic gas (syngas), and; conversion of syngas into biofuels and green chemicals via catalysis. In Enerkem's bubbling fluidized-bed gasifier, a continuous feed of biomass is carried into a reactor where an inert heat carrier (i.e., sand) is "fluidized" in a zone known as "the bed" under relatively low temperatures (700–750°C) and moderate pressures (~2 atm.). In this environment, biomass is converted into gas and small amounts of ash. Oxygen and steam are used as fluidizing gases and gasification agents. Gasification takes place throughout the bed. The injected oxygen and added steam prevent the ash from melting and maintain the injection nozzles at appropriate temperatures, so available steels can be used, thus avoiding exotic and expensive ceramics. The syngas is drawn from the top of the gasifier. In its movement towards the top of the bed, the synthetic gas disengages from the sand and undergoes extensive cracking and reforming in the freeboard stage – key chemical processes, which convert organic intermediates to simple molecules, having the desired molecular characteristics (H2, CO, CH4 and CO2 – the main constitutive molecules in synthetic gas). The superior mixing, which occurs in the bubbling fluidized bed used by Enerkem, generates high rates of heat and mass transfer which subsequently yield stable temperatures, resulting in well controlled reaction kinetics. Thus, Enerkem's reactor is capable of handling feedstock that is not uniform, is geometrically dissimilar and is heterogeneous. Enerkem's reactor is also efficiently sized due to high throughput rates facilitated by the design – a feature that contributes to the commercial viability of the Enerkem platform.

The raw syngas produced in Enerkem's gasifier is subjected to sequential conditioning as follows: (i) cyclones remove particulates (pilot plant capture efficiencies are ~90%); (ii) secondary cracking and steam-carbon reactions reduce tar and carbon loads for the next step, catalytic steam reforming, which is carried out in a fluid bed configuration, using proprietary reforming catalysts; (iii) heat recovery is performed on the reformed syngas, which is quenched and scrubbed through a two stage system, that ensures removal of any residual Cl and S and that recovers the NH3 produced. The latter is recycled back or is separated and sold; (iv) any residual tar or light hydrocarbon that has escaped the fluid bed reforming operation is recovered and, as an emulsion, reinjected into the gasifier where it is converted into syngas, as well as green chemicals, including acetic acid and acetates.

The catalytic conversion into alcohols is the fourth step of the process. The cleaned syngas is a mixture of H2, CO, CO2, residual argon and N2 and residual methane (only traces of other light hydrocarbons are present in the syngas). The molar ratio H2/CO resulting from the reforming is essentially 2 with no necessity of shift conversion. Should a specific molar ratio be desired, the reforming step can be adjusted, or a classical shift conversion can simply be added, either before or after the compression step. Enerkem's extensive gas conditioning process produces a tailored syngas that can then be converted into alcohols using established and readily available industrial catalysts at the proper temperatures and pressures. Enerkem can produce methanol and ethanol, other advanced fuels, such as synthetic diesel, dimethyl ether, and synthetic gasoline, as well as other green chemicals.

Enerkem has developed a modularized compact design of "100,000 metric tons in - 10 million gallons of EtOH out". Its manufacturing system is easily scalable with skidded 10MM gallon-per-year reaction vessels that can be used in an array of larger plants. This gives the company more flexibility in engineering plants of different sizes and allowing it to replicate the design without significant re-engineering.

The use of waste biomass as feedstock is economically attractive given the negative cost of the raw material. However, the presence of contaminants in the waste requires an effective sequence of conditioning steps in the overall process, to: (a) produce a syngas that is clean and; (b) handle contaminants in a manner that they become fully neutralized and thus can be safely disposed of.

Gas- and Aqueous-Phase Hydrogenation of Acetic Acid Over Ru Catalysts

George W. Huber, Chemical Engineering, University of Massachusetts, Amherst, MA

Hakan Olcay, Chemical Engineering, University of Massachusetts, Amherst, MA

Ye Xu, Center for Nanophase Materials Sciences, Oak Ridge National Laboratory, Oak Ridge, TN

Research on biofuel production from biomass has rapidly increased due to the depletion in fossil fuel reserves and to the rising concerns with greenhouse gas emissions. Hydrogenation reactions, and particularly the hydrogenation of organic acids, are critical in most pathways for production of biofuels. This work presents and compares the results of Ru catalyzed acetic acid hydrogenation in both gas and liquid phases in various types of reactor configurations. In this presentation we will show that the aqueous phase does not change the catalytic properties, and that the catalytic activity is a function of the concentration of both the acetic acid and the hydrogen in the reactor. We will compare our experimental results with density functional theory (DFT) studies to gain a better understanding of the reaction chemistry.

All experiments were carried out with a Ru/Carbon catalyst. We studied the effect of water concentration at 175"aC in a flow reactor by keeping the acetic acid and hydrogen concentrations constant (0.0016 M and 0.061 M, respectively), and by varying the water/water vapor concentration in the system (0-0.13 M). We have observed that the presence of liquid water has very little effect on the activity and selectivity of the catalyst. It has been concluded that the hydrogen and acetic acid concentration on the catalyst surface changes the catalytic properties.

We also compared acetic acid hydrogenation in both batch and flow reactors. Again in these experiments the hydrogen and acetic acid partial pressures are the controlling variables. In the flow reactor the hydrogen partial pressure is typically significantly higher than in typical batch reactors. Therefore, flow reactors are significantly better hydrogenation reactors than batch reactors.

These results provide important insights as to how hydrogenation reactions can be improved.

See more of this Session: Production of Fungible Biofuels From Lignocellulose
See more of this Group/Topical: Forest and Plant Bioproducts Division

Commercializing Pyrolysis Oil Into Motor Fuels Leveraging the Existing Refining Infrastructure

Thomas D. Foust, National Renewable Energy Laboratory, Golden, CO

Robert Baldwin, National Bioenergy Center, National Renewable Energy Laboratory, Golden, CO

Jim Frederick, National Bioenergy Center, National Renewable Energy Laboratory, Golden, CO

Fast pyrolysis of biomass shows considerable promise as a conversion route for converting lignocellulosic biomass into fungible motor fuels. Unlike other biomass conversion routes that produce a gaseous or sugar intermediate, pyrolysis produces a liquid intermediate that could potentially be processed at an existing petroleum refinery into conventional motor fuels synergistic with conventional crude oil refining. Processing pyrolysis oil at conventional refineries would significantly reduce costs of processing and facilitate the rapid transition into the market place. However, raw pyrolysis oil has numerous qualities which will prevent it from being processed at conventional refineries such as its corrosiveness, insolubility in hydrocarbons and thermal and temporal instability. Because of these characteristics, raw pyrolysis oils must be upgraded before they would be acceptable to enter a refinery for further processing into conventional motor fuels.

Raw pyrolysis oils typically contain up to 40% oxygen and this high oxygen content is linked to many of the undesirable properties of pyrolysis oils, hence oxygen removal is critical. A detailed market assessment and experimental program has been undertaken to address the challenges of introducing pyrolysis oils into conventional refineries from both a technical and cost barrier perspective. This study looked at various intermediate oxygen levels to determine the physical and chemical properties and the processing costs associated with achieving these intermediate states. Several scenarios were identified where an initial state processing to approximately 10% oxygen would be sufficient to allow entry into a conventional refinery for ultimate processing to conventional motor fuels at approximately 0.2% oxygen. This two stage processing maximizes the advantages of economies of scales and existing infrastructure utilization. Results of this program will be shown that indicate that this approach shows considerable promise for producing conventional motor fuels from biomass at costs highly competitive with petroleum derived fuels.

See more of this Session: Production of Fungible Biofuels From Lignocellulose
See more of this Group/Topical: Forest and Plant Bioproducts Division

Thermal Pretreatment Options for Thermochemical Conversion of Lignocellulosic Biomass

Wei Yan, Chemical Engineering, University of Nevada, Reno, Reno, NV

Charles J. Coronella, Chemical Engineering, University of Nevada, Reno, Reno, NV

Tapas Acharjee, Chemical Engineering, University of Nevada, Reno, Reno, NV

Victor R. Vásquez, Chemical Engineering, University of Nevada, Reno, Reno, NV

Thermochemical conversion of lignocellulosic biomass is a promising technology for production of renewable power, fuels, and chemicals. Both pyrolysis and gasification of biomass have significant technical barriers that must be eliminated for successful widespread commercialization. Feedstock handling is complex due to the diverse nature of important feedstocks, including commercial timber, slash removed due to fire hazard reduction, switchgrass, and agricultural residues as diverse as rice hulls, corn stover and wheat chaff. Seasonal availability and low forest density make feedstock logistics complex and expensive. A process to homogenize the feedstocks that will simultaneously produce a stable energy-dense fuel is needed. Two such processes are considered here: dry torrefaction, and wet torrefaction.

Dry torrefaction is a low-temperature pyrolysis process in which the biomass is heated in a chemically inert environment at temperatures ranging between 200 and 300 °C. The reaction produces two products, a solid and a gas. The solid has about 70% - 80% of the mass, and approximately 90% of the fuel value of the original biomass.

Biomass is treated in hot compressed water in the case of wet torrefaction, also known as hydrothermal pretreatment. This process results in three products, including gases, aqueous chemicals, and a solid fuel. The temperature is in the range of 225° - 275 °C, and the pressures are up to 60 bar. The solid product contains about 50 - 70% of the mass and 70 - 90% of the fuel value of the original feedstock. The gas product is about 10%, and the aqueous chemicals make up the balance.

We have characterized experimentally the solid fuels resulting from both processes, which indicate significant similarity. Solid product from both processes have an enhanced energy density relative to the starting feedstock, and are is friable (and might be easily pelletized.) In both cases, an ultimate analysis of the solid fuels reveals decreased oxygen content and increased carbon content. In both cases, a fiber analysis of the fuel demonstrates that a significant fraction of the original hemicellulose component has been extracted. Proximate analyses reveal fuels suitable for gasification, and possibly for pyrolysis. The solid fuel in both cases exhibits hydrophobic qualities, as demonstrated by equilibrium moisture.

See more of this Session: Production of Fungible Biofuels From Lignocellulose
See more of this Group/Topical: Forest and Plant Bioproducts Division

Pyrolysis investigations and ash melting behaviour of agricultural residues used as a fuel in small furnace installations

Martina Poppenwimmer[1], *Harald Raupenstrauch*[1]

[1]*Chair of Thermal Processing Technology, University of Leoben, Franz-Josef-Strasse 18, 8700 Leoben, Austria (martina.poppenwimmer@unileoben.ac.at).*

Abstract

During thermal combustion of solid fuel in furnace installations the applied solid is first dried, then pyrolysed and finally the gasification and combustion reactions take place. In this process the pyrolysis is an important sub-step, in which the volatiles are released. Especially in case of biomass it is of special interest, since more than 85 % of the solid may be converted.

Unfortunately biomass like agricultural residues involves various problems when used as fuel in small furnace installations. The most important reasons are high ash contents on the one hand side and the low melting points of ash on the other, which are mainly caused by a high rate of alkali metals. For preventing the agglomeration of ash it is necessary to provide ideal combustion conditions and to avoid temperature peaks. Therefore it is essential to get more detailed information of devolatilisation behaviour of the various fuels.

In addition to furnace design and operation conditions it is possible to manipulate the fuel properties by additives in order to reduce ash melting and agglomeration. Concerning the selection of additives attention was paid to the high content of alkaline earth metals, because these elements increase the ash melting point.

The results of the tests show on the one hand the influence of several additives to ash melting behaviour and on the other the effect on the progression of calorific value of formed volatiles.

Introduction

Because of still decreasing fossil reserves it is consequently searched for alternative fuels. One example is wood, but it is essential to minimize the dependence on wood industry. Therefore other biomass like agricultural residues is analysed. However there are some difficulties and problems in use of these residues for generation of energy. In addition to the high ash content and the low ash melting point also the nitrogen content of the fuel is a problem. But it is possible to reduce it fractionally by means of elution. Due to washing of the residues also the unwanted ingredients chlorine and potassium can be reduced and furthermore corrosion damages by condensation on cold plant components can be avoided. However the washing of straw involves high costs.

Concerning the low melting behaviour of ash it is also caused by fuel content (see Fig.1).

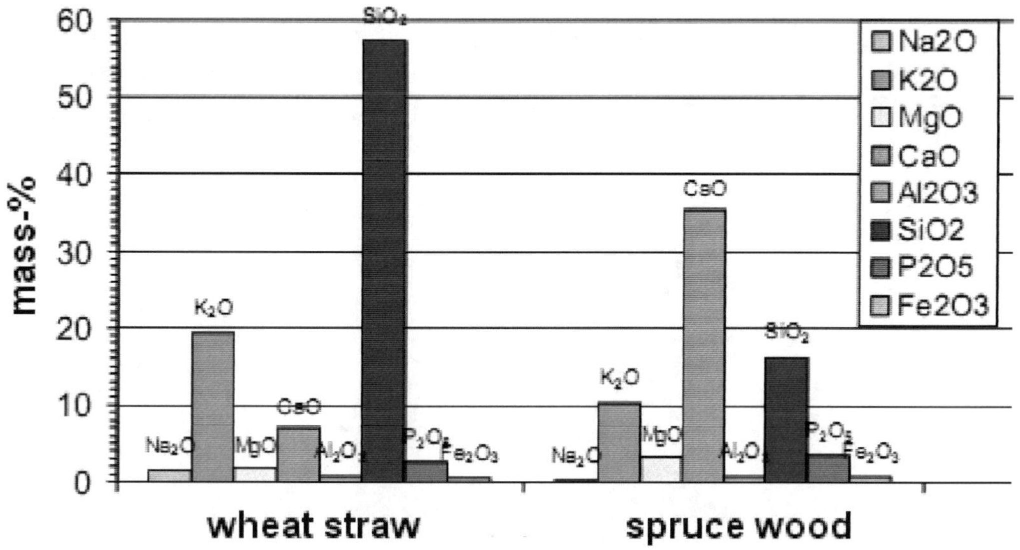

Figure 1: Typical values for ash composition [1]

Compared with spruce wood wheat straw contains high value of SiO_2 and K_2O. Alkaline metals like potassium and sodium lower the ash melting point as well as silicon. On the other hand alkaline earth metals increase the melting point. The content of these metals in wheat straw is rather low. For improving the ash melting behaviour and reducing agglomeration it is possible to manipulate the fuel properties by additives. Concerning the selection of additives attention is paid to the high content of alkaline earth metals. Therefore materials like lime flour, which includes mainly calcium carbonate with a small share of magnesium carbonate, are used. Two other possibilities, which were applied for analyses, are blast furnace slag and molasses. However, it is important to add only a smidgen of an inert substance, because the ash volume and the abrasion resistance get worse with it. Only in case of molasses the abrasion resistance might be possibly improved. [1,2,3]

Additives influence not only the melting behaviour but also the devolatilisation of the solid fuel. The analyses of pyrolysis are operated with an analysis device, which includes a thermo gravimetric scale (TGS) in combination with a differential scanning calorimeter (DSC). Because of the DSC it is possible to investigate the time dependency of the heat of reaction of volatiles and in combination with the measured mass loss the calorific value respectively. So the devolatilisation behaviour of pure wheat straw pellets was compared with them of the mixtures with additives.
With the knowledge of the exact curve progression of calorific value the yield of useable heat can be improved and emissions minimized and consequently low environment pollution will be achieved.

Analysis methods

For analysing the fusibility of fuel ash, first the biomass residues are milled to analytical grade. Then they are incinerated in a muffle oven according to CEN TC335/WG N43 (standard for investigation of ash content of solid biomass fuels). Finally the ash is detected with a LEITZ hot stage microscope according to ÖNORM CEN/TS 15404 (Solid derived fuels - Process for determination of fusibility of fuel ash in using various temperatures). During the investigations the sintering and softening temperature are detected. At sintering temperature the sample reaches only 95 percent of his input size. At this temperature the sintering process of the ash is starting and the particles adhere together. First indications of ash melting occur at the softening temperature.

Pyrolysis investigations are made with the pilot plant of the Chair of Thermal Processing Technology (TPT) at the University of Leoben, which combines a Thermo Gravimetric Scale (TGS) and a Differential Scanning Calorimeter (DSC). A flow sheet of this test facility is given in Figure 2.

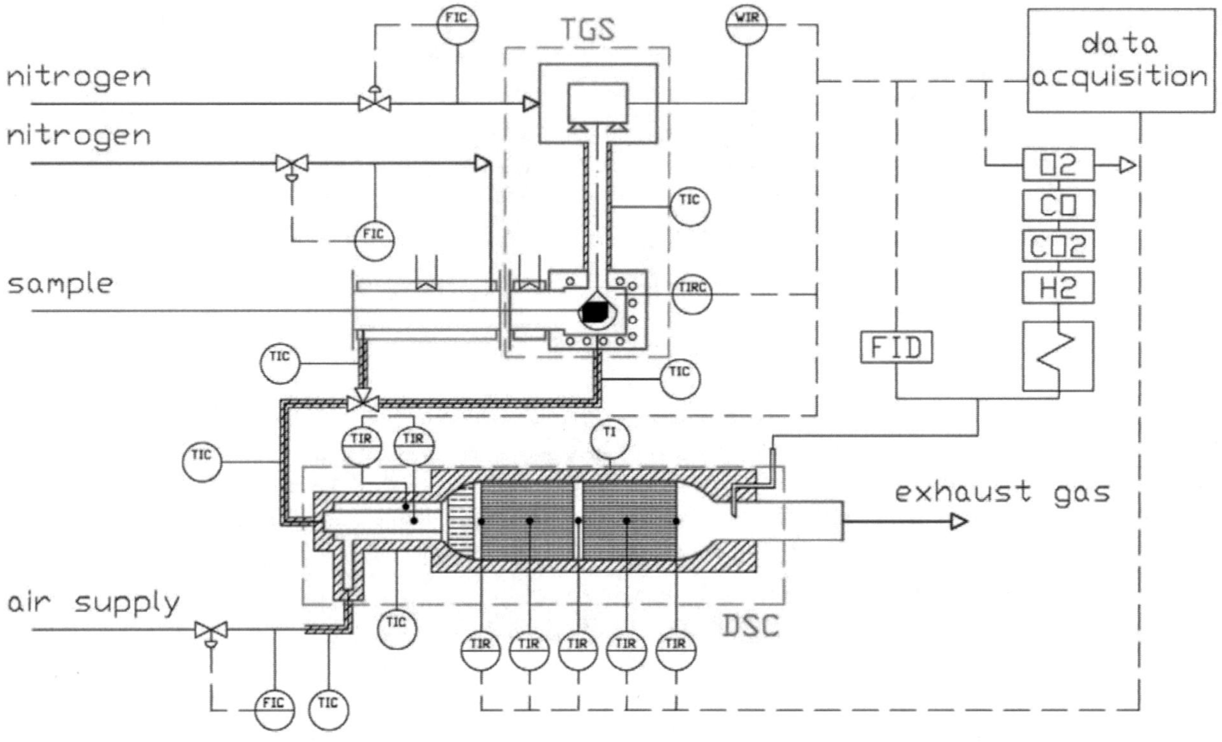

Figure 2: Flow sheet of test facility [4,5]

The fuel is pyrolysed in a muffle oven on a scale basket under nitrogen atmosphere and afterwards the produced pyrolysis gas is mixed with air. In the heat exchanger the pyrolysis gas and the combustion air are equalised and after a mixing comb burned on the surface of a platinum comb catalyst. The occurring difference of temperature in the catalyst during the combustion is quantified and with it the heat of reaction of volatiles is determined.
After the catalyst a sampling unit is implemented, which leads the flue gas to the analysers. For synthesis analyses O_2, CO_2 and CO analysers as well as a flame ionisation detector are installed.
In addition, the calorific value of the synthesis gas is also calculated, simultaneously measuring the mass loss of biomass (cf. 1).

$$H_{u(t)} = \frac{Q_{(t)}}{dm/dt} \qquad (1)$$

For determining the oxygen demand of the volatiles dependent on time, an oxygen mass balance of the DSC was performed. Thereby the oxygen entry is known because of the defined combustion air and the oxygen in the flue gas is measured. The oxygen demand of all volatiles was considered as the difference between these two values (cf. 2).

$$O_{2,dem} = O_{2,in} - (O_{2,out} + 5 \cdot C_3H_{8,eq}) \qquad (2)$$

Additionally the calorific value of the fuel was investigated with a bomb calorimeter in order to close the energy balance of the test evaluation and to get a verification of the established results compared with the theoretical ones. [4,5]

Results

Ash melting behaviour

For the investigations of ash melting behaviour the additives lime flour, molasses and blast furnace slag (BFS) were used.
The results at figure 3 show, that a larger quantity of additive does not automatically mean a further increasing of temperature. The non expectable reduced temperature can be explained with occurring eutectics. Especially in case of addition of lime flour a better increasing of the sintering temperature is noticeable with 3% additive than with 4%. The same way show the mixtures with molasses. 10% molasses involve a better effect on sintering and softening temperature than 20% molasses.

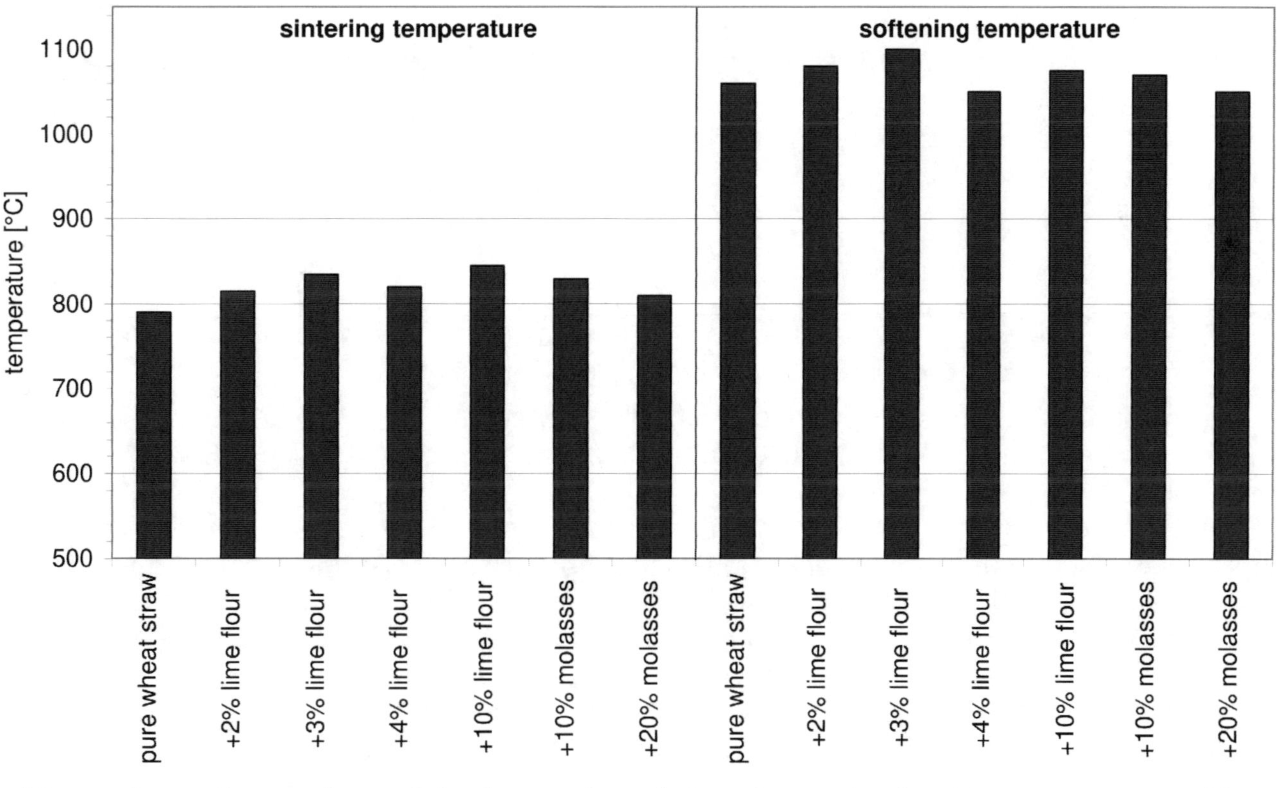

Figure 3: Comparison of the increasing of sintering and softening temperature with additives

The softening temperature of the analysed wheat straw is 1060°C. This value is rather high compared with values from literature. There you will find softening temperatures of wheat straw from 860 to 910°C.

Pyrolysis investigations

Pure wheat straw pellets and in combination with additives like blast furnace slag and molasses were determined about their devolatilisation behaviour at a temperature of 800°C.
For closing the energy balance of the evaluation, the calorific value of pyrolysis coke of fuels is investigated with a bomb calorimeter in addition to the calorific value of the solid.
Always two pellets with a length of 15 millimetres and a diameter of 6 millimetres, which weight is about 1 gram, were put on the scale basket and pyrolysed.

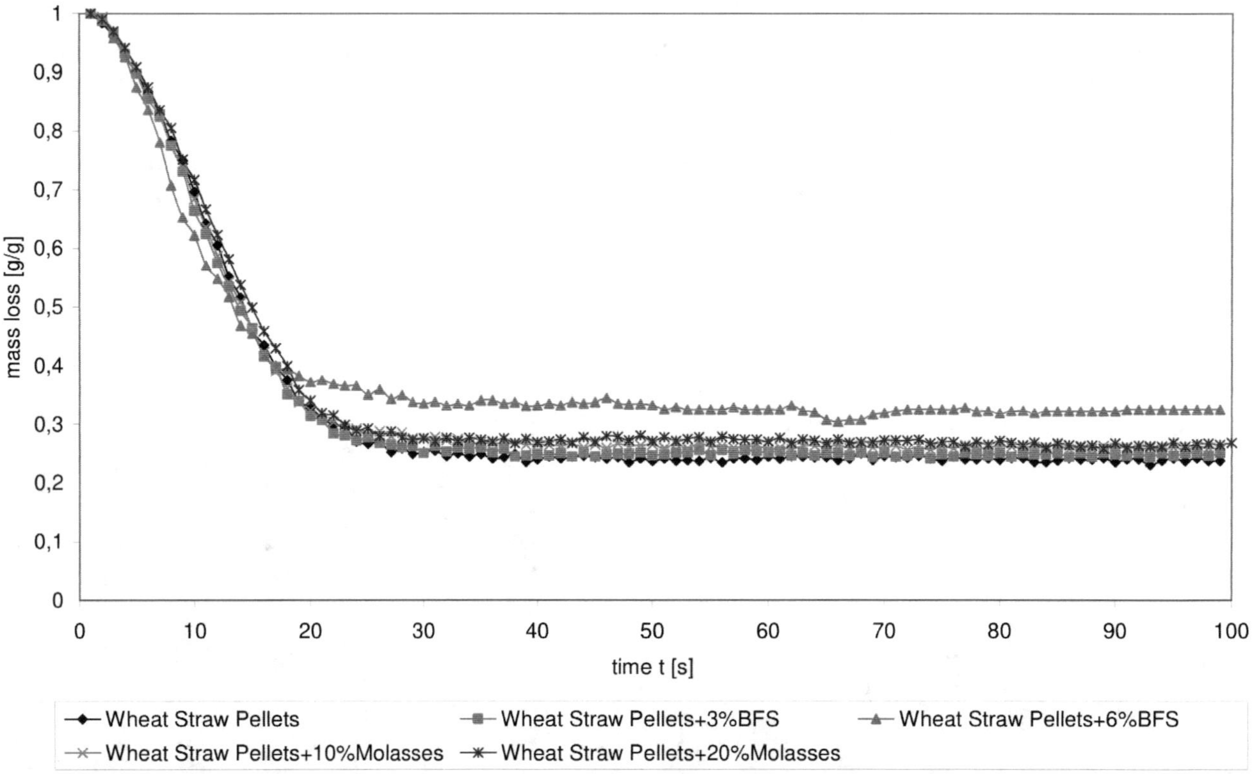

Figure 4: Mass loss of 15mm long wheat straw pellets (m_0 ~1 g) dependent on time (T_{oven} = 800°C, nitrogen atmosphere, $T_{catalyst,start}$ = 480°C)

In figure 4 it is well noticeable that the mass losses of the various samples proceed a similar way. All curves continue linear after about 30 seconds. At this moment the pyrolysis of the fuel is finished. The small variations at the end of the mass curves are effected on the one hand by the vibrations of the scale, which are caused by the perfusion of the furnace chamber and on the other hand probably by the small sample mass, whereby the solid fuel can't stand smooth on the basket cage. But these variations do not influence the following evaluations, because the mass loss curve is fitted by a proper function (cf. 3).

$$m_{loss} = \frac{a + c \cdot t^2 + e \cdot t^4}{1 + b \cdot x^2 + d \cdot x^4} \qquad (3)$$

Considering figure 4 it is observable that the share in pyrolysis coke is different for each sample. The lowest value is measured for pure wheat straw pellets (23%). This means that 77 % of the solid were pyrolysed. For wheat straw with 6% blast furnace slag 32% coke are detected. Because of the inert blast furnace slag the content of ash is much higher than in case of molasses as additive and following also the coke fraction is the biggest. Coke is composed of ash and fixed carbon.

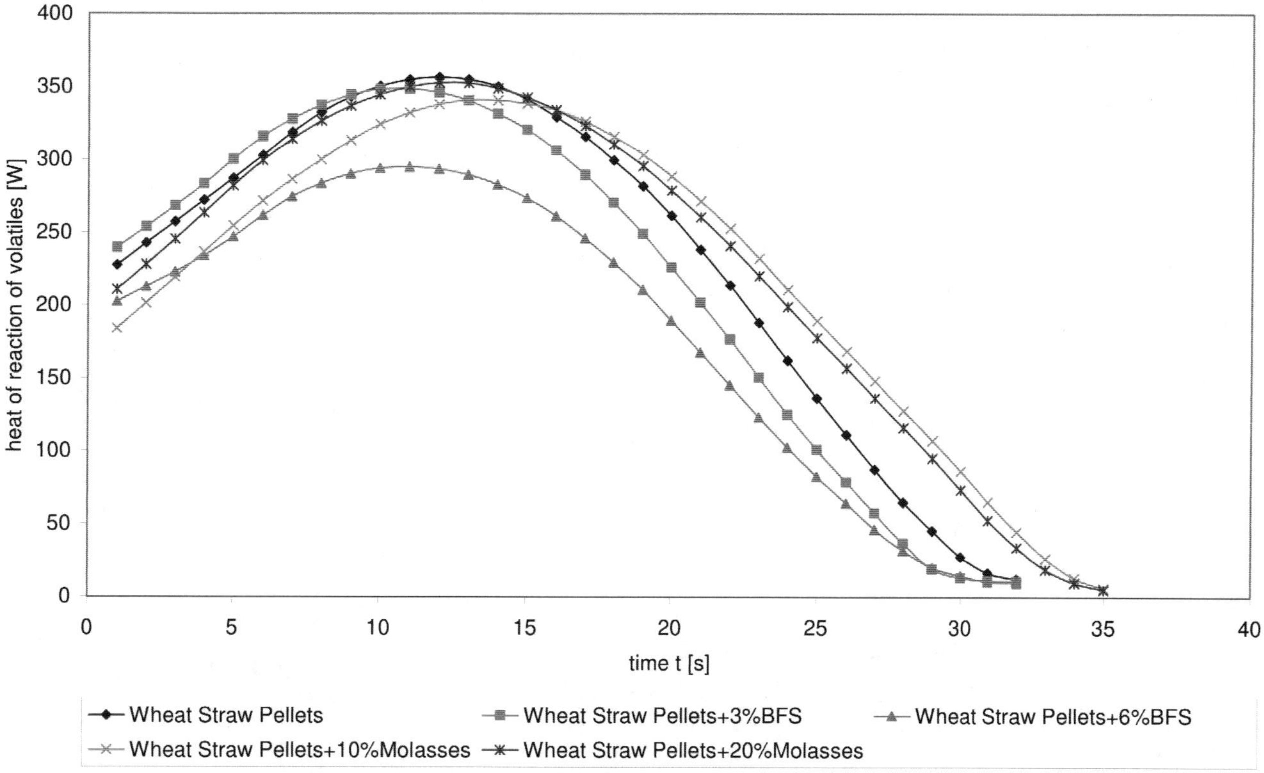

Figure 5: Heat of reaction of all volatiles of 15mm long wheat straw pellets (m_0 ~1 g) dependent on time (T_{oven} = 800°C, nitrogen atmosphere, $T_{catalyst,start}$ = 480°C)

Figure 5 shows the average values of heat of reaction of volatiles from all experiments. The curve of pure wheat straw pellets has got the highest maximum and the total released energy the highest value as well. Due to addition of inert blast furnace slag the maximum and also the total energy decrease. This confirms the results of the measurements with bomb calorimeter. The sum of the heat of reaction of volatiles from wheat straw pellets with 10% molasses and 20% molasses is nearly the same as the one of pure wheat straw. Molasses is unlike blast furnace slag an organic material, which has got a calorific value a little lower than the one of wheat straw.

The courses of all heat of reactions of volatiles proceed in a similar way. At the beginning many volatiles release, because the energy increases rapidly. The end of pyrolysis is given after about 30 seconds for the experiments with blast furnace slag as additive. The other pyrolysis processes of wheat straw with molasses end approximately after 35 seconds. This can be explained with structure of the pellets. Pellets in combination with blast furnace slag are not so compact like those with molasses, which acts as a binding material and thereby volatiles can not exhaust so easily.

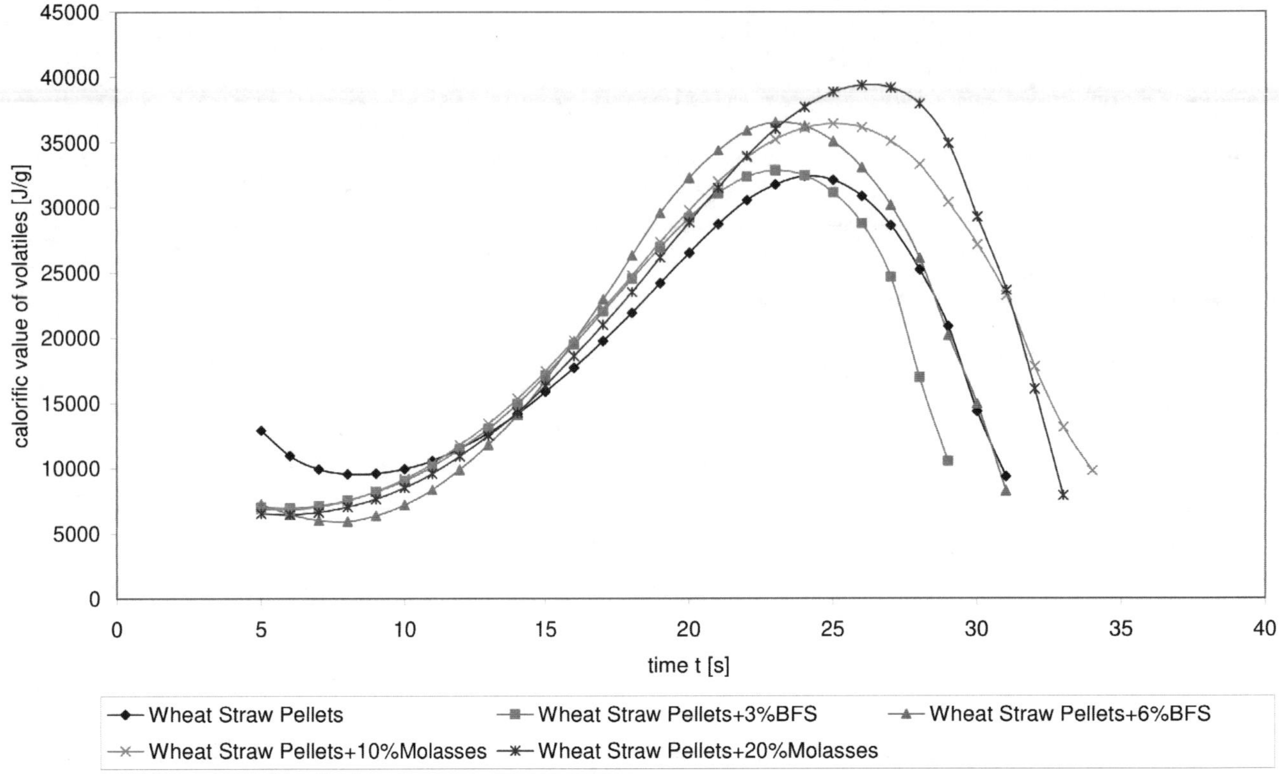

Figure 6: Calorific value of volatiles of 15mm long wheat straw pellets (m_0 ~1 g) dependent on time (T_{oven} = 800°C, nitrogen atmosphere, $T_{catalyst,start}$ = 480°C)

The calorific value of the volatiles (cf. figure 6) rises to a maximum at the end of pyrolysis. At the beginning only volatiles with a low specific calorific value exhaust. The end of the averaged devolatilisation of each sample combination differs a little bit. First the calorific value curve of wheat straw with 3% blast furnace slag ends after about 30 seconds. Second finish the one of pure wheat straw and 6% blast furnace slag. The two with molasses addition have got the longest exhausting time, which could be seen as well in figure 5. The compacter construction announces furthermore the high maxima of the two curves with molasses. Because of this additive the pores at the outside of the pellet are closed and the volatiles have to outgas axial. So the longer condensable hydrocarbons (tar) are easier to split on the surface of the hot ash-coke construction into shorter compounds with a high rate of hydrogen.

Blast furnace slag has also got a positive effect on cracking reactions and degassing of components with a high calorific value.

Conclusions

The results of the ash melting experiments showed that it is possible to raise the sintering and softening temperatures with addition of suitable substances. However the increasing was not very high with small amount of additives. But a larger quantity would worsen the abrasion resistance and the ash volume. Certainly it is also necessary to enhance the furnace installations for avoiding temperature tops and following ash agglomerations.

Therefore the information of pyrolysis investigation can be applied. Concerning the devolatilisation, all samples showed a similar characteristic. At the beginning the heat of reaction increased rapidly, just as the mass loss fell off. Also the length of pyrolysis was nearly the same except for the samples with molasses, which lasted a bit longer. Molasses closed the outer pores and supported the axial emission of pyrolysis gases. Due to this fact also the characteristics of calorific value of volatiles differed from the samples without additives. Towards the end of pyrolysis the specific calorific value increased higher. This points to a better tar cracking tendency into shorter components with a high share in hydrogen. Also the addition of blast furnace slag showed a rising of the maximum of calorific value curve and consequently a possible positive effect on cracking reactions.

Table of symbols

SiO_2	silicon oxide
K_2O	potassium oxide
O2	oxygen
H_u	calorific value [J/g]
m	masse [g]
Q	energy [J]
t	time [s]
dem	demand
eq	equivalent

References

1 Ottmann M., Verbrennung biogener Brennstoffe in stationären Wirbelschichtfeuerungen. PhD thesis, München University of Technology, 2006

2 Bensmann M., Stroh vergolden. Neue Energie - Erneuerbare Energie/ Biomass, Volume 03, 2006

3 Kiesewalter S. and Röhricht Ch., Pelletproduktion aus halmgutartiger Biomasse. Waemre 2004, 2004

4 Poppenwimmer M., Reaktionswärmestrom von Flüchtigen und Teer in Abhängigkeit der Zeit bei der Pyrolyse von festen Brennstoffen. Master's thesis, Graz University of Technology, 2007

5 Maurer Ch., Development of a Differential Scanning Calorimeter - Experimental Investigations of Wood and Plastics. PhD thesis, Graz University of Technology, 2006

Dissolution and Fractionation of Wood and Straw Using Ionic Liquids

Miguel A. Tavares Cardoso, Delft ChemTech, Delft University of Technology, Delft, Netherlands

Gianluca Marcotulio, Process & Energy Laboratory, Delft University of Technology, Delft, Netherlands

Jaap van Spronsen, Process & Energy Laboratory, Delft University of Technology, Delft, Netherlands

Geert-Jan Witkamp, Process & Energy Laboratory, Delft University of Technology, Delft, Netherlands

W. de Jong, Process & Energy Laboratory, Delft University of Technology, Delft, Netherlands

J. Ruud Van Ommen, Delft ChemTech, Delft University of Technology, Delft, Netherlands

The decrease in the resources of fossil fuels together with the global warming alarms have led to an intensified search for alternative fuels to supply the modern society. Among them, the concept of biorefinery is becoming increasingly important since it presents the potential to be a real alternative for large scale production of fuels with a neutral carbon dioxide balance. In this biorefinery context, biomass resources such as wood and straw have not yet been properly valorised since these materials are mostly incinerated to produce heat. In addition, production of fuels from lignocellulosic biomass, especially wood, minimizes the competition with the food chain and presents higher yields when compared to first generation biofuels. The challenge now is to make this process attractive from an economical point-of-view. One of the major difficulties when dealing with these raw materials is related with the need of pre-treatments to dissociate the biomass structure. Lignocellulose consists mainly of cellulose (35 -50%), hemicellulose (20 -35%) and lignin (5-30%). These components are assembled in a complex three-dimensional structure remarkably resistant against chemicals and microbial attacks that makes them very difficult to hydrolyze, which is key for its future utilization. An effective dissociation of these components and their separation can lead to the production of several high value products prior to the production of fuels. This integrated approach increases the competitiveness of the biorefinery industry and can make it attractive even considering the actual oil prices (2009). As previously reported, ionic liquids offer a new and attractive route for the dissolution of lignocellulosic biomass. Besides the amount of biomass that can be dissolved in some ionic liquids (that can reach 8 % w/w), their complete recycling and significant environmental advantages make this class of compounds the best possible choice for a new pre-treatment to dissolve and fractionating the lignocellulosic biomass prior to biorefinery processing.

In this work, the complete dissolution of both wood and wheat straw in ionic liquids was followed by the fractionation of the several components that constitute the biomass and the analysis of each fraction was performed to close the mass balance. Two different ionic liquids, 1-ethyl-3-methylimidazolium chloride and 1-butyl-3-methylimidazolium chloride, and their mixtures were tested. The results were compared both in terms of yield and selectivity for the different constituents of both types of biomass. The obtained results show that this is a promising process for the pre-treatment of lignocellulosic biomass, regardless the type (wood or straw). Moreover, since the ionic liquids can be almost 100% recycled, this is a low cost, fast, simple and environmentally friendly pre-treatment that can increase significantly the competitiveness of modern biorefineries.

See more of this Session: Integrated Biorefineries for Lignocellulosic Biomass
See more of this Group/Topical: Forest and Plant Bioproducts Division

An Assessment of The Recovery and Recycle of Ionic Liquids (ILs) Following Lignocellulosic Biomass Pretreatment

Thehazhnan K. Ponnaiyan, Chemical & Environmental Engineering, University of Toledo, Toledo, OH

Ananth P. Dadi, SuGanit Systems, Inc., Toledo, OH

Fei Zhao, Chemistry Department, University of Toledo, Toledo, OH

Christa M. Graham, Chemistry Department, University of Toledo, Toledo, OH

Constance Schall, Chemical & Environmental Engineering, University of Toledo, Toledo, OH

Jared Anderson, Chemistry Department, University of Toledo, Toledo, OH

Sasidhar Varanasi, Chemical & Environmental Engineering, University of Toledo, Toledo, OH

Many technical challenges exist in the development of commercially viable biorefining processes for plant (lignocellulosic) biomass, although it is an abundant, inexpensive and sustainable source of organic carbon for the production of fuels and chemicals. Pretreatment of biomass and enzymatic hydrolysis are critical steps, and most of the current pretreatment methods fail to effectively disrupt the crystallinity of cellulose, which is a major barrier for efficient enzymatic hydrolysis. Ionic liquids (ILs) being non volatile, non-flammable, recyclable and designer friendly, are gaining wide recognition as green solvents, and their unique solvating properties make them ideal for pretreating lignocellulosic substrates. IL pretreated lignocellulosic substrates like corn stover (softwood) and poplar (hardwood) are shown to produce enhanced enzymatic hydrolysis yields of glucan to glucose and xylan to xylose. However, economic viability of this promising pretreatment technique materializes only when the process solvent (IL) is completely recovered and recycled. After incubation with biomass, the IL is separated from the latter via dissolution in a suitable anti-solvent (water/ethanol). Subsequently, IL can be recovered from the anti-solvent via evaporative and/or membrane based separations. It is of importance to quantify the residual anti-solvents in the recovered IL, to establish their permissible levels.

This study reports: (1) the use of ILs as non-derivitizing green solvents for the pretreatment of lignocellulosic biomass for subsequent bio-ethanol production, (2) the recovery of ILs using an evaporative separation scheme, along with a concomitant assessment of the efficiency of the recycled IL for biomass-pretreatment over several cycles of recovery and reuse, and (3) the establishment of acceptable levels of anti-solvent in the recycled IL (without compromising on its effectiveness as a pretreatment agent), and the development of analytical methods to accurately measure these concentrations.

See more of this Session: Integrated Biorefineries for Lignocellulosic Biomass
See more of this Group/Topical: Forest and Plant Bioproducts Division

Acetic Acid Removal from Pre-Pulping Wood Extract

Aymn Abdulrahman, Byung-Hwan Um, Audrey Polifka
Adriaan van Heiningen, G. Peter van Walsum,

University of Maine, Department of Chemical and Biological Engineering
Orono, Maine 04469 USA

Abstract

Pre-pulping extraction is a means of deriving a hemicellulose-rich process stream from the front end of a Kraft pulp mill. When the extraction is carried out using green liquor, pulp quality and quantity can be retained while still releasing hemicelluloses and acetic acid for recovery as bioprocessing feedstock or chemical products. The acetic acid that is present in the wood extraction is inhibitory to microorganisms and can inhibit fermentation. It is also a commodity chemical that may provide retain sufficient value to justify recovery and purification.

In this study, a liquid-liquid extraction method is applied to extract acetic acid from a green liquor pre-pulping hardwood extract. The acetic acid removal process takes place as an initial step prior to the fermentation process. An organic solution, such as Trioctylphosphine oxide (TOPO) in Un-decane, is prepared and mixed with wood extract. Next the extract phase is centrifuged in order to separate the aqueous and organic phases. The aqueous phase is sent on to fermentation while the organic phase is sent on to distillation to separate acetic acid from the organic phase. Finally, the organic solvents are recycled back to the extraction. Results present the extraction and recovery efficiencies. Preliminary comparisons are made with other potential separation technologies.

Introduction

Fossil fuels, generated from crude oil, coal and natural gas, have been reported to be one of main causes of the global warming, greenhouse gas emissions and associated environmental problems. Moreover, increasing crude oil prices and petrochemical fuel demands, and decreasing fuel supplies and foreign oil imports lead researchers to search for renewable, reliable and alternative energy sources. As a renewable, widespread, cheap and sustainable fuel, biomass-derived fuels such as ethanol are more attractive because of their potentially plentiful supply and relatively low cost. Biomass conversion is also utilized to produce valuable and important by-product chemicals such as acetic acid [1,2,3].

Biomass Conversion

Biomass has been used for a long time as a primary source of energy all over world. Today approximately 12% of the world's energy supplies are produced from biomass and in many developing countries the percentages are reaching to 40% and 50%. Biomass has been used or converted to produce renewable fuels and energy sources in solid, liquid and gas phases.

The United States is one of many countries trying to use biomass conversion processes to produce bio-fuels and energy in an effort to reduce fossil energy consumption. As an example of that, it has been proposed by the U.S. department of Energy [5] that about 512 million dry tons of biomass residues in the United States could be converted to liquid fuels and 10% to 50%

of transportation fuels could be replaced by biomass fuels between 2010 and 2050.

Lignocelluloses Composition

Wood lignocelluloses mainly consist of a mixture of polymerized carbohydrates (cellulose and hemicelluloses), lignin, extractive components and ash. Wood can be generally categorized as hardwood or softwood and the physical and chemical properties between them vary [4].

Hemicelluloses are linear and Y branched polymers that are shorter than cellulose with a degree of polymerization of 50-200 units. They consist of three hexoses (D-glucose, D-galactose, and D-mannose) and two pentoses (D-xylose and L-arabinose). Hemicelluloses work as a linkage between cellulose and lignin [3]. There are two common types of hemicelluloses in wood: Xylans and galactoglucomannans, (Figures 1 and 2). These two are varying in terms of compositions and properties from one species to another.

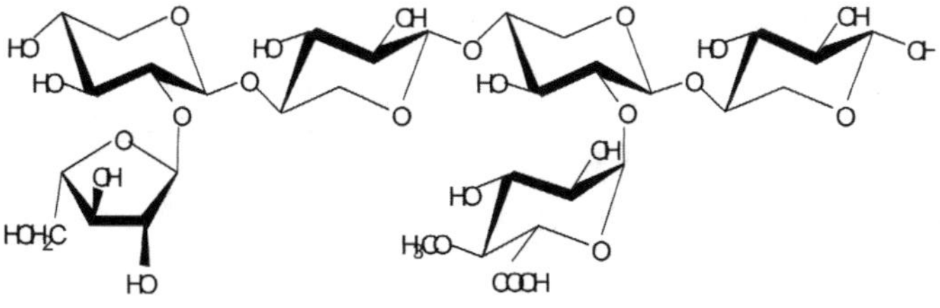

Figure 1: Structural representation of softwood xylan. Containing acetyl groups in 4-O-methyl α-D glucopyranosyluronic acid (bottom right) [9]

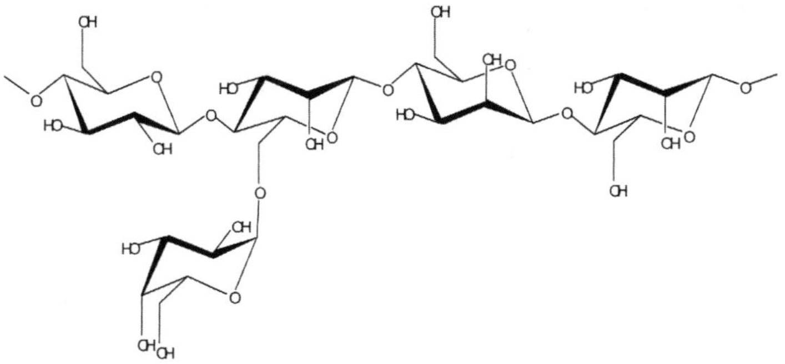

Figure 2: Structural representation of softwood galactoglucomannans [9]

Kraft Pulping Process

As an important process for pulp and paper technology, the Kraft pulping process is used widely all over the world and especially in the state of Maine. For more than 100 years, it has to be the most economical chemical pulping process. In 1998, about 3.8 million tons of products were produced from pulp manufacturing [4]. In the Kraft process approximately 50% of the wood weight is dissolved into pulping liquor. This solution is primarily composed of hemicelluloses and lignin [7].
About half of wood hemicelluloses can be removed in the Kraft pulping process because hemicelluloses have a lower degree of polymerization (50-200) that can be more easily dissolved into the aqueous solution than other lignocellulosic components [7]. Hemicelluloses are usually burned to produce energy during the chemical recovery process [7], but because they have a heating value of about 13.6MJ/kg, which is only about half the heating value of lignin [9], they may find a more optimal use as a feedstock for chemical conversion.

Pre-pulping extraction process

The primary processes to produce ethanol from extracted hemicellulose are shown in Figure 3. These processes are: hemicelluloses extraction, hydrolysis to release sugars using sulfuric acid, filtration to extract lignin, liquid-liquid extraction to recover acetic acid and furfural, gypsum separation and fermentation of C5-C6 sugars to produce ethanol [6].

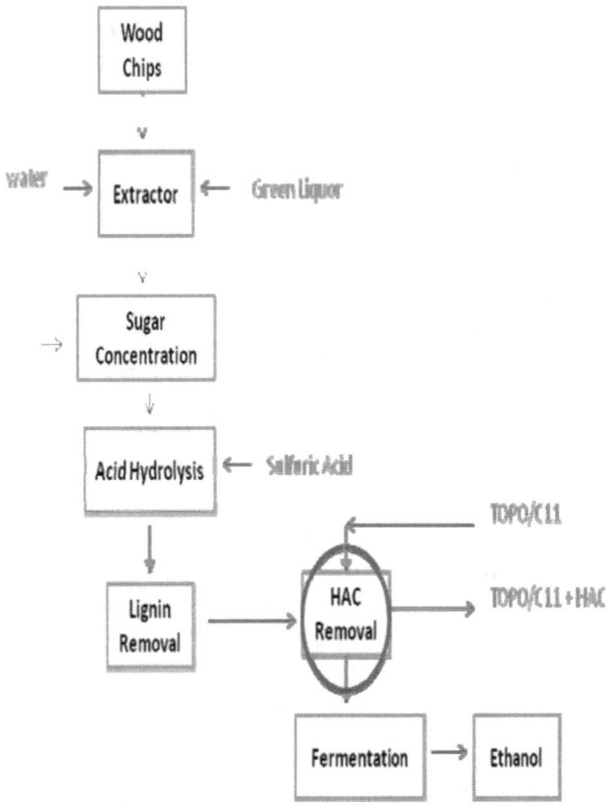

Figure 3: Overall processes to produce ethanol

There are three common processes to convert lignocelluloses to ethanol: hydrolysis, either by acid or by enzymes, and fermentation. The hydrolysis process is used to break the crystalline structure of the lignocellulosic material (cellulose and hemicelluloses) and release monomer sugars such as glucose and xylose in order to make them more digestible in the fermentation process. Hydrolysis can be accomplished using acids, such as sulfuric acid, or cellulolytic enzymes. These sugars are then fermented using microbes such as Baker's yeast to produce ethanol [3]. Degradation products from the hydrolysis process, such as furfural, hydroxymethyl furfural (HMF) and acetic acid could be harmful for some fermentation organisms. Xylose is converted to furfural while glucose is converted to HMF.

Fermentation affected by acetic acid

Acetic acid, CH_3COOH, is an important biomass product which has a high marketing selling price (approximately double the ethanol price) [6]. It is released when the acetyl groups of hemicelluloses, especially hardwood xylans, are hydrolyzed. Acetic acid can be used in many industrial applications such as chemicals, light industry, textiles, pharmaceuticals, printing/dyeing, rubber, pesticides, photographic chemicals, electronics, and food processing (as vinegar) [1].

Acetic acid is a typical weak acid and is partially ionized in its solution according the following reaction.

$$CH_3COOH \rightleftarrows H^+ + CH_3COO^-$$

The thermodynamic equilibrium constant, K_a can be defined by:

$$K_a = \frac{[H^+][CH_3COO^-]}{[CH_3COOH]}$$

Acetic acid is one of the components that could inhibit fermentation organisms, which affects ethanol production efficiency. Acetic acid can penetrate the bacteria cell walls and acidify the cytoplasm. As a result of that, the protein gradient across the cell membrane is disrupted [10]. Therefore, it is recommended to extract acetic acid early before reaching the fermentation stage. Liquid- liquid extraction and distillation are two common processes to extract acetic acid from hemicellulose extracts.

Method

In the University of Maine's pilot plant, hemicellulose extracts are prepared by loading 2 kilograms of commercial freshly mixed northeastern hardwood chips sized between 16 and 22.6 mm, green liquor (contain 1 - 4% of equivalent Na_2O) and water (including wood moisture) at overall ratio of 4:1 liquor/dry weight into a 20 liters rocking digester. The system is then agitated at a speed of 2 rpm for 110 minutes and heated at 160 °C. Next, the system is cooled below 100 °C and the freely drained liquor is collected. The composition of the green liquor and hot water extract are shown in Table 1 [7].

Table 1: Composition of 3 % Green Liquor (GL) and Hot Water (HW) Extracted Northern Hardwood

Composition of Extract	GL Extract[2] [g/L]	GL Extract[3] free monomers [g/L]	HW Extract[2] [g/L]	HW Extract[3] free monomers [g/L]
Glucose	0.36	0.09	1.01	0.09
XMG[1]	1.62	0.11	9.06	0.65
Arabinose	0.24	0.02	1.37	1.07
Acetic Acid	9.40	8.32	3.47	1.27
Total Sugars	2.22	0.22	11.44	1.81

[1] xylose, mannose, and galactose, [2] secondary hydrolyzed with 4 % sulfuric acid, [3] analyzed without secondary hydrolysis.

Materials

The chemicals used were: Trioctylphosphine oxide (TOPO) ($[CH_3(CH_2)_7]_3PO$, Mol W. 386.63, 99% purity, Sigma) , Un-decane ($CH_3(CH_2)_9CH_3$, Mol W. 156.3, 99% purity, Sigma), Trioctylamine (TOA) ($[CH_3(CH_2)_7]_3N$, Mol W. 353.67, 98% purity, Sigma) , Octanol ($CH_3(CH_2)_7OH$, Mol W. 130.23, 99% purity, Sigma), and acetic acid (CH_3CO_2H, Mol W. 60.05, 99.99+% purity, Sigma).

Liquid-Liquid Extraction

The liquid-liquid extraction method is applied first to clean aqueous solution (acetic acid and water) and then to real extract (green and hot water) to extract acetic acid.

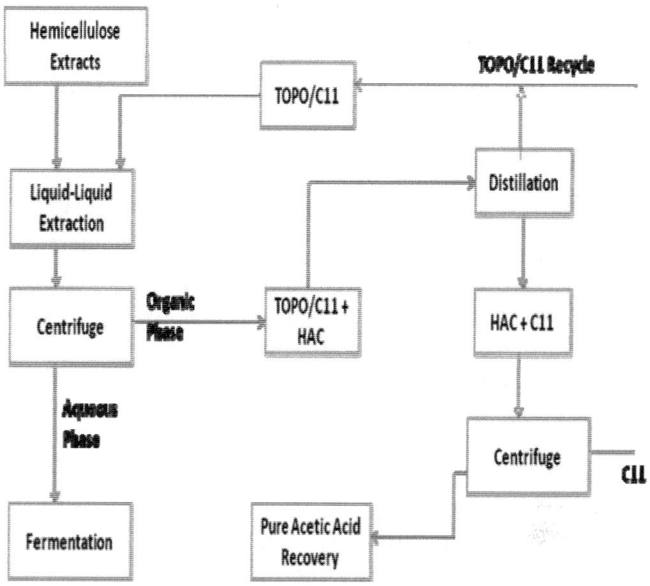

Figure 4: Liquid-liquid extraction for acetic acid followed by distillation system

TOPO has a high molecular weight and works as an extraction agent to remove fatty acids like acetic acid from aqueous solutions. TOPO has strong hydrogen bonding properties that induce the carboxylic acid to transfer into the organic phase. In addition, TOPO is characterized by its excellent stability, high boiling point and low solubility in water [11].

An organic solution of TOPO is prepared by mixing TOPO with undecane ($C_{11}H_{24}$). Then a typical extraction procedure is as follows: 10.0 mL of the aqueous phase (green liquor- or hot water-generated hemicelluloses extract) which contain a certain amount of acetic acid was shaken vigorously with 10.0 mL of the organic phase (undecane as a diluent containing a certain amount of TOPO) for 36 minutes at various temperatures. Samples of both phases were analyzed immediately after the phase separation. The concentration of acetic acid in the organic phase was determined by titration using 0.05 M NaOH with phenolphthalein as indicator while the concentration of acetic acid in the aqueous phase was determined by High Performance Liquid Chromatography (HPLC). Gas Chromatography (GC) was used to detect quantities of organics in the organic phase.

Distillation

The organic phase after extraction process is taken to a distillation system to separate acetic acid from the organic phase. The TOPO in the organic phase after distillation is recycled again to reduce the cost.

Results

The following tests have been done for this system using both solvents TOPO and TOA:

- Effect of varied concentrations of acetic acid in aqueous solution
- Effect of varied concentrations of solvent in organic solution
- Effect of varied PH
- Effect of varied ratios of organic to aqueous phase
- Recycle efficiency of solvent

This extraction is used for a clean solution (acetic acid and water) with varied of acetic acid concentrations from 10 to 100 g/L (figure 5). The organic to aqueous ratio was 1:1 and PH was 2 at reaction temperature was 70 C°.

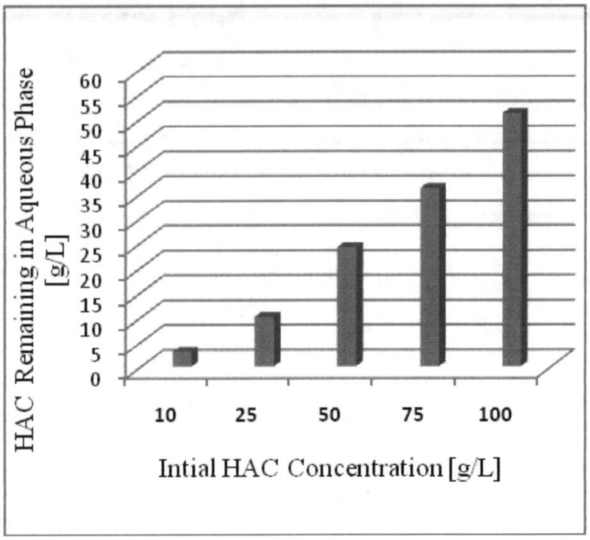

Figure 5: Effect of varied concentrations of acetic acid in aqueous solution

The extraction efficiency is shown to decrease as the initial acetic acid concentration increases. This is because the organic solution reaches a saturated condition with no more capacity to bind with acetic acid.

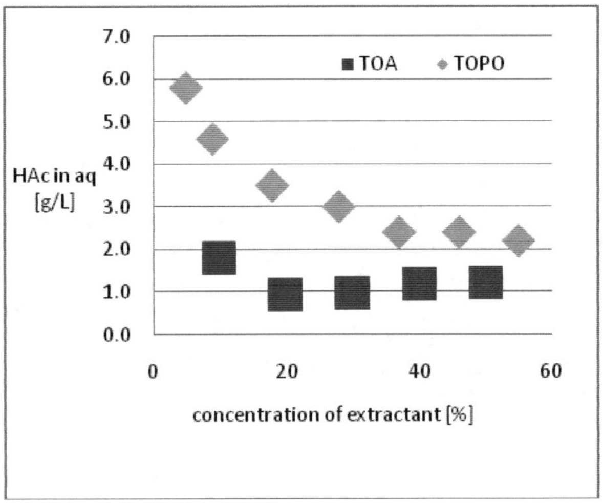

Figure 6: Effect of varied concentrations of solvent in organic solution

Figure shows the effect of using different concentrations of TOPO and TOA. The system is tested for a simple water and acetic acid system. Both the solvents are varied from 10 to 50 g/L in their diluents.

TOPO is shows good extraction when it has a concentrationof 37% (w/v) of or more in Un-decane, while TOA shows high efficiency down to 20%in its solvent. In addition, the maximum extraction percentage of TOA is 90% while TOPO is 70%. Therefore, TOA superior to TOPO at extracting acetic acid .

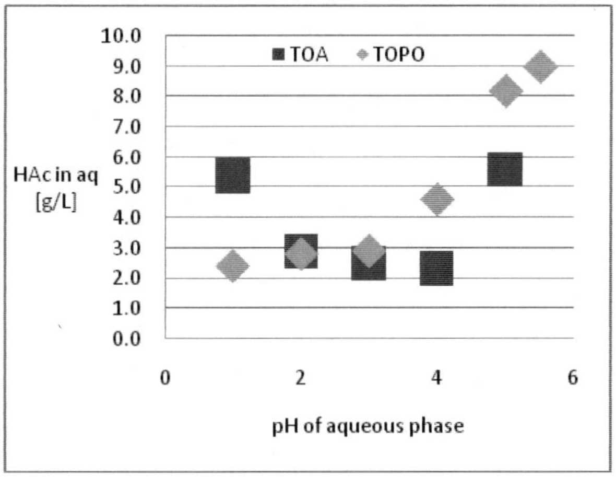

Figure 7: Effect of varied pH

Figure 7 is shows a good extraction when the pH is low. Therefore, TOPO has optimal extraction when pH is at or below 3 and shows less efficient extraction when pH is above 3. On other hand, TOA is shows good extraction up to a pH of 4 and only deteriorates above pH 4. Since the green liquor extract can have an approximate pH of 4, using TOA solvent may eliminate the need to lower pH to get a good extraction while using TOPO does require reduction in pH..

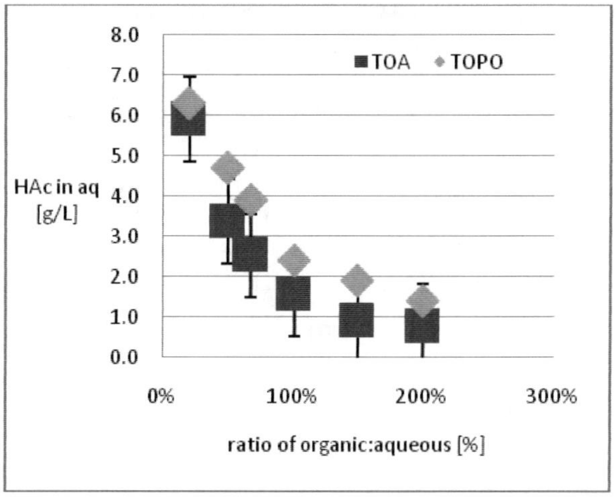

Figure 8: Effect of varied ratios of organic to aqueous phase

The extraction efficiency is affected with organic to aqueous ratio. When the ratio increases, the extraction efficiency also increases. However extraction efficiency appears to plateau near a 100%.ratio, Therefore, the best ratio is found to be 1:1.

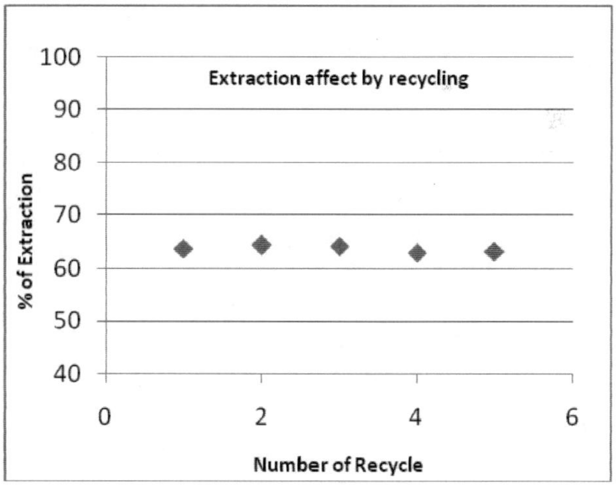

Figure 9: Recycle efficiency of solvent

This test has been done for TOPO solvent is very expensive and therefore requires good recovery. Figure 9 illustrates the extraction performance through 5 recycles of the solvent through distillation. This demonstrates that on a clean system (water and acetic acid) TOPO in undecane appears to be readily recoverable. At the time of this writing, the recyclability has not yet tested on real extract such as green liquor extract.

Table 4: Comparisons of using TOPO and TOA for Liquid-Liquid extraction

	TOPO	TOA
Max Extraction	70 %	90 %
Acetic acid concentration in aqueous	10 g/L	10 g/L
Solvent concentration in organic	37 %	20 %
PH	≤ 3	≤ 4
Ratio	1:1	1:1
Reaction Temperature	70 C^0	70 C^0
Reaction Time	1 hr	1 hr
Recyclability	Stable	-

Note: The extraction parameters are used for a clean system

Future Works

- Test the extraction and recyclability of TOA on model solutions and wood extracts.
- Test the recyclability of TOPO solvent for real extract system

Conclusion

Liquid-Liquid extraction is used to extract acetic acid from aqueous solution. The extraction was done first for a clean system (acetic acid in water) and then for wood extracts (green liquor extract). The recyclability of TOPO solvent was done for the clean system to reduce the cost of the solvent.

TOPO shows 70% extraction for clean system and 64% for wood extract system. While TOA is shows 90% acetic acid removal in the clean system. The best extraction parameters for this system are :

- The concentration of acetic acid in aqueous solution is to be 10 g/L.
- The concentration of TOPO in organic solution is to be 37%, while TOA is to be 20%.
- PH is to be ≤ 3 for using TOPO and ≤ 4 for TOA
- The ratio of organic to aqueous is to be 1:1

Finally recyclability of TOPO for the clean system is showing stable performance with no affect to extraction efficiency.

References

[1] X. Xu, Direct Conversion of Carboxylate Salts to Carboxylic Acids Via Reactive Extraction, M.S. Thesis, Texas A&M University, (2008).

[2] A. van Heiningen, Converting a Kraft Pulp Mill into an Integrated Forest Biorefinery, PAPTAC. 107 (2006) 141-146.

[3] F. Agbogbo, Anaerobic Fermentation of Rice Straw and Chicken Manure to Carboxylic Acids, PhD Thesis, Texas A&M University, (2005).

[4] M.F Demirbas, Current Technologies for Biomass Conversion into Chemicals and Fuels, Energy Sources Part A. 28 (2006) 1181-1188.

[5] B. Han, W. Carvalho, L. Canilha, S.S. da Silva, J.B.A. e Silva, J.D. McMillan, S.R Wickramasinghe, Adsorptive Membranes vs Resins for Acetic Acid Removal from Biomass Hydrolysates, Desalination. 193 (2006) 361–366.

[6] H. Mao, J.M. Genco, A. van Heiningen, H. Pendse, Technical Economic Evaluation of a Hardwood Biorefinery Using the "Near-Neutral" Hemicellulose Pre-extraction Process, J. Biobased Mater. Bioenergy. 2 (2008) 1-9.

[7] S.L. Walton, Biological Conversion of Hemicelluloses Extracted from Hardwood,

Unpublished Results, University of Maine, (2008).
[8] B.J.W. Cole, Wood Chemistry, Class Notes, University of Maine, (2009).
[9] E. Sjöström, Wood Chemistry: Fundamentals and Applications, second ed., Academic Press, California,1993, pp.65.
[10] C.M. Takahashi , D.F. Takahashi, M.L.C. Carvalhal, F. Alterthum, Effects of Acetate on the Growth and Fermentation Performance of *Escherichia coli* KO11, Appl. Biochem. Microbiol. 81 (1999) 193-203.
[11] H.F. Al-Mudhaf , M.F. Hegazi, A.I Abu-Shady, Partition Data of Acetic Acid Between Aqueous NaCl Solutions and Trioctylphosphine Oxide in Cyclohexane Diluent. *Sep. Pur. Technol.* 27 (2002) 41–50.

Key Metric Comparison of Five Cellulosic Biofuel Pathways

by B. A. Thorp [1]

Purpose

Cellulosic biofuels are relatively new products except for a handful of sulfite pulp mills that have fermented their red liquor to obtain ethanol as early as 1909.[2] Sulfite pulping is obsolete and the modern pathways for cellulose are pyrolysis, gasification, acid hydrolysis and enzymatic hydrolysis. While the pathway from pyrolysis to transportation fuel is not complete, the pathway to fuel for industrial boilers will be included for reference. Commercial efforts in cellulosic biofuels are new and data to compare a specific offering to others within a class or to compare one pathway to another do not exist. Data are lacking for a complete comparison. This paper will start building a comparison metrics framework for cellulosic biofuel pathways. Others will need to further build on and update this comparison. The metrics for ethanol from corn and biodiesel from transesterification are well known and can be added as needed. Later, advanced biofuels like those derived from barley, sorghum , algae, sugar cane and hybrid pathways can be added.

The metrics chosen are capital in dollars per annual gallon, yield in gallons per ton, the co-production of MAJOR secondary products, operating cost per gallon, and the selling price of the biofuel as a percentage of gasoline. These key metrics focus on conversion to liquid transportation fuels. Conversion to chemical products can improve overall economics. The desired yield is BTUs sold divided by BTUs purchased but there is not yet enough published to determine this key variable. <u>It is shocking to find so little data or understanding of this metric as biorefineries are fundamentally a BTU transforming facility</u>. All petrochemical refineries know, track and optimize this metric. The operating cost per gallon is highly dependent of feedstock cost and that is a function of time, location and type. By establishing a selling price as a percentage of gasoline the time variation of commodity oil pricing is eliminated.

There are a large number of technology approaches. Grouping of technologies will be required to simplify metric development shown in Figure 1 below. Gasification can be used with two very different Gas-to-Liquid (GTL) pathways so each will be discussed.

Figure 1

Metrics of Conversion Pathways

*data not available for the desired BTU yield

Conversion Pathway	Capital Cost ($ per ag)	Yield* (gal per d ton)	Secondary Products	Op. Cost ($ per gal)	Selling Price (% gasoline)
Pyrolysis					
Gasify then convert to Ethanol					
Gasify then F-T to Diesel					
Acid Hyd. then Fer.					
Enz. Hyd. then Fer.					

Introduction

Most qualified forecasters conclude that global energy prices will increase as demand stresses supply. The only difference in the forecasts are time and intensity of the increase. Some forms of energy like electricity have alternate production sources. Those sources include wind, solar, hydro and nuclear. The only non-petrochemical alternatives for liquid transportation fuel are biofuels or switching engines from internal combustion to electric power. Hydrogen may appear as a fuel in the future but most predict that it is decades away. Now, biofuels become one of the factors which are critical to the economic health and even survival any oil importing nations so that biofuels can supplement oil for transportation needs and industrial process needs.

Understand the relationship of raw materials, markets and technology is critical. For any given location the quantity of raw material that exists or can be grown within an economical harvest radius is fixed. This makes raw material an independent variable which must be quantitavely understood. Location can also have a major impact on the type of biofuel that can be effectively marketed. For example, some regions have an excess of gasoline and ethanol while others have shortages or higher logistics costs. Considerations such as these make markets an independent variable which must be quantitavely understood. If it is possible to define the raw material and the market, then conversion technology becomes a dependent variable and should be selected because of its suitability to convert the selected raw material to the biofuel selected for a market.

After identifying raw materials and markets, the next step is identifying the most promising pathway, then identifying the promising technology within a pathway. The last step is to validate all choices. This means that there will not be a single technology of choice but several technologies of choice for combinations of raw materials and markets.

Figure 2 below is an illustration of the optimum area among raw materials, markets and technologies.

Figure 2

Optimum Area

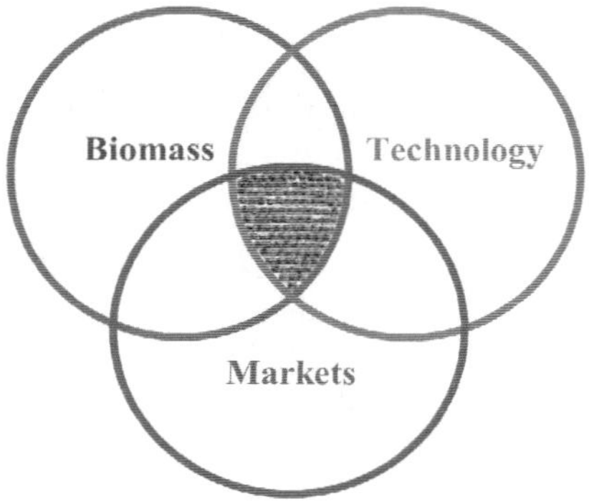

Pyrolysis

The first pathway is pyrolysis of biomass. Producing bio-oil is an old technology in which biomass is volatized by high temperatures, the condensed into a liquid. Pyrolysis oil is acidic having a pH in the range of 2 to 3, has a shelf life as it continues to slowly react and has a lower BTU content because it contains low BTU components like water. The process is outlined in Figure 3 below.

Dynamotive, UOP/Honeywell and others are developing new and improved processes to convert bio-oil into transportation fuels. While progress has been made and reported, these developments are not far enough along to establish key metrics. However, it is important to know the key metrics of the base process.

Figure 3

Pyrolysis of Biomass to Bio-oil

- Green biomass is harvested and sized. For sawdust , no preparation is needed.
- Rapid thermal pyrolysis volatizes the mass.
- Char is rejected with a simple cyclone.
- The volatiles are quenched/condensed and collected.
- Products from sawdust are ~70% Bio-oil (hi yield), ~15% Char and ~15% non-condensable gasses.
- This is an old process that has been used in laboratories for decades.

There has been ample commercial activity using the pyrolysis technology. A summary of commercial Plants in North America include those shown in Figure 4.

Figure 4

Pyrolysis Plants Include

- There are several pilot plants in Canada including Guelph and West Lorne, Ontario.
- The Ensyn technology is used in 4 Red Arrow plants where flavors are extracted from the Bio-oil which then becomes a by-product.
- The first commercial plant announced *2009/10* in Willow Springs, MO by Dynamotive. It has a cost of $24 million and will produce 12 million gpy Bio-oil sold for commercial boilers.
- Overseas plants have also been announced.

Since pyrolysis is an older technology a number of the metrics are documented. Advantages include that is a known process with demonstrated costs and produces a biofuel with known properties. The major advantages and disadvantages include those shown in Figure 5 [3].

Figure 5

Pyrolysis Major Strengths and Weaknesses

- **Major strengths include:**
  Low capital: ~$2.00 per annual gallon
  Any tree or crop species
  Simple preparation of raw material
  Hi yield: ~170 gpt, and low operating cost

- **Major weaknesses include:**
  Unstable as reactions slowly continue; shelf life issues
  Low BTU content as it contains ~15% water
  Very acidic, cannot be used in all boilers
  Therefore, lower selling price: ~50% of gasoline

Gasify Biomass & Convert Syngas to Ethanol

Gasification, or heating biomass with controlled oxygen, is well known process. There are numerous commercial industrial processes in North America.[4] The resulting syngas must be cleaned to specifications determined by the downstream process which will convert the syngas to liquid. After cleaning the syngas will consist of C, CO_2, H_2 and only traces of other compounds. The syngas can be converted to ethanol chemically via Mixed Alcohol Synthesis (MAS) (Range Fuels), packed bed catalytic reactors (Fulcrum), or fermentation (INEOS and Coskata). In addition to gas cleanup, the GTL processes can require compression to achieve the needed temperatures and pressures. Fermentation does not require compression but does require an "incubation tank farm" sized with enough units to match gas production capacity. A generalized process description is shown in Figure 6. Diverse raw materials can be gasified and include forest slash, spent crops, energy crops, and municipal solid waste (MSW) and sorted MSW (SMSW). Regulation prohibits sorting inside the landfill. Diverse feedstocks can impose additional requirements for storage, preparation and drying prior to gasification.

Figure 6

Biomass Gasification Followed by Conversion to Ethanol

- Biomass is harvested, sized and dried as required by gasification technology. MSW is sorted and prepared outside the landfill as required by regulation. Sorted MSW goes to plant at no/low cost.
- Feedstock is then gasified and the syngas cleaned. For SMSW some prefer downdraft partial oxidation followed by plasma arc. Heat can be recovered as syngas is cooled and cleaned.
- The syngas is converted to ethanol by MAS or proprietary packed bed catalytic reactors. Alternately it can be fermented. Heat can be recovered from some processes. Fermentation requires little/no additional heat.
- Distillation/dehydration is required for most processes. Membrane separation has lower costs but requires demonstration.

Recent Department of Energy (DOE) grant solicitations have augmented commercial activity. There are several pilot plants and at least six announced and partly-funded commercial plants which are shown in Figures 7 and **7a.**

Figure 7

Biomass Gasification Plants Include

- Range Fuels is constructing a phase 1 plant in Soperton, GA which is scheduled to make 8 million gallons ethanol and 2 million gallons methanol.
- The first SMSW plant may be Sierra BioFuels (Fulcrum Energy), McCarran NV, *scheduled to start in 2011.* This plant will produce 10.5 million gpy ethanol + some solid waste for landfill. Permits have been obtained and engineering complete. Cost estimated to be ~$120 million.
- Enerkem has announced financing to gasify SMWS and produce 9.5 million gpy and 20 million gpy respectively in Edmonton, Canada and Pontotoc, MS.
- Coskata is building a demo plant in Warren County, PA which will ferment syngas to produce 40,000 gpy ethanol. They are actively looking for a commercial site.

Figure 7a

Biomass Gasification Plants Include *(continued)*

- Powers Energy has announced a 32 million gpy plant in Indiana converting MSW to ethanol using INEOS technology.
- Gulf Coast Energy has a 20,000 gpy demonstration plant in Livingston, AL based on Pearson technology funded for years by DOE. Higher yields reported in electronic media have not been validated.

The major strengths and weaknesses that have been identified to date from proposed installations of gasifying biomass for the production of ethanol include those shown in Figure 8.

Figure 8

Biomass Gasification then Ethanol Major Strengths & Weaknesses

- **Major strengths include:**
  - Feedstock diversity
  - SMSW feedstock operating cost can be as low as $0.24/gal
  - Lower capital cost claimed by those using fermentation
  - Yield (80 gpt to claims of 111 gpt)
  - Smaller units can be self-financed
- **Major weaknesses include:**
  - Some approaches lack extended pilot line runs
  - Pilot lines are propriety, not always available
  - Sierra BioFuels will be a pilot/demo site
  - Plasma arc gasifiers are difficult to operate
  - Some syngas conversion technology unproven
  - Product selling price ~70% of wholesale gasoline (BTU basis)
  - Many plants are stand alone and appear under optimized for heat recovery

Gasification followed by Fischer-Tropsch (F-T)conversion

This process was first used commercially by the German military in WWII to produce ~5 million gallons per day when oil supplies were curtailed. The raw material was coal. The next commercial use was in the 1960's by South African Synthetic Oil Ltd (SASOL) as a way for South Africa to survive under the embargo. The Secunda facility still produces ~6,720,000 gallons per day. The US challenge is to replace coal with biomass and use a gas cleanup process which is optimum for biomass. Figure 9 shows the general process.

Figure 9

Biomass Gasification then Fischer-Tropsch Conversion to Diesel & Wax

- Size and dry biomass
- Gasify (Choren, GTI, TRI & Rentech)
- Gas cool down, heat recovery and cleanup
- Compress and pass through F-T
- Burn naphtha and F-T tail gas
- Generate green power for facility or grid, recover steam for a user-host
- Separate diesel and wax. Wax can also be converted to diesel

There are several plants being designed for the U.S. and the major ones are shown in Figures 10 and 10a. Choren is also shown because their plant in Germany is being commissioned and is an important benchmark.

Figure 10

Biomass then F-T Plants Include

- Choren has a 4.2 million gpy stand alone plant in Freiberg Germany with a yield of ~60 gpt. Thermal Eff. ~41%.
- NewPage won a $30 million DOE grant to build an integrated 5.5 million gpy D&W facility in Wisconsin Rapids WI. The mill also gets 200 million BTU/hr stream, 123 million BTU/hr hot water and 14 million BTU/hr gas for the lime kiln. Thermal Eff. ~70%. Cost ~$90 million. *Start date not announced.*
- Flambeau River Biofuels won a $30 million DOE grant to build an integrated 17 million GPY D&W facility in Park Falls WI. The mill also gets 5.8MW power and 190 million BTU/hr steam. Thermal Eff. ~60%. Cost $260 million. *Start date not announced, but 2012 likely.*

Figure 10a includes configurations which have not been discussed so far. That is some mixture of biomass and fossil fuel which is available in the area. The reason this will be important in the future is that biomass availability may limit biomass biorefineries to a size of ~50 to 100 million gpy. By incorporating 50% fossil fuel the size could be increased to 100 to 200 million gpy, thereby gaining economy of scale. Capital intensity is decreased 34% by increasing plant size from 25 to 100 million gpy and 20% by increasing capacity from 50 to 100 million gpy.[(5)] Some of these configurations could possibly meet the Green House Gas (GHG) standards in the 2007 Energy Independence and Security Act (EISA). This approach has technical, environmental and social challenges.

Figure 10a

Biomass then F-T Plants Include

(continued)

- Rentech will build a 1.5 million gpy plant in California and has a signed contract with airlines to use the fuel in ground operations.
- Rentech proposed a 95% petroleum coke and 5% biomass for Phase 1, 24 million gpy plant in Natchez, MS by 2011. Phase 2 is 412 million gpy.
- Rentech - Mingo proposed to make 441 million gpy in Mingo County, WV from coal and wood waste.
- Rentech - Solena proposed to make 220 million gpy jet fuel from BOIMASS in northern CA using the Solena gasifier.

The strength and weaknesses are well known due to commercial experience and recent biomass design. These systems can have greatly improved economics when the heat from syngas cool down/clean-up and from F-T tubes can be recovered and sold.

Figure 11

Biomass Then F-T
Major Strengths & Weaknesses

- **Major strengths include:**
  Process commercial in "components"
  Uses the whole plant/tree
  Can use mixed biomass
  Can sell waste heat as well as diesel & wax
  Sulfur free diesel can sell for ~150% gasoline
  Pilot and demo plants proprietary but available

- **Major weaknesses include:**
  High capital cost
  Lower yield but high thermal eff., if integrated
  Has high power requirement for F-T unit

<u>Strong Acid Hydrolysis</u>

Like pyrolysis, this is an old technology. In this pathway there can be mild acid hydrolysis first stage treatment if the raw material has hemicellulose in quantity. The remaining biomass is subjected to concentrated sulfuric acid to convert the cellulose to glucose which is then fermented to ethanol or chemical products. Sulfuric acid is recovered, purified and reused. Lignin will be a residue and can be collected and sold. It can also be burned to generate process heat if the capital for the boiler can be justified by the savings. This has not been part of the projects listed. Figure 12 shows a generic outline of the process.

Figure 12

Strong Acid Hydrolysis

- Can have "mild acid" pretreatment to remove hemicelluloses if raw material has them in quantity
- Reclaims concentrated sulfuric acid, applies it to the raw material converting cellulose to glucose, then uses chromatographic separation to isolate the sugars.
- The sugars are then fermented to ethanol with conventional technology, including distillation and dehydration.
- If the raw material contains lignin there is discussion about using it to produce energy. Otherwise natural gas is used.
- Yield is low in the 60 to 70 gpt range.

The most modern facility is the demonstration plant outlined in Figure 13 below. This is followed by a very old plant in Russia and a modern plant targeted for California.

Figure 13

Acid Hydrolysis Plants Include

- There is a 9,000 gpy demonstration plant in Izumi Japan (southern tip). Various yields reported but 64 gpt seems reasonable. SEKAB has a pilot plant in Ornskoldsvik, Sweden.
- Tavada, Siberia has produced 3.3 million gpy of ethanol from 70% wood, 20% wheat straw and 10% beet molasses since 1943. The 1997 cost estimate was $5.67 per gallon with raw materials and energy (coal) totaling 70% of costs.
- In 2007 BlueFire Ethanol won a $40 million dollar DOE award to build a 100 million dollar plant using sorted MSW. Yield reported to be 68 gpt. BFE has applied for a DOE Loan Guarantee to build a 3.2 million gpy plant in California.
- There is discussion of a plant in South Korea.

Since this technology has been studied for decades the strengths and weaknesses are well understood and shown in Figure 14. A major strength is the ability to use Sorted Municipal Solid Waste and perhaps get a tipping fee from urban areas where landfill capacity is critical. This is an example of locating the raw material then selecting the most appropriate technology for conversion.

Figure 14

Acid Hydrolysis
Major Strengths & Weaknesses

- **Major strengths include:**
  Known process, over 100 years old
  Excellent pilot/demo plant in Japan
  Can use low cost raw material like SMSW

- **Major weaknesses include:**
  Capital cost high due to use of acid
  Operating cost high
  Yield on the low side of "average"

<u>Enzymatic Hydrolysis followed by Fermentation to Ethanol or Other Chemicals</u>

This is a newer technology with a great deal of emphasis, research and federal funding over the last decade or so. So far, lignin has not responded to enzymatic hydrolysis so it becomes a by-product. Lignin can also slow the enzymatic process so many companies have focused on low-lignin-content biomass like corn cobs, corn stover, spent sugar cane and various straws. This minimizes issues with lignin. Figure 15 below outlines a generalized process.

Figure 15

Enzymatic Hydrolysis

- Adequate feedstock storage for ~10 months (for crops)
- Feedstock preparation
- Mild acid hydrolysis for separation/removal of hemicellulose
- Fermentation of 5 carbon sugar to ethanol
- Enzymatic hydrolysis of cellulose to glucose
- Fermentation of glucose to ethanol
- Removal of lignin for sale or energy use
- Distillation
- Dehydration

Consistent with the emphasis there are a growing number of pilot, demonstration facilities and proposed commercial facilities. The major ones are shown in Figures 16 and 16a. It is likely that there will be additional announcements, sometimes on a weekly basis.

Figure 16

Enzymatic Hydrolysis Plants Include

- Verenium, Jennings, LA demo plant converting spent sugar cane and switchgrass to 1.4 million gpy ethanol. Also has pilot plant.
- KL Energy has a 1.5 million gpy demonstration plant in Upton, WY since 2008 using pine chips
- Mascoma recently started a 200,000 gpy pilot plant in Rome, NY.
- Verenium announced a ~$275 million commercial plant sized for 36 million gpy for Highlands County, FL. Partners include Lykes Bros. and BP Oil. Received $7 million grant from Florida.

Figure 16a

Enzymatic Hydrolysis Plants Include
(continued)

- Mascoma received a \$30 million DOE grant and a \$24 million Michigan grant to build a 40 million gallon per year woody biomass to ethanol plant in Michigan. Estimated cost was ~\$300 million.
- Poet has a self funded 20,000 gpy pilot line in Scotland, SD using a proprietary process to convert corn cobs to ethanol. The 30 million GPY commercial facility is planned for Emmetsburg, IA for which Poet has a DOE grant.
- Abengoa constructed a 20,000 gpy pilot line in York, NE using enzymatic hydrolysis of corn stover. Salamanca, Spain has a 1.3 million GPY demonstration plant. The 12 million gpy commercial plant is planned for Hugoton, KS for which they have a DOE grant.

Since this technology is relatively new and much is still proprietary the major strengths and weaknesses shown in Figure 17 below are more subjective.

Figure 17

Enzymatic Hydrolysis Major Strengths & Weaknesses

- **Major strengths include:**
  Have operating demo plants with good scale to commercial
  Have strong partners
  Moderate capital cost

- **Major weakness include:**
  Must accumulate and store crop feedstock for long periods
  Moderate yield, reported at 80 gpt
  Enzyme cost typically high
  Selling price ~70% gasoline

There are also a number of hybrid technologies and two are highlighted in Figure 18 below. When demonstrated, Value Prior to Pulping (VPP) could add value to pulp mills because there is more value in converting the hemicellulose to ethanol or other chemicals than burning sugar in thermally inefficient recovery boilers.

Figure 18

Hybrid Technologies

- VPP-Value prior to HW pulping removes ~60% of hemicellulose before pulping, lowering pulping and bleaching cost. Hemicelluloses are then fermented to ethanol or other chemicals.
- Another hybrid combines enzymatic hydrolysis and gasification. The syngas derived from lignin is reacted with CO_2 from fermentation to produce more ethanol. The unique use of lignin increases yield.

The proposed plants are shown in Figure 19.

Figure 19

Hybrid Technologies Plants Include

- Old Town Fiber and Fuel, Old Town, ME won a 30 million dollar DOE grant to install the process at Old Town. Reports indicate a potential switch to a butanol process.
- Applied Biorefinery Sciences, Syracuse, NY is commercializing processes with higher extraction.
- ZeaChem, is constructing a 1.5 million gpy demonstration pilot plant in Boardman, OR.

The major strengths and weaknesses are shown in Figure 20. The VPP process can also be used on feedstock prior to pelletizing or combusting in a solid fuel boiler.

Figure 20

Hybrid Technologies
Major Strengths and Weaknesses

➤ Major strengths include:

Low/no cost raw material for VPP

VPP can be used on processes other than pulping

Highest reported yield for ZeaChem

➤ Major weaknesses include:

Both lack pilot/demo plants

Selling price ~70% gasoline

Conclusions

There is sufficient publically available data and in the above presentation graphics to complete the Key Metrics boxes for Figure 21[(6)]. This figure is useful when studied for implications. Since a corn-ethanol facility can be completed for $2.00 to $2.50 per annual gallon, all pathways have relative high capital cost at this point in their development. At this time commercial success depends on capital reduction instruments like grants and/or unique synergies like tipping fees for raw materials or the sale of co-products like steam and power to host facilities.

Figure 21

Metrics of Conversion Pathways

*data not available for the desired BTU yield

Conversion Pathway	Capital Cost ($ per ag)	Yield* (gal per d ton)	Co-Products	Op. Cost ($ per gal)	Selling Price (% gasoline)
Pyrolysis	2	170	Char	~1	~50
Gasify then convert To Ethanol	4-6	80-100	TBD	1-2	70
Gasify then F-T to Diesel&Wax	6-8	50-70	Power & steam	1-2	~150
Acid Hyd. & Fer.	6-8	60-70	TBD	1.5-2.5	70
Enz. Hyd. & Fer.	4-7	70-80	TBD	1.4-2.5	70

Figure 22 shows general conclusions. Specific conclusions require a raw material availability analysis, a market analysis and a thermal balance for all facilities involved.

Figure 22

Conclusions

- Capital costs are the major barrier at current oil prices. Partnerships can be beneficial.
- Hydrolysis and pyrolysis pathways (and corn-ethanol) are more logical for facilities who have excess steam.
- Gasification pathways are more logical for facilities which buy fossil fuel due to the use of recovered biorefinery heat
- Commercial success is dependent on capital reduction (like grants) and/or unique synergies.
- Feedstocks and markets are the "independent" variables, technology is the dependent variable.
- There may be different technologies of choice depending upon feedstocks and thermal host needs.

Another conclusion is that capital cost should receive more attention from Federal agencies. Neither DOE or USDA have an announced goal in this area and no funding opportunity to date has focused on capital intensity reduction.

References:

(1) Strategic Energy Consultant and Renewable Energy Expert, *email:* bathorp@comcast.net; or phone 804-873-6561.

(2) Presentation at SEKAB, July 2007.

(3) References include web sites for all companies listed, papers they presented at major conferences and private discussion with most

(4) Thorp, Ben, "Paper Industry Must Protect its Status in Cellulosic Innovation", Pulp and Paper, May 2007.

(5) Figure 14, "Advanced Biofuels, Promises, Challenges & Sustainability Issues", Dr. Helena Chum XtL Symposium Montreal Canada, 26 Aug 09

(6) Capital cost based on fall 2009 prices. Operating costs assume $50 per delivered dry ton for slash, spent crops or energy crops.

Preliminary Studies of pH and Char Particle Content On Bio-Oil Aging

Caitlin D. Naske, Dave C. Swalm School of Chemical Engineering, Mississippi State University, Mississippi State, MS

Sarah E. Crosby, Dave C. Swalm School of Chemical Engineering, Mississippi State University, Mississippi State, MS

Andrew McMaster, Dave C. Swalm School of Chemical Engineering, Mississippi State University, Mississippi State, MS

Keisha B. Walters, Dave C. Swalm School of Chemical Engineering, Mississippi State University, Mississippi State, MS

Bio-oil produced by biomass pyrolysis has been shown to degrade over time and at elevated temperatures resulting in increased viscosity that makes processing and handling difficult. It has been theorized that char particles remaining in the bio-oil after pyrolysis and condensation catalyze polymerization reactions leading to the viscosity increases during bio-oil aging. Additionally, pyrolysis bio-oil has a low pH as produced (pH ~2-3) and the effect of the low pH on the polymerization reactions is unknown. In this study, the effects of particulate filtration and pH adjustment are investigated on pyrolysis bio-oil samples produced from ground clear wood cottonwood and pine biomass using an auger reactor under vacuum. The bio-oil control samples were aged up to 21 days at 80 °C and 28 days at 50 °C. For the filtration experiments, pine whole tree fractionated bio-oil was filtered in series by vacuum filtration at 20 μm followed by centrifugal filtration at 0.2 μm and then aged at 80 °C up to 21 days. For the pH adjustment experiments, the pH of pine clear wood fractionated bio-oil was adjusted by the addition of KOH to obtain nominal pH values of 5, 7 and 9 and these pH-adjusted samples were aged at 80 °C up to 14 days. All bio-oil samples were characterized with Fourier transform infrared (FTIR) spectroscopy, gel permeation chromatography (GPC), and combined gas chromatography-mass spectroscopy (GC/MS); select samples were analyzed by light microscopy and electrospray ionization mass spectroscopy (ESI-MS). GPC analysis showed an increase in molecular weight and PDI during aging. The ESI-MS results revealed a polymeric pattern in several of the bio-oil samples, including the aged samples. Light microscopy verified that a large amount of the char particles were removed by centrifugal filtration.

See more of this Session: Integrated Biorefineries for Lignocellulosic Biomass
See more of this Group/Topical: Forest and Plant Bioproducts Division

Production of Fisher-Tropsch liquids from Biomass-derived Syngas

Weihua Deng, Zheng Li, Justinus Satrio and Robert Brown

Center for Sustainable Environmental Technologies (CSET)
Iowa State University
Ames, IA 50011

1. Introduction

Depletion of fossil oil resources, global climate change and national security concerns have made the development of alternative energy increasingly important. Compared to other alternative energy choices, the production of energy and chemicals from biorenewable resources, such as biomass, has attracted increasing research interests in recent years due to the vast potential availability of biomass materials and the availability of existing technologies that potentially enable the production of energy and chemicals in commercial scales in shorter time period. Center for Sustainable Environmental Technologies (CSET) is a research organization at Iowa State University (ISU) that has been working on promoting, developing and demonstrating thermochemical technologies such as gasification and fast pyrolysis, and catalytic upgrading processes, for the production of fuels, chemicals, and energy from biomass. While for many years focuses have been given to gasification and fast pyrolysis technologies for converting biomass into synthesis gas and bio-oil, respectively, in the past two years, CSET researchers have been working on the development of catalytic processes for upgrading syngas and bio-oil into value-added products. This paper is to report the results of research work that has been performed at CSET on the production of liquid fuels from biomass-derived syngas by using Fischer-Tropsch Synthesis (FTS) technology [Dry et al., Anderson et al.]. The work included synthesis and characterization of an FTS catalysts and the demonstration of FTS fuel production from a fixed bed reactor.

One FTS catalyst used for this study is a cobalt-based catalyst which was prepared by using zirconium and γ-alumina as the promoter and support, respectively. Previously, the catalyst has been successfully synthesized, characterized and tested for its performance by using a high pressure slurry reactor constructed at CSET. In order to improve efficiency and liquid selectivity for FTS process, a novel catalyst containing the same active components but supported on SBA-15 based material was developed recently. Since SBA-15 mesoporous silica has large surface area and well defined nanostructures, reaction capacity and shape selectivity of this catalyst is expected to be much better. In this work, the relationships between catalyst property and hydrocarbon fuel production will be thoroughly investigated. The catalysts will be engineered to improve thermal dissipation in the catalyst bed since FTS reaction is highly exothermic. In addition, a comparison between the results of FT synthesis on a slurry reactor and those on a fixed-bed reactor at similar operating conditions will be studied as well. Finally, the effect of impurities in synthesis gas on the catalyst performance will be addressed. This final discussion will relate to the need addressing a compromise between development of robust FTS catalyst and on the development of synthesis gas technology in order to obtain an FTS process that is most sounds techno-economically.

2. **Experimental**

2.1. Catalyst preparation

Catalysts were prepared by means of impregnation with a total metal content of 20wt% Co and 5wt% Zr, respectively. Gamma-alumina and SBA-15 mesoporous silica were used as catalyst supports. The as synthesized catalysts were calcined at 500°C for 6 hours with a temperature increasing rate of 1°C/min form room temperature to the target.

There are two major problems associated with the usage of catalyst powder on the fixed bed reactor: (1) local heating inside catalyst bed resulted from the heat generated from this exothermic reaction, and (2) large pressure drop along catalyst bed which may create gas shortcuts and consequently lower catalyst utilization. To resolve these problems, the catalyst powder was pelletized to make small particles with certain size following the procedure:

- Catalyst powder was pelletized using a Carver press and a set of die to make 0.3 cm thick and ~0.75 cm in diameter pellets. The powder was pressed and held under 5000 lb for 5 minutes.
- Then the pellets were broken down into small pieces.
- A set of sieves was used to separate catalyst particles with different size, and particles with certain size were chosen for this project.
- Selected catalyst was diluted by silica carbide spheres (with similar size to catalyst pellets) and loaded into isothermal zone of the fixed bed reactor.

2.2. Catalyst testing

Figure 1. Fixed bed reactor system constructed at CSET (left: reactor and the right: whole system).

Catalysts were tested on the fixed bed reactor (as shown in Figure 1) constructed at CSET, which is a continuous system with precise flow and pressure control. This system uses circulated oil to provide or remove heat from reaction zone so heat generation of FTS reaction, which probably a perturbation for temperature control if a electrical furnace were used as heating source, will not affect temperature control significantly. Before each FTS reaction, the catalyst was reduced at 350 psi, 330°C under flowing hydrogen (100 mL/min) for 5 hours. Tail gas was sampled and analyzed every one hour throughout the entire reaction period. Liquid products were collected in a condenser after reactionperiod and analyzed by a Varian GC with a FID detector installed, as well as by a GC-MS.

3. **Preliminary results**

A comprehensive investigation of the performance of two above mentioned FTS catalysts under various reaction conditions has not been completed. Some preliminary results obtained so far are discussed in the following.

The activity of homemade FTS catalysts is sensitive to reaction temperature. It was observed that there was almost no activity for the catalysts when reaction temperature was below 210°C. During a temperature range of 220°C to 250°C, FTS catalysts showed good activity as evidenced by a CO conversion between 30% and 70%. On the other hand, total gas selectivity increased significantly as reaction temperature went up indicating less CO was transformed to hydrocarbon fuels. GC analysis of FTS liquid products showed that a broad range of hydrocarbons were generated although quantitative determination of each compound has not been achieved.

Testing of SBS-15 based FTS catalyst under different reaction conditions is underway. By comparing performance of FTS catalysts using different supports, more mechanistic insights into material property-performance relationship can be obtained. The finding will provide useful information in designing the new generation FTS catalyst.

4. **References**

- Dry, M.E., In *Catalysis Science and Technology*; Anderson, J.R., Boudart, M., Eds.; Springer-Veerlag: New York, 1981.
- Anderson, R. B. The Fisher-Tropsch Synthesis; Academic Press: Orlando, FL, 1984.

Simultaneous Infrastructure Investment and Operation Decisions for Bio-Refinery Networks

Jinkyung Kim, School of Chemical & Biomolecular Engineering, Georgia Institute of Technology, Atlanta, GA

Matthew J. Realff, School of Chemical & Biomolecular Engineering, Georgia Institute of Technology, Atlanta, GA

Jay H. Lee, School of Chemical & Biomolecular Engineering, Georgia Institute of Technology, Atlanta, GA

Craig Whittaker, Research & Development, Weyerhaeuser NR

Ludwig Furtner, Research & Development, Weyerhaeuser NR

Amar Neogi, Research & Development, Weyerhaeuser NR

There is a growing interest and technological advances being made in converting biomasses into bio-fuels. Even though we are still years away from its full commercialization, it is important to begin to assess the economic feasibility of various emerging technologies and evaluate the advantages and disadvantages of different design and logistics options. In this presentation, we propose an optimization model that enables the selection of fuel conversion technologies, processing capacities and locations, and the logistics of transportation from forestry resources to conversion and from conversion to final markets. A mixed integer linear program (MILP) model is built and implemented in a commercial software package (GAMS) using databases built in Excel.

The MILP solves for (1) the optimal numbers, locations, and sizes of various types of processing plants and (2) the amounts of biomass, intermediate products, and final products to be transported between the selected locations, with the goal of maximizing the overall profit under present constraints. It also outputs the cost and profit data associated the selected network in a convenient form for analysis. The model is tested using a industry-representative data set that contains information on the existing wood resources, final product market locations and demands, candidate locations and sizes for different types of processing plants, and costs associated with the various processing units and transportation of materials, within the Southeastern region of the United States. We investigate: 1) Which parameters have major effect on the overall economics, and 2) benefits of going to more distributed types of processing network, in terms of the overall economics and the robustness to demand variations.

See more of this Session: Thermochemical Biorefineries II
See more of this Group/Topical: Forest and Plant Bioproducts Division

INTEGRATING BIOMASS TORREFACTION WITH THERMO-CHEMICAL CONVERSION PROCESSES

Dr. Sudhagar Mani, Faculty of Engineering, University of Georgia, Athens, GA
Email: smani@engr.uga.edu

Introduction

Agricultural and forest residues are the major feedstock for a biorefinery that can process and convert them into valuable chemicals, heat, fuels and materials. Biomass in its original state have wide range of moisture content (25-60%), low bulk density (60-100 kg/m^3) and low energy density (8-14 MJ/kg) and is difficult to transport, handle, store and feed into the existing combustion and gasification systems (Sokhansanj et al., 2002; Mani et al., 2006b). The uneven and high moisture content of biomass feedstock make it difficult to transport and store for a long time before its use in the biorefinery. Energy required for grinding of biomass is very high to produce a powdered biomass supply at the throat of the conversion chamber (Mani et al. 2004). In addition, gasification of lignocellulosic biomass with high volatiles and moisture content yields heavy tar formation that makes the process more expensive and harder to operate, .e.g. tar contamination in catalysts used in downstream conversion of syngas into methanol and ethanol (Dayton, 2002; Boerrigter et al., 2005). Major power companies are interested in co-firing of biomass with coal, but face several operating issues due to imbalance in the energy density of coal and biomass, low burning rate and low flame temperature. Biomass also creates additional problem due to poor flowability, poor blending and reduces the thermal efficiency and capacity of existing boiler units (Tillman, 2000). Pyrolysis of lignocellulosic biomass yields high molecular weight organic liquid called bio-oil and a carbon rich material called bio-char. Bio-oil is recognized as a major feedstock for producing chemicals, fuels and heat. However, application and use of bio-oil have several technical challenges due to low pH, high oxygen and water contents, high molecular weight tar compounds and high viscosity. Several of the above problems on the thermo-chemical conversion biomass were partly or mostly originated from feedstock properties. Thus, thermal pretreatment of biomass or torrefaction process can provide a promising solution for thermo-chemical conversion technologies.

Torrefaction is a process of thermal degradation of biomass at relatively low temperatures (200-300°C) under anoxic conditions resulting in an energy dense, stable char or coal-like material that also eliminates or minimizes coke forming precursors, such as acetic acid. Removal of these coke forming precursors could significantly improve catalytic upgrading of the bio-oil generated by pyrolysis. It is anticipated that integrating biomass torrefaction with thermo-chemical conversion processes will not only improve the energy density and quality of bio-crude oil/syngas, but also minimizes preprocessing and storage costs. Co-firing torrefied biomass with coal can be possible without compromising the energy efficiency and plant production capacity. This paper present the recent progress on the torrefaction characteristics of forest biomass and the effect of torrefaction on the pyrolysis and gasification processes studied using a thermal gravimetric analyzer with mass spectrometry (TG-MS).

Materials & Methods

Materials

Southern pine forest residue was selected as a biomass feedstock for this study. Large volume of forest residue was collected from Macon, GA and was frozen for the torrefaction experiment. The physical properties of forest residues such as moisture content, mass distribution, bulk density, particle density and particle size were determined based on the procedures used by Mani et al. (2004 & 2006a). The proximate and ultimate analyses and heating value of the biomass were determined as per standard procedures. To study the effect of torrefaction with various thermo-chemical conversion processes, southern pine chips were used.

Torrefaction Experiment

A batch torrefaction reactor (Figure 1a) with a capacity to hold 3-5 lbs of biomass samples was used to study the torrefaction characteristics of forest residue. The batch reactor has three ports as 1) for temperature sensor; 2) for carrier gas inlet and 3) for carrier gas outlet. We used nitrogen as a carrier gas for all the torrefaction experiment. For each torrefaction experiment, about one lb (0.454 kg) of forest residue was placed in the cast iron container and was sealed and placed in the temperature controlled oven. The oven temperature and sample temperature data were data logged in the computer and the temperature inside the reactor was controlled accurately. Figure 2b shows a typical temperature profile during torrefaction of forest residues. A sweep gas flow rate of 2 L/min of nitrogen was used to maintain anoxic condition inside the reactor and the exhaust gas line was passed through a series of condensers to collect all the condensable compounds generated from the torrefaction experiment. The non-condensable gases were passed though a water column to remove any dirt and were collected using gas bags for further analyses. Once the torrefaction temperature and holding time were reached, the sample will be cooled using liquid nitrogen or by forced convection until the sample reaches close to room temperature. Figure 2(b) shows the typical time –temperature profile of a batch torrefaction reactor. Then, the torrefied samples were removed from the reactor, weighed and stored for further analysis. The condensed liquids were collected in sample tubes for GC-MS analysis.

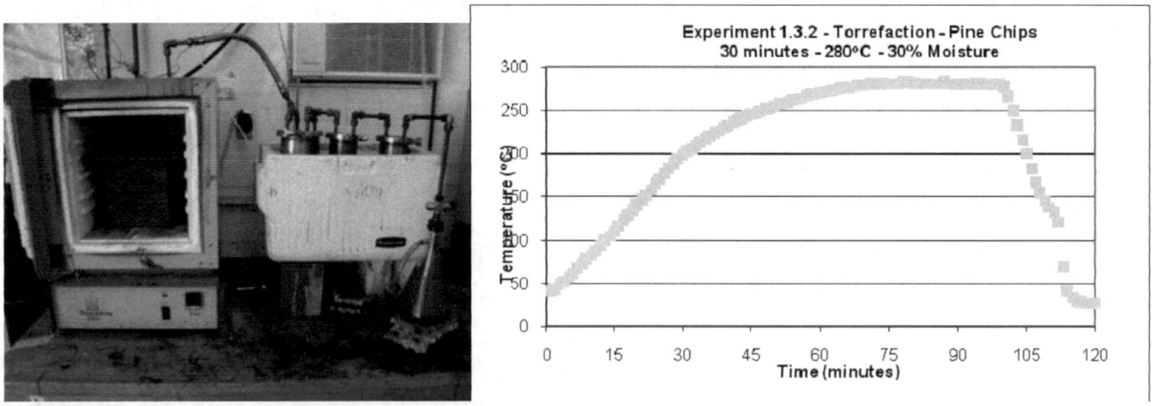

Figure 1. Schematic diagram of a batch torrefaction reactor (a) with a typical temperature profile (b)

The quality of torrefied biomass depends on the number of process variables, which are chemical composition of the biomass, particle size, moisture content, reaction temperature, reaction time, heating rate and types of carrier gases. Among various parameters, torrefaction temperature has a significant effect on the yield and composition of biomass. Therefore three different torrefaction temperatures: 220, 250 & 280°C were selected for this study. Torrefied biomass collected at the end of each experiment was analyzed for chemical compositions, heating value, and structural characteristics of biomass. Finally, the process variables responsible for producing high energy density torrefied biomass (biocoal) were optimized.

Integration of torrefaction with thermo-chemical conversion processes

In order to study the effect of biomass torrefaction with thermo-chemical conversion processes such as combustion, gasification and pyrolysis processes, a Thermal Gravimetric Analyzer with Mass Spectrometry was used. Southern pine chips were ground using a knife mill and passed through 0.25 mesh size was used for this study. For each experiment, about 15 mg of sample was torrefied at four temperatures (225, 250, 275 & 300°C) and the solid samples were quenched with liquid nitrogen. Then, the samples were used for combustion, gasification and pyrolysis experiments respectively using air, partial oxygen mixture and nitrogen as sweep gases. For each type of thermo-chemical conversion processes, mass degradation rate and the off-gas compositions monitored using selective ion were recorded and analyzed. Each test combinations were repeated trice.

RESULTS & DISCUSSIONS

Forest residues characterization

Forest residues were fractionated into clean chips, bark, small stems, leaves, and small particles. More than half (51%) of the samples were chips, followed by 19% of small stems and 13% of bark. Small stems and particles contributed about 8% and 9% respectively. Most of the small particles were pine needles ground up during the chipping operation. The particle size distribution of each mass fraction is given in Table 1. The average bulk density of the whole forest residues was (on dry basis) 149.17 kg/m^3 (SD = 4.47), compared to sawdust of 142.2 kg/m^3 and clean pine chips of 209.8 kg/m^3 (Gigler et al. 2000a). The lower bulk density of forest residues was mainly attributed to uneven particle size distribution of small stems and leaves.

Table 1. Particle Size Distribution of Components of Forest Residues

Residue fractions	Thickness[a] (mm)	Length[a] (mm)	Width[a] (mm)
Chips	6.3 (0.8)	26.1 (8.3)	18.5 (0.0)
Barks	4.4 (1.2)	32.0 (0.9)	15.0 (0.8)
Small stems	4.7 (0.5)	124.9 (93.3)	4.6 (0.2)
Leaves	0.5 (0.1)	93.0[b] (28.2)	___
Small particles[c]	<4.8		

[a] Numbers in the parenthesis are standard deviation with five replications.
[b]The length of the leaves ranges from 36.1 to 124.1 mm.
[c]The size of small particles represents particles that are smaller than 4.8 mm screen size.

Biomass Torrefaction Experimental Data

Figure 2 shows the example of torrefied products obtained from each experiment. Figure 3 shows the yield of torrefied biomass (in dry basis) at various temperatures and initial residue moisture content with 30 min residence time. In overall, lower torrefaction time in the batch reactor results in higher yield as the solid hydrolysis of hemiellulose takes longer time to decompose into vapors. Among the variables studied, torrefaction temperature and residence time had significant yield on the total yield of torrefied biomass. Torrefaction temperature had a significant effect on the torrefied biomass yield, chemical compositions and the heating value of biomass. The results were comparable with experimental data obtained by Prins et al. (2006a). Tables 3 & 4 show the changes in chemical composition of forest residue at various torrefaction temperatures. It was evident that an increase in torrefaction temperature increased the heating value of biomass to a level closer to that of lignite or low grade bituminous coal. A further increase in torrefaction temperature will result in low yield of biomass and an increase in the volatile oil (bio-oil) yield. Therefore, a torrefaction temperature of 280°C is optimal for forest biomass based on the heating value and the yield of biomass obtained from this process.

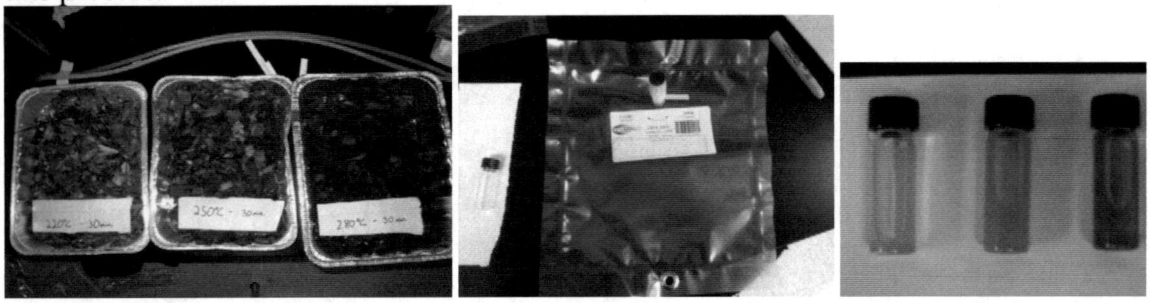

Figure 2 shows the products of torrefied biomass (a) torrefied biomass; (b) non-condensable gases and (c) condensable liquids

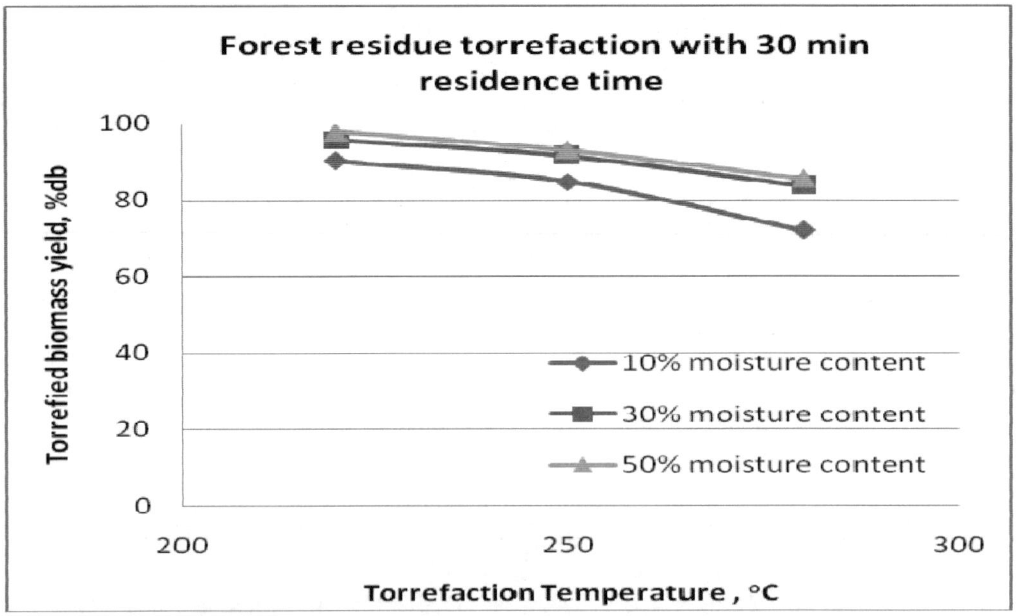

Figure 3. Yield of torrefied biomass at 30 min torrefaction time in the reactor

High initial moisture content of biomass is not preferred as more time and energy required removing moisture from the biomass before starting the torrefaction process. If low temperature gas sources are available in the power plant, this process can be integrated to the power plant for efficient utilization of low heat energy sources such as exhaust gases. A significant reduction in the composition of hemi-cellulose was noted with an increase in torrefaction temperature indicating that hemi-cellulose is thermally hydrolyzed during torrefaction (data not shown). Some degree of cellulose was also decomposed during torrefaction and the rate of cellulose decomposition increased with increase in torrefaction temperature.

Table 3. Elemental analysis of torrefied forest residue samples

Torrefaction temperature °C	C (%)	H (%)	O (%)	N (%)	S (%)
Non-torrefied	45.3	5.9	48.5	0.3	0.1
220	49.4	5.5	44.1	0.3	0.0
250	50.5	5.4	41.8	0.4	0.0
280	56.4	5.4	36.1	0.9	0.1

Table 4 Proximate analysis of torrefied forest residue samples

Torrefaction temperature °C	Moisture (%)	Volatile C (% wt)	Fixed C (% wt)	Heating value (Btu/lb)	Ash (%)
Non-torrefied	10	75.3	16.3	7,774	0.67
220	3.2	76.8	19.1	8,474	1.1
250	2.3	74.9	20.6	9,376	1.2
280	2.1	70.8	25.6	10,167	1.4

Figure 4 shows the concentration of various chemicals present in the non-condensable portion of gases collected during torrefaction of forest residues at 280°C. Three major compounds: carbon monoxide, carbon dioxide and methane were emitted in large quantities during the early stage of torrefaction and their concentrations were decreased as the residence time increases. Other compounds such as hydrogen, propylene, ethane, ethylene, n-butane and methyl acetylene were also present in the gases at low concentrations. The rate of emission of above compounds varies significantly as the torrefaction temperature changes and can be potential used as a carrier gas during torrefaction process instead of nitrogen as an inert gas.

Figure 5 shows the partial concentration of condensable liquids collected during torrefaction of biomass at various residence times. Although several hemi-cellulose and extractives degraded compounds were noticed from the GC-MS data, higher concentrations of alpha-pinene, camphene, beta-pinene and beta-phellandrene were present.

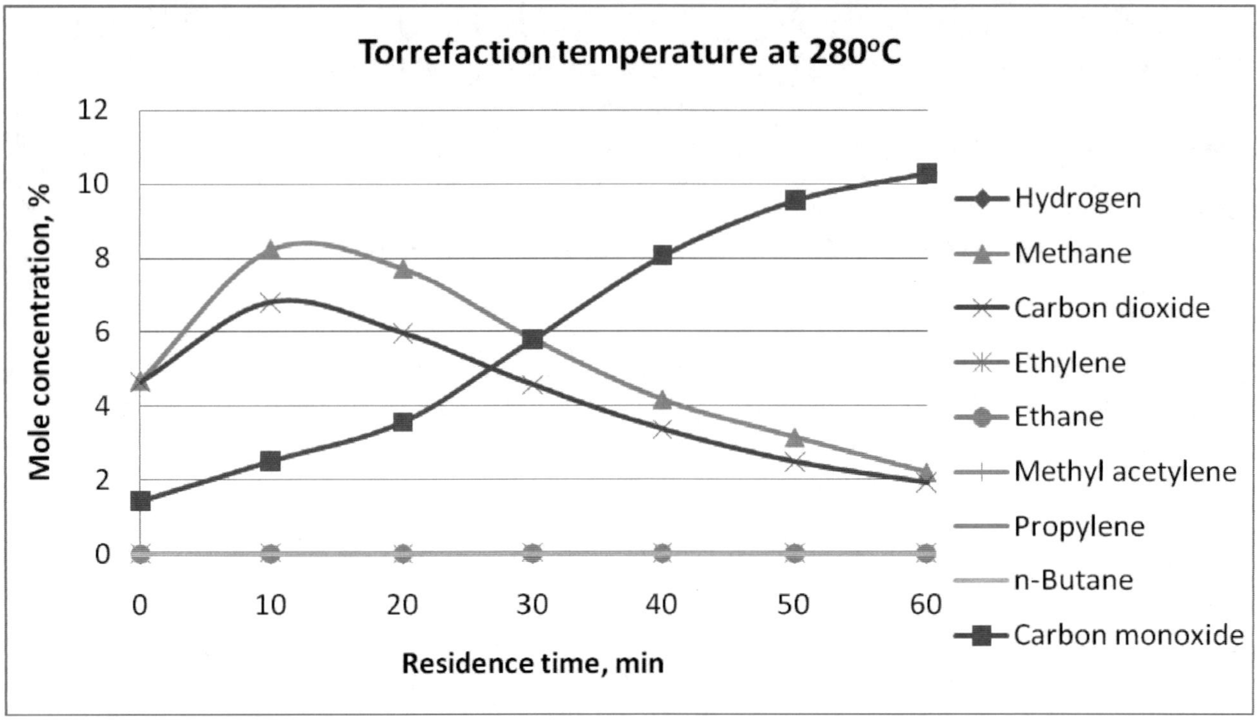

Figure 4. Concentration of various compounds present in the non-condensable gases generated during torrefaction of forest residues.

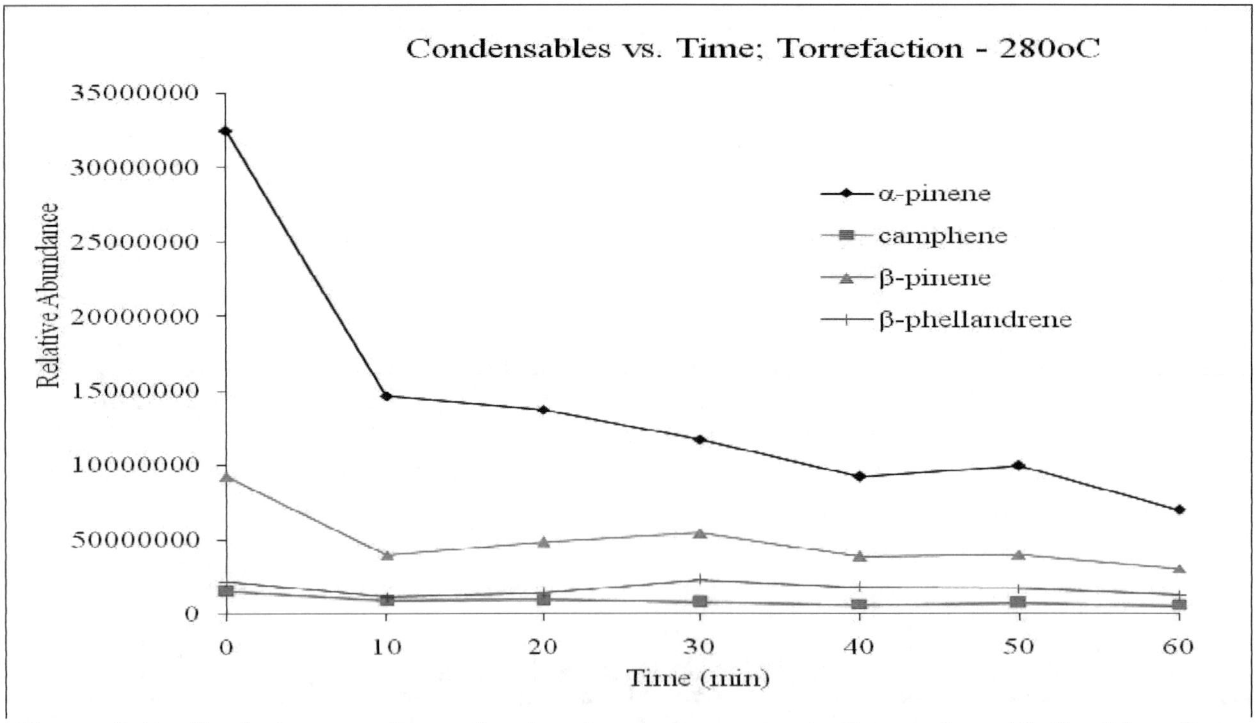

Figure 5. Qualitative composition of major chemicals present in the condensable liquids collected at various torrefaction times.

Integrating Torrefaction with Combustion, Gasification & Pyrolysis Processes (preliminary data)

Combustion: Based on the mass decay data from TG-MS experiment, both the control and torrefied biomass was more reactive than coal. The rate of mass loss was higher for both control and torrefied biomass compared to coal. As the torrefaction temperature of biomass increases, little or no effect on the rate of mass loss was noticed. This indicates the there were no significant difference among the control and torrefied biomass burning rate. The preliminary data from the combustion study will be useful for conducting the lab scale combustion study on both torrefied biomass and coal.

Gasification: Mass spectrometry (MS) data collected using selective ion monitoring from torrefaction with gasification experiment showed that there were significant reduction in carbon dioxide and acetol production when pine was torrefied at 275°C for 20 min compared to untreated (control) biomass. Furthermore, a marked reduction in peak acetic acid production relative to baseline signal was observed for torrefied samples compared to control biomass. There was a statistically significant change in the amount of guiacol (One of the critical pre-cursor compounds for tar formation during biomass gasification) detected in one sample (torrefied at 225°C) compared to the control. The results were partially comparable with results obtained from Prins et al. (2006b). Further research on this topic is required to critically evaluate the effect of torrefacion on the gasification of biomass.

Pyrolysis: MS data from torrefaction with pyrolysis experiment indicated that higher torrefaction temperatures resulted in a significant decrease in acetol formation. Acetol was considered as the primary tar precursor during pyrolysis of biomass and lower concentration of acetol from torrefied biomass may generate lesser tars in the bio-oil. Torrefaction of biomass at 225°C followed by pyrolysis increased the amount of CO_2 formed relative to the control. However, higher torrefaction temperatures reduced the amount of CO_2 formed during pyrolysis. Concentration of water formed during torrefaction followed by pyrolysis was significantly less compared to control biomass.

Conclusions

Biomass torrefaction is the promising thermal pretreatment method for lignocellulosic biomass for thermo-chemical conversion platform. Torrefaction process significantly increases the energy density of biomass close to coal in addition to improve the storability and grindability of biomass. Higher torrefaction temperature had a significant effect on the increased energy density and decreased hemi-cellulose content (data not shown). Identification and separation of volatile compounds generated during torrefaction can become potential value added compounds in the future. Effect of torrefaction on the combustion, gasification and pyrolysis processes showed promising preliminary results on the final product compositions. Further research in this area at UGA is directed towards better understanding on the structural and functional modification of torrefied biomass and evaluating thermo-chemical reaction kinetics of both gasification and pyrolysis processes.

Acknowledgement

The author acknowledges the Traditional Industrial Program (TIP) of Georgia for Forest Resources and Products for funding this project and the invaluable research support from Dr. Jim Kastner & Mr. Gerry Lindo at the University of Georgia, Athens (UGA).

References

1. Boerrigter, H., S. V. B. Van Paasen, P. C. A. Bergman, J. W. Konemann, R. Emmen, A. Wijnands. (2005). "OLGA tar removal technology". *ECN Report # C-05-009*, The Netherlands: Energy Research Center of the Netherlands (ECN).
2. Dayton, D. (2002). "A review of the literature on catalytic biomass tar reduction". *NREL/TP report # 510-32815.* National Renewable Energy Laboratory, Golden, CO.
3. Mani, S., L. G. Tabil and S. Sokhansanj. (2004). "Grinding performance and physical properties of wheat and barley straws, corn stover and switchgrass". *Biomass and Bioenergy* 27(4): 339-352
4. Mani, S., L. G. Tabil and S. Sokhansanj. (2006a). "Effects of compressive force, particle size and moisture content on mechanical properties of biomass pellets", *Biomass and Bioenergy,* 30(7): 648-654.
5. Mani, S., S. Sokhansanj, X. Bi and A. Turhollow. (2006b). "Economics of producing fuel pellets from biomass", *Applied Engineering in Agriculture*, 22(3): 1-6
6. Prins, M. J., K. J. Ptasinski, F. J. J. G. Janssen. (2006b). "More efficient biomass gasification via torrefaction". *Energy* 31: 3458-3470.
7. Prins, M.J.; K.J. Ptasinski.; F.J.J.G. Janseen. (2006a). "Torrefaction of wood: Part 1. Weight loss kinetics". *J. Anal. Appl. Pyrolysis.* 77: 28-34.
8. Sokhansanj, S., A. Turhollow, J. Cushman and J. Cundiff. (2002). "Engineering aspects of collecting corn stover for bioenergy". *Biomass and Bioenergy* 23: 347-355.
9. Tillman, D. A. (2000). "Biomass co-firing: the technology, the experience, the combustion consequences". *Biomass and Bioenergy* 19(6): 365-384.

Simulation of Combustion and Gasification of Biomass Including Agricultural and Forest Residues and Pulp Mill Black Liquor Using FactSage

Hua-Jiang Huang, Department of Bioproducts and Biosystems Engineering, University of Minnesota, Saint Paul, MN

Shri Ramaswamy, Department of Bioproducts and Biosystems Engineering, University of Minnesota, Saint Paul, MN

There is an increasing interest to convert biomass such as agricultural waste, forest residue, and black liquor solids from pulp mills into heat, power and biofuels and biochemicals.

In conventional Kraft pulp mills, the black liquor solids containing organic chemicals such as hemicellulose and lignin, and inorganic chemicals (Na2S and Na2CO3 used for pulping) is currently burnt in the Tomlinson recovery boiler for recovery of the inorganic chemicals and generation of energy via combustion. The combustion with CHP (combined heat and power) system is not very energy-efficient. Instead, IGCC (Integrated Gasification Combined Cycle) with gasification replacing combustion can bring higher efficiency and environmental benefits. In addition, part of syngas produced from gasification can be used to produce biofuels and chemicals by syngas synthesis (e.g., Fisher -Tropsch synthesis) and conversion.

The combustion/gasification of black liquor is a very complicated process as it involves multiphase reactions including inorganic solid phases. In this study, the thermochemical software Factsage is used to simulate the combustion/gasification of biomass including black liquor solids and other agricultural or forest biomass, in order to better understand the combustion/gasification process and predict its performance or efficiency. Potential chemical emissions were also predicted to evaluate the environmental impact. The results can be useful for techno-economic and environmental analysis of the whole biomass-based biorefinery.

See more of this Session: Thermochemical Biorefineries II
See more of this Group/Topical: Forest and Plant Bioproducts Division

Catalytic Steam Reforming of Water Extracted Bio-Oil Fractions

Satya K. Jujjuri, Center for Sustainable Environmental Technologies, Iowa State University, Ames, IA

Justinus A. Satrio, Center for Sustainable Environmental Technologies, Iowa State University, Ames, IA

Robert C. Brown, Center for Sustainable Environmental Technologies, Iowa State University, Ames, IA

Steam reforming of bio-oil over a Ni-catalyst mainly generates syngas which can further be converted into a range of fuels and chemicals. Moreover, under suitable conditions and through catalyst modifications, product selectivity toward highly desirable hydrogen can be significantly improved by promoting water-gas shift reaction. Production of syngas from bio-oil is highly endothermic requiring higher temperatures while water-gas shift reaction that produces hydrogen is exothermic preferring lower temperatures. Thus optimum reaction conditions are required for improving yield toward syngas/hydrogen. Although steam reforming of petroleum-derived feedstocks is well established but because of the higher reactivity and thermal instability of bio-oil several challenges are faced in the fixed bed reformer wherein catalyst deactivation is rapid with bio-oils. Typically this condition is minimized by the utilization of higher steam to carbon ratios. The focus of this work is to study the catalytic conversion efficiency of different fractions of bio-oil that were obtained during the fast pyrolysis of biomass mainly in terms of the yield of syngas/hydrogen from wheat straw bio-oil. One of the main goals of this work is to identify the most desirable bio-oil fraction for the reforming step. A commercial Ni/Al2O3 catalyst is employed for testing. High steam to carbon ratios of 10 or higher are maintained by selectively extracting water soluble compounds from the four different bio-oil fractions produced during fast pyrolysis of wheat straw. The temperature range studied is between 350°C and 550°C at atmospheric pressure. Effluent gases are analyzed on a Micro-GC.

See more of this Session: Thermochemical Biorefineries II
See more of this Group/Topical: Forest and Plant Bioproducts Division

Review and Analysis of Transportation Fuel Production From Biomass-Derived Syngas

Christopher M. Kinchin, National Renewable Energy Laboratory, Golden, CO

Thermochemical conversion of biomass to synthesis gas (syngas), followed by catalytic conversion of syngas to hydrocarbons is one potential avenue for producing renewable liquid transportation fuels. The Biomass Program at the National Renewable Energy Laboratory (NREL) has performed techno-economic analyses for producing clean syngas and converting the syngas to mixed alcohols. However, the syngas can be catalytically converted to other hydrocarbons with higher energy density and potentially better infrastructure compatibility.

The Massachusetts Institute of Technology School of Chemical Engineering Practice recently worked with NREL to compile and compare a list of liquid transportation fuels that can be produced from syngas. The fuels are compared and ranked according to criteria such as energy density, infrastructure compatibility, technology maturity, and environmental impact. An initial screening process reduced an extensive list to the four most promising options, which were selected for further evaluation. A literature review and techno-economic analysis was thus performed for each of these four options.

The resulting scoping analysis includes mass and energy balances for each process as well as economic evaluation using assumptions and cost data consistent with previous NREL techno-economic analyses on biomass to mixed alcohols processes. Yields and costs for each process are estimated and compared. Additional results of the project are identification of critical knowledge gaps and recommendations for future research and development.

See more of this Session: Thermochemical Biorefineries II
See more of this Group/Topical: Forest and Plant Bioproducts Division

Chemistry Modeling of Biomass Pyrolysis: Application to the CFD Simulation of a Laminar Entrained Flow Reactor

Perrine Pepiot, National Renewable Energy Laboratory, Golden, CO

Mark W. Jarvis, National Renewable Energy Laboratory, Golden, CO

Mark R. Nimlos, National Renewable Energy Laboratory, Golden, CO

A current limitation to the efficiency of biomass gasification is the high level of tars usually found in the product gas, causing fouling and hazardous waste. Costly cleaning treatments to control the tar content of the exiting gas are usually applied downstream. Another appealing approach would be to directly optimize the gasifier design and operating conditions to the biomass properties for minimal tar production. However, such an approach requires a better understanding of the detailed chemical and physical processes occurring simultaneously in the reactor, including mixing, heat and mass transfer, and chemical reactions. Computational Fluid Dynamics (CFD) offers a unique opportunity to incorporate fundamental results, for example from detailed chemical modeling, into a larger scale framework, allowing to gain some valuable insight on tar formation pathways.

In this work, an extensive experimental investigation of biomass pyrolysis in a laminar entrained flow reactor is used as a basis for the development and validation of a reactive CFD model. The three-dimensional, cylindrical configuration remains simple enough to allow for a relatively detailed description of the chemical processes occurring in the reactor. Particles considered in the experiments are finely milled to a maximum size of 180 µm, which is below the size of a typical cell structure. Therefore, intra-particle transport processes are neglected in the CFD model. Particles are assumed to remain spherical in shape as they shrink during conversion, allowing the use of well-characterized correlations for the drag and heat transfer models. Biomass density is kept constant, and a volumetric approach is adopted to evaluate the chemical source terms for the solid phase.

A preliminary chemical mechanism describing biomass volatilization and further conversion of the products into permanent gases and polycyclic aromatic hydrocarbons has been assembled from our previous work on soot formation and data available in the literature. To reduce the computational cost associated with solving the chemistry in CFD, automatic reduction techniques have been applied to this detailed mechanism containing several hundreds of species to obtain a low-order scheme able to correctly reproduce the dynamics of the full chemical system. Since tar formation is the focus of this study, emphasis in the reduced model has been placed on the accurate description of aromatic rings formation from small hydrocarbon species and their subsequent molecular growth. The obtained chemical scheme has been integrated into CFD, and simulation results have been compared to the experimental data collected in the LEFR for a wide range of reactor temperatures and residence times. An improved description of the early stages of biomass volatilization based on carefully chosen model compounds has been derived from the analysis of the computations and the principal component results obtained in the experimental work.

See more of this Session: Thermochemical Biorefineries II
See more of this Group/Topical: Forest and Plant Bioproducts Division

Estimation of Lithium-Ion Battery Condition Using Reduced Electrochemical Cell Modeling

Sungwoo Cho, School of chemical and biological engineering, Seoul National University, Seoul, South Korea

Hyunseok Chung, School of chemical and biological engineering, Seoul National University, Seoul, South Korea

Hoi-in Jeong, School of chemical and biological engineering, Seoul National University, Seoul, South Korea

Chonghun Han, School of chemical and biological engineering, Seoul National University, Seoul, South Korea

Lithium-ion polymer batteries (LiPB) are extensively used in hybrid electric vehicles (HEV), plug-in hybrid electric vehicle (PHEV) and consumer portable electronics. In order to operate these batteries more efficiently, accurate estimation of battery conditions such as state-of-charge (SOC), state-of-health (SOH) is necessary. Electrochemical modeling of the battery is important for accurate estimation of these battery's conditions. The reduced dynamic equivalent circuit battery model with modified discrete time concept is suggested. This model is based on first order RC circuit model which represents the polarization phenomenon in battery. The model parameters are estimated with adaptive parameter estimation algorithm using least square estimators based on experimental cell data with pulse patterns. The electrochemical model and the parameter estimation algorithm are validated with urban driving pattern data for HEV. Using proposed battery model, estimation of SOC algorithm is developed. From these models, current-based SOC is calculated. Also, voltage-based SOC is estimated by open-circuit voltage calculation from battery model. The current-based SOC and the voltage-based SOC are compensated for each other by PID control system with current-based criteria. The model parameters and the controller parameters are estimated by neural-network and adaptive filters. These models are validated by experimental data with various SOC and temperature environment. The result is that proposed battery model and estimation algorithms are appropriate for estimation of SOC to various use of LiPB.

See more of this Session: Lithium Battery Technology and Materials
See more of this Group/Topical: Fuels and Petrochemicals Division

Thermal Phase Behavior of Ionic Liquid-Lithium Salt Electrolytes

Wesley A. Henderson, Department of Chemical & Biomolecular Engineering, North Carolina State University, Raleigh, NC

Qian Zhou, Department of Chemical & Biomolecular Engineering, North Carolina State University, Raleigh, NC

Ionic liquids have become a focal point for electrolyte research in recent years due to their potential to replace the volatile and flammable aprotic solvents now used. Many ionic liquids exhibit high ionic conductivity, high thermal and electrochemical stability, negligible vapor pressure at ambient pressure, low flammability and other desirable properties for electrolytes. The ions from the ionic liquid do not, however, participate in the battery electrochemical reaction. Thus, a lithium salt with electroactive lithium cations must be added. Although the ionic liquid may be a liquid at ambient temperature and below, dramatic differences in the thermal properties of the binary salt mixtures (relative to the neat ionic liquid) are frequently found. The phase behavior of binary ionic liquid-lithium salt mixtures is therefore reported to examine the useful liquid range of such electrolytes and to aid in explaining electrolyte transport properties such as ionic conductivity and viscosity.

See more of this Session: Lithium Battery Technology and Materials
See more of this Group/Topical: Fuels and Petrochemicals Division

Uncertainty Analysis in Battery Packs

Shriram Santhanagopalan
Celgard LLC
13800 South Lakes Drive
Charlotte NC 28273
shriramsanthanagopalan@celgard.com

November 12th 2009 AIChE Fall 2009 Annual Meeting

Experimental Data

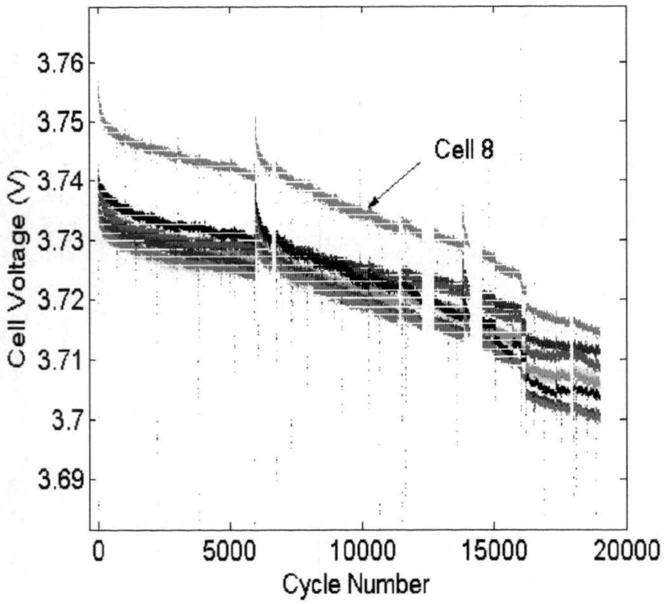

Individual End of Discharge Cell Voltages in a string of 8 cells in series

- **Charge:**

  - CC Charge:
    Constant current until one individual cell reaches 4.1V
  - CV Charge:
    Constant voltage charge at the last battery voltage recorded when first individual cell hit 4.1V.
  - Total charge time of 27.5 minutes

- **Discharge:**

  - CC discharge:
    Constant current to 20% DOD
  - Total discharge time of 17.5 min
  - Temperature: 30°C
  - No rest between charge and discharge steps

Equivalent Circuit Cell Model

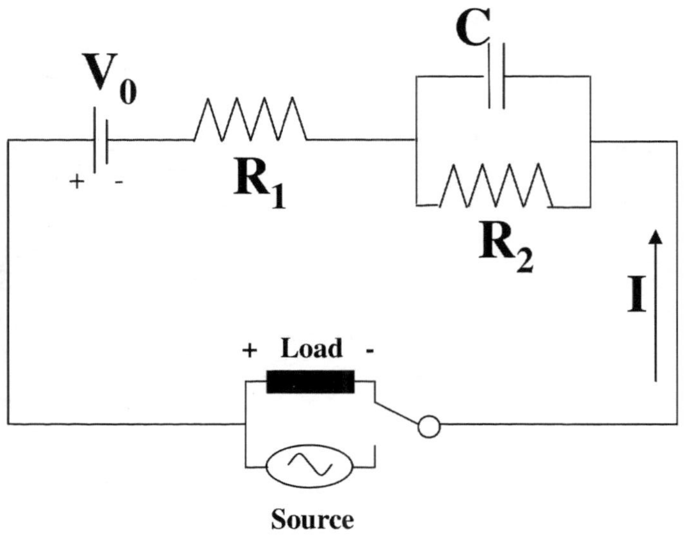

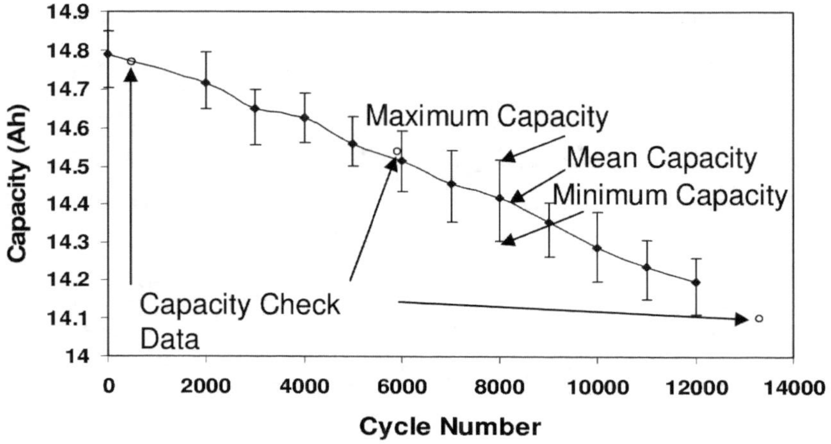

V_0 - Open Circuit (Steady-state) voltage

R_1 - Ohmic drop from contact resistances, SEI film, etc.

R_2 - Charge transfer resistance

C - Time-delay (double-layer effects, diffusion, etc.)

•The model was used to predict decay in capacity Vs cycle number by estimating the circuit parameters.

•Overall, the simple model did a good job!

AIChE 2009 Annual Meeting

Divergence among Cell Voltages

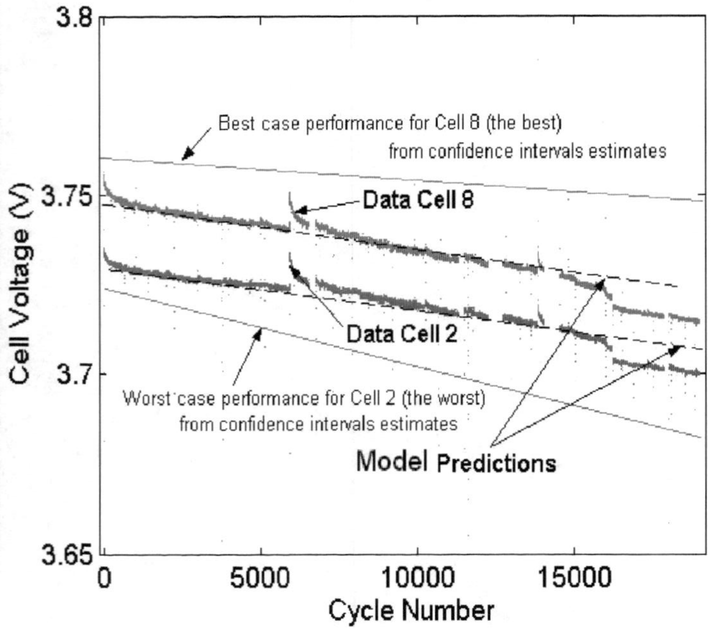

Model predictions for individual cells and the overall confidence interval for the battery

- Discharge time was constant and end of charge was determined by the weakest cell.
- Cell-to-cell variations made predictions difficult: parameters had to be estimated for each cell
- Making predictions with best and worst case parameter sets lead to wide confidence intervals for battery performance.
- With cycling the interval-width increased, leading to poor predictions of end of life for the battery.
- **Solution:**

  \- Consider cell to cell variations within the Cell model, from cycle 1.

AIChE 2009 Annual Meeting

Nominal Plant Transfer Function

- Analytical Solution for the Circuit Model:

$$V = \frac{Q_0}{C} e^{-t/R_{ct}C} + V_0 + IR_1 + IR_2\left(1 - e^{-t/R_{ct}C}\right)$$

- Laplace transform:

$$G_0(s) = \frac{\tilde{V}(s)}{I(s)} = \frac{(R_3 - R_2)}{s + 1/\tau} + \frac{(R_1 + R_2)}{s} \quad \text{or} \quad G_0(s) = K_0 \frac{s + \left(\frac{R_1 + R_2}{R_3 + R_1}\right) 1/\tau}{s\left(s + 1/\tau\right)}$$

- Definition of the lumped parameters:

$$\tilde{V}(s) = V(s) - V_0 \qquad R_3 = \frac{1}{\mathrm{C_{rate}} C} \qquad \tau = R_2 C \qquad K_0 = \left(\frac{1}{R_3 + R_1}\right)$$

Uncertainty Analysis

- Nominal Transfer Function:

$$G_0 = K_0 \frac{s + \left(\frac{R_1 + R_2}{R_3 + R_1}\right) \frac{1}{\tau}}{s\left(s + \frac{1}{\tau}\right)} \equiv K_0 \frac{s+a}{s(s+b)}$$

- Perturbed Transfer Function:

$$G = G_0 K_0 (1 + w_I \Delta_I), \quad \Delta_I(j\omega) \le 1, \forall \omega$$

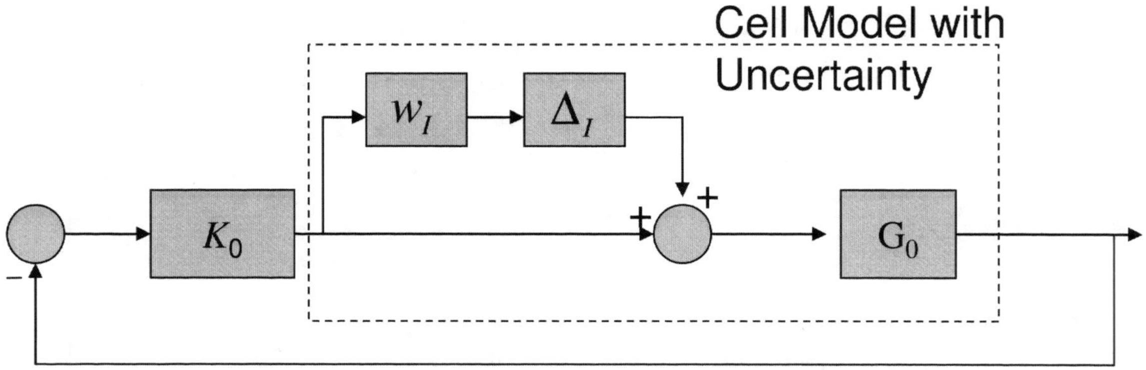

AIChE 2009 Annual Meeting

Robust Stability

- Calculate uncertainty functions w_I from variance in cell capacities among different batches, and confidence intervals for the different parameters.
- Check for Nyquist robust stability criteria:

Nyquist Response of the Cell Model with Uncertainty in:

i) the Cell Capacity (± 10%)

ii) Internal Impedance (R_1 was perturbed by 20 % in this case)

Plot shows that the battery will meet end of life requirement of cell voltage divergence < 50 mV.

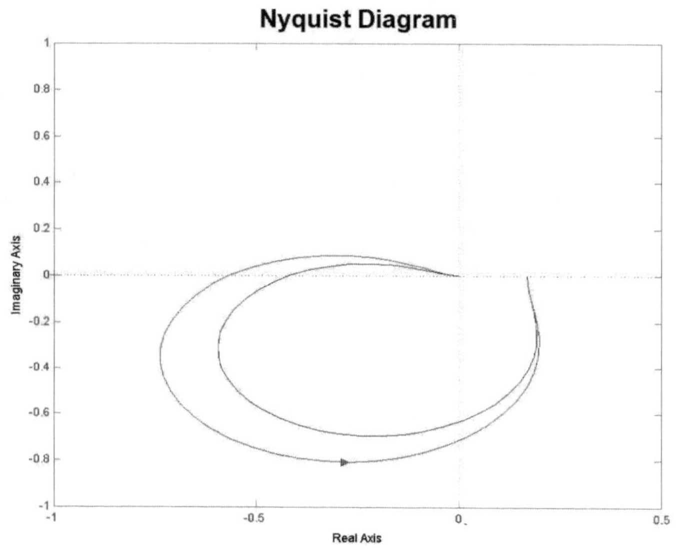

AIChE 2009 Annual Meeting

Summary

- A battery model comprised of cell model with parameters for individual cells was developed: however predictions based on parameter estimates for the different cells gave rise to wide confidence intervals and unreliable predictions of end of life.
- Incorporation of uncertainty in parameters such as cell capacity and internal impedance enables one to make more definitive predictions.
- This approach is particularly useful when a large number of cells are used to build the battery pack (individual estimates are impractical for this case).
- The proposed approach provides statistically representative estimates for parameters that characterize the cell.

Acknowledgements: Prof. Ralph E. White and Prof. Antonello Monti, University of South Carolina, Columbia, SC.

Analysis of Electrode Materials for Use in Lithium Ion Cells for Automotive Applications

Rajeswari Chandrasekaran, School of Chemical & Biomolecular Engineering, Georgia Institute of Technology & Center for Fuel Cell and Battery Technologies,Georgia Tech Research Institute, Atlanta, GA

Jagjit Nanda, Research and Advanced Engineering, Ford Motor Company & (currently at)Materials Science and Technology Division, Oakridge National Laboratory, Oakridge, TN., Dearborn, MI

Jeffrey Remillard, Research and Advanced Engineering, Ford Motor Company, Dearborn, MI

Ken Nietering, Research and Advanced Engineering, Ford Motor Company, Dearborn, MI

Ann O'Neil, Research and Advanced Engineering, Ford Motor Company, Dearborn, MI

Ted Miller, Research and Advanced Engineering, Ford Motor Company, Dearborn, MI

Abstract:

Understanding the individual electrode materials used in lithium ion cells is of prime importance for their long term use in hybrid electric vehicles. Several analytical techniques have been useful in this regard. The In-Situ Raman spectroscopy technique has been employed before[1],[2] to understand the behavior of negative and positive electrode particles, individually in real time. Efforts also can be seen in the literature that discuss space and time-resolved Raman spectro-microtopography techniques[3] applied to a working battery with $LiCoO_2$/graphite electrode materials. As mentioned in Ref.3, this technique was qualitative at that stage and these authors have suggested that this technique could be used to generate dynamic images of lithium transport within the electrodes by mapping the state of charge as a function of time and space. The authors also identified the rate of spectral acquisition as the rate limiting step. If in-situ Raman techniques can be successfully employed in dual insertion lithium ion cells to understand solid-phase lithium concentration gradients in both electrodes in real time during a charge-discharge process, one can come up with better engineering solutions to overcome problems at the cell level that pose limitations to their performance at the system level. This presentation will attempt to discuss efforts along those lines that have been carried out at Ford Motor Company with electrode materials other than $LiCoO_2$, which was used in Ref.3. In the future, experimental results can be compared with modeling predictions as well.

To begin with, the open-circuit potential data of the positive electrode material (proprietary) is needed. This information can be obtained from slow-rate charge-discharge data (in our case C/27 rate) as shown in Figure 1 and then converting the specific capacity to intercalation coefficient of Li in the positive electrode material. These data were obtained using the cell configuration shown in Figure 2, which is termed the "hockey- puck" cell. Similar experiments were performed to characterize the negative electrode (carbon) material as well. These electrodes were cycled a few times both in half-cell as well as in full-cell configurations to make sure that they have good cyclability before performing any in-situ Raman experiments. In-situ Raman experiments were then carried out in an "edge cell" configuration (figure not shown here) in which, as the name suggests, one can observe the progress of experiments across a cell sandwich. Some of the preliminary results and methods to obtain state of charge estimations from Raman data will be discussed at the meeting.

Figure1. Slow rate charge-discharge data for the positive lithium intercalation electrode

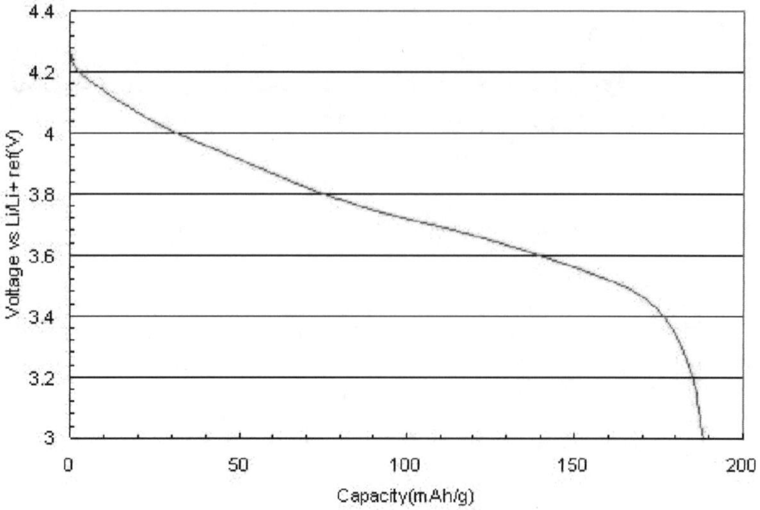

Figure 2 Hockey-puck cell used for half cell and non-optical experiments

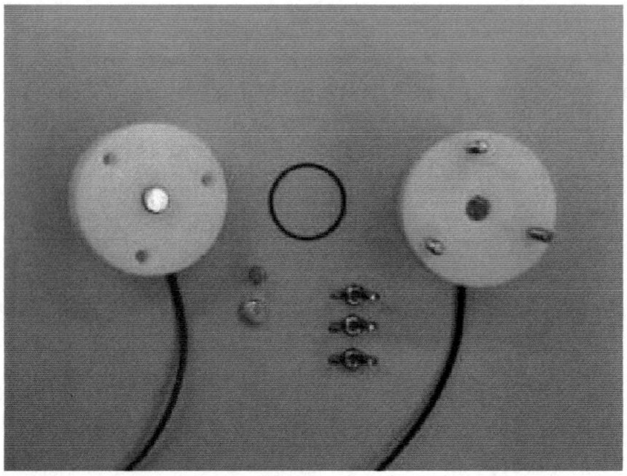

Acknowledgements:

Ford Motor Company is acknowledged for financial support of this research work and Dr.Dawn Bernardi for helpful discussions. Dr. Thomas F Fuller, Professor, Georgia Tech is acknowledged for his support.

References:

[1] Jinglei Lei, Frank McLarnon, Robert Kostecki, *J. Phys. Chem. B, 2005, 109 (2), pp 952-957.*

[2] Qingfang Shi, Kaoru Dokko, Daniel A. Scherson, *J. Phys. Chem. B, 2004, 108 (15), pp 4789-4793.*

[3] Yu Luo, Wen-Bin Cai, Xue-kun Xing, Daniel A. Scherson, *Electrochemical and Solid-State Letters*, **7** (1) E1-E5 (2004).

See more of this Session: Lithium Battery Technology and Materials
See more of this Group/Topical: Fuels and Petrochemicals Division

Battery Design for Large Lithium and Zinc Batteries

Galen J. Suppes, University of Missouri-Columbia, Columbia, MO

Bryan Sawyer, University of Missouri-Columbia, Columbia, MO

Michael Gordon, University of Missouri-Columbia, Columbia, MO

A battery design is being evaluated which is particularly useful for larger battery applications such as in PHEV and electrical grid applications. Preliminary studies validate the approach. Parametric studies are discussed which define performance curve.

See more of this Session: Lithium Battery Technology and Materials
See more of this Group/Topical: Fuels and Petrochemicals Division

Virtual Screening for Lithium Ion Battery Electrolyte Additives

Mathew D. Halls *, Materials Science Division, Accelrys, Inc., San Diego, CA

Ken Tasaki, Technology Research Division, Mitsubishi Chemical, Inc., Redondo Beach, CA

Nick Reynolds, Materials Science Division, Accelrys, Inc., Burlington, MA

The use of high-throughput quantum chemistry to analyze and screen a materials structure library is demonstrated for fluoro- and alkyl-derivatized ethylene carbonate (EC). The fundamental properties of these systems are evaluated for potential application as graphite anode solid-electrolyte interface (SEI) additives in lithium ion battery electrolyte formulations. The effect of added substituents increases the electron affinity (EA) relative to EC which is a critical requirement for improved additive activity. The results presented here highlight fluorination as a promising chemistry for improving anode SEI additive properties and merits further computational and experimental examination.

See more of this Session: Lithium Battery Technology and Materials
See more of this Group/Topical: Fuels and Petrochemicals Division

Volume Expansion and Phase Transformation in Single Lithium Insertion Silicon Electrode Particle

Rajeswari Chandrasekaran[1, 2], Thomas F.Fuller[1, 2]
(1) Center for Innovative Fuel Cell & Battery Technologies, Georgia Institute of Technology, Atlanta, GA 30332. (2) School of Chemical & Biomolecular Engineering, Georgia Institute of Technology, Atlanta, GA 30332.

Introduction: Silicon is being considered for negative electrodes in lithium ion batteries because of their high specific energy, 4200 mAh/g[i]. Recent literatures[ii],[iii] suggest that alloying of lithium in silicon proceeds through an amorphous phase transition. Isotropic volume expansion is suggested to be associated with this amorphous phase transition [iv]. It should be noted that, however, even at room temperature, towards the end of lithiation, a-Li_xSi suddenly crystallizes to $Li_{15}Si_4$[ii]. Few literatures have suggested that amorphous silicon phase is only a metastable phase and arises due to selective equilibrium at room temperature [v] and that the order-disorder transition is reversible [vi].Moreover at high temperatures [vii], the lithiation proceeds through distinct crystalline phases. So, overall, it will be worthwhile to pursue a unique modeling approach for silicon electrodes among lithium insertion electrodes since both phase transformation[i] and associated huge volume expansion [iii] occur in silicon electrodes. If one wants to consider lithiation through distinct crystalline phases, the interface between the inner phase and the outer phase moves towards the center of the particle as lithiation occurs and simultaneously the overall radius of the particle increases. Thus we have two moving boundaries. Literature[viii] suggests that a shrinking core model to capture the interface motion is applicable only when there is a three-dimensional growth mechanism which need not be the case always. Moreover, it is also suggested[ix] that one needs to consider phase transformation kinetics along with solid state diffusion to study phase transformation electrodes.

Current work will help in a better understanding of the solid phase transport for the case of increasing particle size in silicon electrodes during lithiation, which could then be used in cell sandwich model for practical lithium ion cells with silicon negative electrodes. Thus, this work attempts to develop a particle model to capture the growing particle size considering amorphous lithiation and associated solid state transport processes in silicon electrodes. In future, to model both volume expansion and phase transformation, one could adopt the framework provided by Harrison *et al*[x] to model phase change in lithium ion batteries for an arbitrary number of phases using diffuse interface method considering phase transformation kinetics as done by Kasavajjula[ix], while considering growing particle size as suggested in this work for silicon electrodes. While doing so, it should be noted that diffusion co-efficient of silicon-lithium alloy is at least two orders of magnitude more than that of the silicon[xi]. Current work considers a particle size such that fracture can be assumed to not occur[xii]. But for a complete picture, stress generation and fracture as done by Christensen and Newman[xiii] can also be incorporated in future.

Single Particle Level Model: Schematic of a lithium-silicon composite electrode /lithium foil electrode cell and single particle of lithium-silicon electrode is given below.[xiv]:

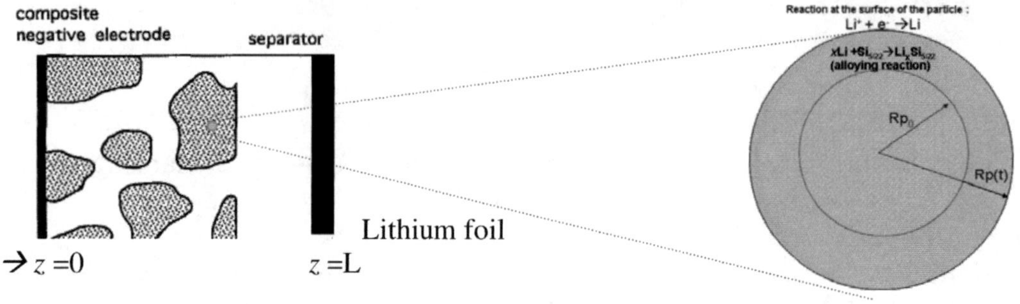

Consider the reaction, $Li^{+} + e^{-} + Si_{5/22} \rightarrow LiSi_{5/22}$

It can be thought of two reactions:

$Li^{+} + e^{-} \rightarrow Li$

$Li + Si_{5/22} \rightarrow LiSi_{5/22}$

So concentration of the species $LiSi_{5/22}$ is the same as Li.

Material balance equation for species '$LiSi_{5/22}$' is given by

$$\frac{\partial c_{LiSi_{5/22}}}{\partial t} + \frac{1}{r^2}\frac{\partial}{\partial r}(r^2 N_{LiSi_{5/22}}) = 0$$

$$N_{LiSi_{5/22}} = y_{LiSi_{5/22}}(N_{LiSi_{5/22}} + N_{Si_{\frac{5}{22}}}) - D_{LiSi_{5/22}}\frac{\partial c_{LiSi_{5/22}}}{\partial r}$$

where

$(N_{LiSi_{5/22}} + N_{Si_{\frac{5}{22}}}) = c_{solid}v^{*}$ At any location, $c_{LiSi_{5/22}} + c_{Si_{5/22}} = c_{solid}$

To evaluate v^{*} (molar average velocity),momentum balance in solid phase (stress) has to be considered. As an approximation, particle growth alone will be considered neglecting stress for now

$$\frac{\partial c_{LiSi_{5/22}}}{\partial t} - D_{LiSi_{5/22}}\frac{1}{r^2}\frac{\partial}{\partial r}\left(r^2\left(\frac{\partial c_{LiSi_{5/22}}}{\partial r}\right)\right) = 0$$

<u>Initial conditions and boundary conditions:</u>

$c_{LiSi_{5/22}} = 0 \quad @\ t = 0$ (lithiation)

$c_{LiSi_{5/22}} = \dfrac{N_{tot}}{\frac{4}{3}\pi\left(4.33^{\frac{1}{3}}R_{p_0}\right)^3}$ $@\ t = 0$ (de-lithiation) (N_{tot} is defined later)

Note: Usually it is said $V_{final} = 4V_i$; here we have $V_{final} = 4.33V_i$

$D_{LiSi_{5/22}}\dfrac{\partial c_{LiSi_{5/22}}}{\partial r} = \dfrac{I_{app}}{\left(4\pi R_p(t)^2\right)F}$ $@\ r = R_p(t)$

$I_{app} = 1.107\times10^{-15}A$ (lithiation) ; $I_{app} = -1.107\times10^{-15}A$ (de-lithiation)

Axial symmetry $@\ r = 0$

Diffusion co-efficient in current work (10^{-18} m^2/s) is based on Pell et.al. Recent literatures suggest a value of 10^{-16} m^2/s.

<u>Expression for the change in particle radius (with time):</u>

Volume of the particle = Average Molar volume of the lithium-silicon electrode × Number of initial moles of silicon

$$\frac{4}{3}\pi R_p(t)^3 = V_{avg} \times \frac{\frac{4}{3}\pi R_{P0}^{\ 3}}{V_{Si_{5/22}}}$$

$$R_p(t) = \frac{R_{P0}}{\sqrt[3]{V_{Si_{5/22}}}} \times \sqrt[3]{V_{avg}}$$

In the lines of Obrovac et.al.

Molar volume of $Li_xSi_{5/22}$ electrode, (where $0 \le x \le 1$)

$$V_{avg} = \overline{V}_{Si_{5/22}} + \overline{V}_{Li} \times x_{Li_{avg}}$$

where $\overline{V}_{Si_{5/22}}$ and $\overline{V}_{Li}$ are partial molar volumes of the host silicon and lithium respectively and $x_{Li_{avg}}$ is the average moles of Li in the particle per mole of $Si_{5/22}$.

Pure molar volume of silicon is used.

For lithium, value provided by Obrovac et.al. is used.

$$V_{avg} = \left(12\ mL/mol\right) + 9\ mL/mol \times 4.4 \times \frac{c_{LiSi_{5/22\ avg}} \times \frac{4}{3}\pi\left(R_p(t)\right)^3}{N_{tot}}$$

where N_{tot} is the maximum number of moles of lithium that can be lithiated in silicon

$$N_{tot} = \frac{1}{\text{Avogadro's Number}}\left(\frac{\frac{22}{5}}{20A^{0^3}}\right)$$

(The assumption is the final lithiated state is $Li_{22}Si_5$ and the volume of pure silicon atom is $20A^{0^3}$)

Change in radius of the particle:

$$\frac{R_p(t)}{R_{P0}} - 1 = \frac{\sqrt[3]{V_{avg}}}{\sqrt[3]{V_{Si_{5/22}}}} - 1 = \frac{\sqrt[3]{V_{Si_{5/22}} + V_{Li} \times 4.4 \times \frac{c_{LiSi_{5/22\ avg}} \times \frac{4}{3}\pi\left(R_p(t)\right)^3}{N_{tot}}}}{\sqrt[3]{V_{Si_{5/22}}}} - 1 = \frac{\sqrt[3]{V_{Si_{5/22}} + V_{Li} \times 4.4 \times \frac{\frac{\int_0^{R_p(t)} c_{LiSi_{5/22}}\,dr}{R_p(t)} \times \frac{4}{3}\pi\left(R_p(t)\right)^3}{N_{tot}}}}{\sqrt[3]{V_{Si_{5/22}}}} - 1$$

Note: $R_p = R_{P0}$ @ $t = 0$

While de-lithiating,

$$R_p = R_{p_{x_{Li_{avg}}=1}} = R_{p_0} \times (4.33)^{\frac{1}{3}} \ @ \ t=0$$

$$\frac{R_p(t)}{R_{p_{x_{Li_{avg}}=1}}} - 1 = \frac{\sqrt[3]{V_{avg}}}{\sqrt[3]{\left(V_{avg}\right)_{x_{Li_{avg}}=1}}} - 1 = \frac{\sqrt[3]{V_{Si_{5/22}} + V_{Li} \times 4.4 \times \dfrac{\dfrac{\int_0^{R_p(t)} c_{LiSi_{5/22}}\,dr}{R_p(t)} \times \dfrac{4}{3}\pi\left(R_p(t)\right)^3}{N_{tot}}}}{\sqrt[3]{V_{Si_{5/22}} + V_{Li} \times 4.4 \times 1}} - 1$$

Simulations:

Total capacity of the particle of initial size of 60 nm (in diameter) is calculated to know the C-rate for lithiation. Lithiation is stopped when total number of moles that is lithiated in the particle reaches the maximum number of moles that can be lithiated based on the initial radius of the silicon particle. Delithiation is stopped when the concentration at the boundary reaches zero. As mentioned before, stress is neglected due to particle size being lesser than the critical size. Hence interior is allowed to displace as dictated by the boundary movement (which implies in the interior mesh points, one may not see exact correlation between concentration and displacement).

Results and discussions:

Figure 1 below gives the concentration profile in the particle when lithiated at 1.1076×10^{-15} A (C- rate). Figure 2 gives the concentration profile at 10C-rate. Figure 3 is a plot of the final radius of the growing particle vs. capacity for different lithiation rates. Alternatively, the stop conditions for lithiation can be changed such that the process stops when the surface concentration reaches maximum concentration of lithium .This is equivalent to stopping the simulations when the potential at the surface of the particle vs. Li/Li^+ ref. electrode reaches a pre-determined value. In such a scenario, the rate capability effects are more pronounced.

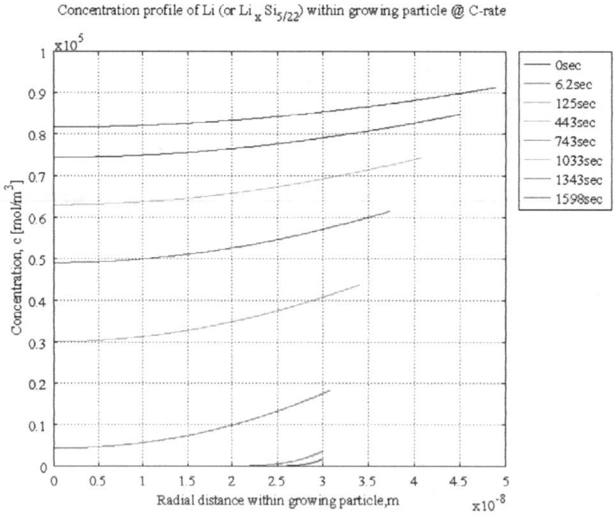

Figure 1: Concentration profile during lithiation at C-rate

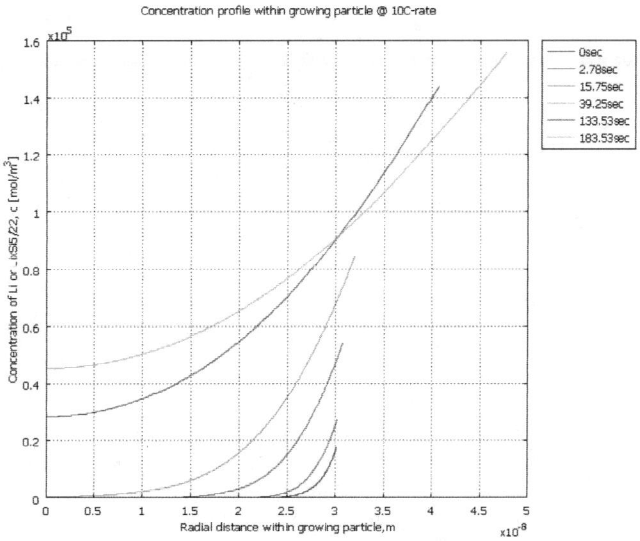

Figure 2: Concentration profile during lithiation at 10C rate

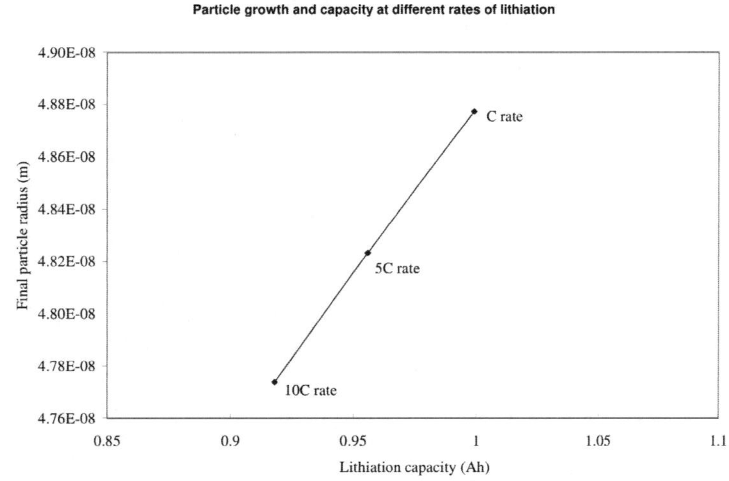

Figure 3: State of charge and final particle radius at different lithiation rates

Figure 4 gives the concentration profile within the particle when de-lithiated at C rate from the maximum concentration.

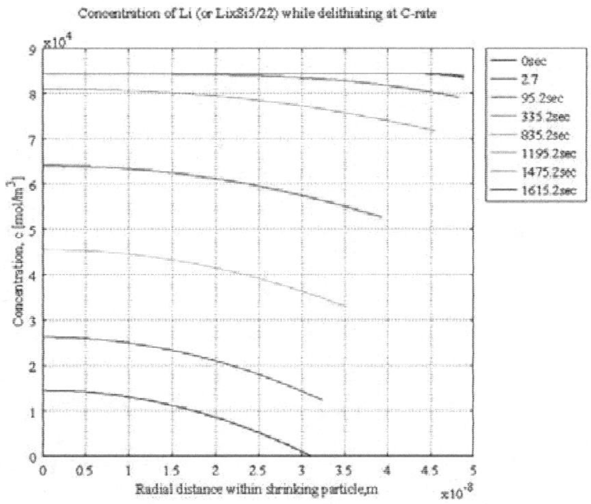

Figure 4: Concentration profile in a shrinking particle when delithiated at C-rate

Conclusions: Single particle model for the case of amorphous lithiation/de-lithiation of silicon electrode that can account for volume changes is finalized. Detailed discussion will follow during the meeting.

References:

[i] Huggins, R.A., *J.Power Sources*, 1999 82, 13-19.
[ii] V.L. Chevrier, J. R. Dahn, *J.Electrochem.Soc.*, 156 (6) A454-A458 (2009).
[iii] L. Y. Beaulieu, K. W. Eberman, R. L. Turner, L. J. Krause, J. R. Dahn, *Electrochem Sol State Letts.*,4 (9) A137-A140 (2001).
[iv] A. Timmons and J. R. Dahn, *J.Electrochem.Soc.*, 154 (5) A444-A448 (2007).
[v] Robert A. Huggins, *J.Power Sources,* 81–82 (1999) 13–19.
[vi] Hong Li, Xuejie Huang, Liquan Chen, Guangwen Zhou, Ze Zhang, Dapeng Yu, Yu Jun Mo , Ning Pei, *Solid State Ionics,* 135 (2000) 181–191.
[vii] Ram A. Sharma and Randall N. Seefurth, *J.Electrochem.Soc.,* 123(12), 1976, 1763-1768.
[viii] Jan L. Allen, T. Richard Jow, and Jeffrey Wolfenstine, *Chem. Mater.* 2007, 19, 2108-2111.
[ix] Uday S. Kasavajjula,Chunsheng Wang and Pedro E. Arce, *J.Electrochem.Soc.*, 155(11), A866-A874 (2008).
[x] Katharine L. Harrison, Jeremy P. Meyers, Abstract #1176, 214th ECS Meeting, 2008.
[xi] C. John Wen , Robert A. Huggins, *J. Sol. State Chem.*, 37, 271-278 (1981).
[xii] Huggins, R.A. and W.D. Nix, *Ionics*, 6, 2000, p.57.
[xiii] John Christensen, John Newman, *J.Solid State Electrochem.* 10(5) (2006), 293-319.
[xiv] Thomas F Fuller, Marc Doyle, John Newman, *J.Electrochem.Soc.* 141(1), 1994, 1-10.

Mathematical Modeling of Lithium Ion Batteries: A Paradigm Shift

Martin Z. Bazant, Chemical Engineering and Mathematics, Massachusetts Institute of Technology, Cambridge, MA

Damian Burch, Mathematics, Massachusetts Institute of Technology, Cambridge, MA

Liam Stanton, Applied Mathematics, Northwestern University, Evanston, IL

Gerbrand Ceder, Materials Science and Engineering, Massachusetts Institute of Technology, Cambridge, MA

The standard approach to modeling Li-ion batteries, based on porous electrode theory, assumes Butler-Volmer reaction kinetics, isotropic diffusion of intercalated ions, and an open circuit voltage (OCV) vs. composition fitted to experiments. Although not fully predictive, this model can fit the discharge of some solid-solution electrodes, but it fails to describe phase-separating materials, which are currently the most promising for high power density. In this talk, we present a new modeling paradigm based on statistical thermodynamics, which coherently describes the OCV, intercalation, and phase separation. The homogeneous free energy in the Cahn-Hilliard equation is related to the OCV, and a new boundary condition is proposed for intercalation kinetics, which depends on the full chemical potential, including the gradient term. For phase-separating materials, such as lithium iron phosphate, the model predicts, for the first time, many features of recent experiments: (i) the phase boundary aligns with the crystal axis of greatest lattice-mismatch strain; (ii) intercalation proceeds as a nonlinear wave along the crystal surface, filling it layer by layer; (iii) a composite electrode in the miscibility gap undergoes a ``mosaic instability'', where reservoir crystals spontaneously phase separate and fill with ions, one by one; and (iv) ``ultrafast'' battery charging, in seconds rather than hours, can be attained with surface coatings that distribute ions across each active crystal facet. [This work is supported by the National Science Foundation via contract DMS-0842504.]

See more of this Session: Lithium Ion Battery Technology
See more of this Group/Topical: Fuels and Petrochemicals Division

Phase Transformation Kinetics of LiFePO4 Cathodes

Yujie Zhu, Chemical and Biomolecular Engineering, University of Maryland, College Park, MD

Chunsheng Wang,Department of Chemical & Biomolecular Engineering, University of Maryland, College Park, MD

Recently, it has been reported that phase transformation materials (such as LiFePO4, Li4Ti5O12) have superior power densities compared to the traditional LiCoO2 cathodes and are the promising electrode materials for high power Li-ion batteries. However the mechanism of the exceptional rate performance is still a controversial issue undergoing debates, because there is no accurate electroanalytical technique to probe ion transport phenomena in the phase transformation region. All current electroanalytical methods, including cyclic voltammetry (CV), potentiostatic intermittent titration (PITT), galvanostatic intermittent titration (GITT) and electrochemical impedance spectroscopy (EIS), can only be used to analyze the ion transport in solid solution electrode, because these techniques do not consider the effect of inter-phase boundary movement on the ion transport in phase transformation materials. We developed novel GITT and PITT techniques to study the phase transformation electrode materials by integrating our mixed-control phase transformation theory into traditional GITT and PITT techniques. Using these phase transformation GITT/PITT techniques, Li diffusion coefficients of LiFePO4 in a single phase and two phase regions were precisely determined. The dependence of Li diffusion coefficients on the Li content in LiFePO4 is demonstrated in Figure 1. The true Li diffusion coefficient in alpha phase measured using novel GITT and PITT is stable in two phase region and it is slightly higher than the diffusion coefficient in beta phase. This is consistent with the fact that the charge rate capability of LiFePO4 is better than the discharge rate capability. During Li insertion, the Li diffusion coefficient in beta phase gradually decreases with the increasing Li concentration due to the increases of accommodation energy, while the interface mobility (M) of LiFePO4 gradually increases, reaching the maximum value at 50% state of discharge, and then decreases (Figure 2). The details of these novel PITT/GITT techniques will be presented in the meeting.

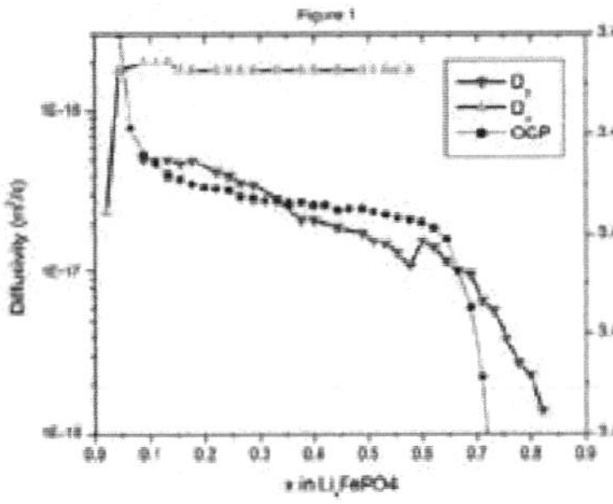

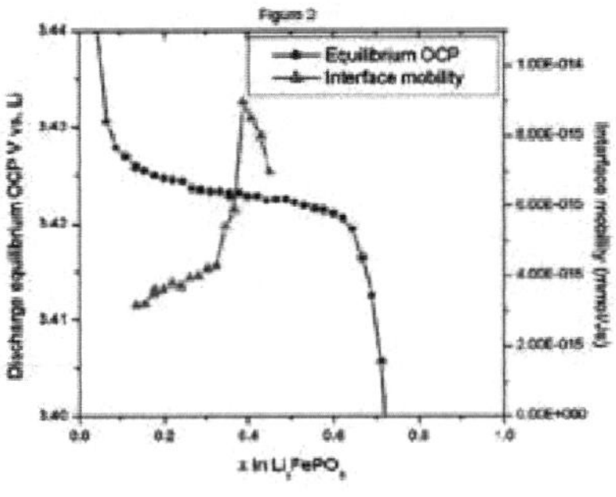

See more of this Session: Lithium Ion Battery Technology
See more of this Group/Topical: Fuels and Petrochemicals Division

A Novel Scaffold Binder Structure for High Capacity Silicon Anodes of Lithium-Ion Battery

Juchen Guo, Department of Chemical and Biomolecular Engineering, University of Maryland, College Park, MD

Chunsheng Wang, Department of Chemical & Biomolecular Engineering, University of Maryland, College Park, MD

A sodium carboxymethyl cellulose (CMC) scaffold as a binder structure for silicon anodes in Lithium-ion battery will be presented. This three-dimensional (3D) porous CMC matrix provides free volume for Si expansion, thus improving the structural stability. The porous CMC scaffold can also enhance the lithium ion transport in the electrodes, as well as being able to improve the electrochemical reaction kinetics. Our results demonstrated that this noval 3D binder structure remarkably improved the cycling stability and rate performance of the silicon anodes comparing to the ones prepared using the tradional sulrry casting technique. For instance, silicon nano-particles in this scaffold can retain 1678 mAh/g capacity (78% retention) after 150 charge/discharge cycles between 0 and 1.5 V at discharge rate of 0.1 C.

The electrochemical performance of the Si-based anodes with polymer scaffold binder can be further improved by carbonization of the polymer scaffold. The mechanism is currently under investigation. The scaffold binder structure works well for electrodes based on both Si nano-particles and micron-sized Si particles.

See more of this Session: Lithium Ion Battery Technology
See more of this Group/Topical: Fuels and Petrochemicals Division

Polymer Nanocomposites: Advanced Polymer Electrolyte Materials for Li Ion Batteries

Tirtha Chatterjee, Materials Research Laboratory, University of California, Santa Barbara, CA

Ramanan Krishnamoorti, Chemical Engineering, University of Houston, Houston, TX

The high degree of poly(ethylene oxide) (PEO) crystallinity (leading to low ionic conductivity) at ambient temperature is unfavorable and excludes many practical applications of PEO as electrolyte in lithium ion batteries. Moreover, in presence of high Li+ ion concentration (optimally [Li+]/[O-] ~ 0.04), PEO degrades fast which in turn reduces the battery lifetime. We have developed carbon nanotubes reinforced PEO nanocomposites as an alternative polymer electrolyte. The successful dispersion of single walled carbon nanotubes in poly(ethylene oxide) using an anionic surfactant (lithium dodecyl sulfate) as compatibilizer is reported here. The geometrical percolation threshold of nanotubes, as revealed by melt state rheological measurements, is found to be at ~ 0.09 vol % loading which corresponds to effective tube anisotropy ~ 650. Significant change in melting and crystallization behavior of poly(ethylene oxide) along with a decrease in fractional crystallinity has been observed in these nanocomposites. Incorporation of nanotubes acts as solid confinements to the PEO chains which not only make the chains stiffer but also provide steric hindrance to the chain transport. This effect coupled with high crystallization activation energy barrier leads to dramatic slowing down in PEO growth kinetics and makes the system more amorphous. The observed changes in the SWNTs based nanocomposites far exceed those observed for equivalent Li+ ion concentration (in this nanocomposites [Li+]/[O-] ~ 0.001) mixture and indicate that the SWNTs and LDS synergistically affect the crystallization of PEO. Finally, nanotubes are successfully aligned (to break their electrical conductivity) to develop this polymer nanocomposite as an alternative polymer electrolyte for Li ion batteries.

See more of this Session: Lithium Ion Battery Technology
See more of this Group/Topical: Fuels and Petrochemicals Division

Inexpensive Oxidative Desulfurization of Middle Distillate Fuels for Portable Power Applications

Michael T. Timko, Aerodyne Research Inc, Billerica, MA
Lino A. Gonzalez, Aerodyne Research Inc, Billerica, MA
Richard C. Miake-Lye, Aerodyne Research Inc, Billerica, MA
Peter E. H. Kracke, Chemical Engineering, Massachusetts Institute of Technology, Cambridge, MA
Jefferson W Tester, Chemical Engineering, Massachusetts Institute of Technology, Cambridge, MA
William H. Green, Chemical Engineering, Massachusetts Institute of Technology, Cambridge, MA

Sulfur compounds contained in petroleum fuels pose an environmental threat. Combustion of fuels containing sulfur releases SO2, degrading vehicles-emissions-control performance and damaging the environment by producing acid rain. Another difficulty that fuel sulfur compounds introduce is the incompatibility of fuel cell catalysts with sulfur. The high sulfur content of jet fuel poses a problem for the U.S. Army as they attempt to adopt fuel cell technology to meet their portable power demands in the 1 kW range. Military jet fuel (JP-8) is the U.S. Army logistics fuel, and reforming the fuel can produce a hydrogen rich syngas that is compatible with solid-oxide fuel cell (SOFC) power supplies. Fuel sulfur compounds rapidly degrade both reformer and fuel cell catalysts, and the sulfur content of the JP-8 fuel must be reduced to less than 10 ppm mass prior to feeding it to the power supply.

Hydrodesulfurization (HDS) is the accepted desulfurization technology applied at the refinery level. The high temperatures and pressures required by HDS, however, make it incompatible with portable power applications. HDS also fails to remove the most refractory sulfur compounds — heterocyclic aromatic compounds such as benzothiophene (BT) dibenzothiophene (DBT) and their derivatives — and therefore cannot reach the ultra-low sulfur levels required to protect the catalysts. Selective oxidation of the sulfur compounds may permit use of small quantities of inexpensive adsorbents to reach the ultralow sulfur levels needed for fuel cell power supplies. The objective of this work is to develop a compact, robust oxidative desulfurization (ODS) process for pretreatment of jet fuel containing up to 3,000 ppm mass of sulfur prior to its reformation and use in SOFC's. We used a novel ODS process that combines activated carbon, formic acid, and ultrasound to oxidize jet-fuel sulfur compounds to sulfones:

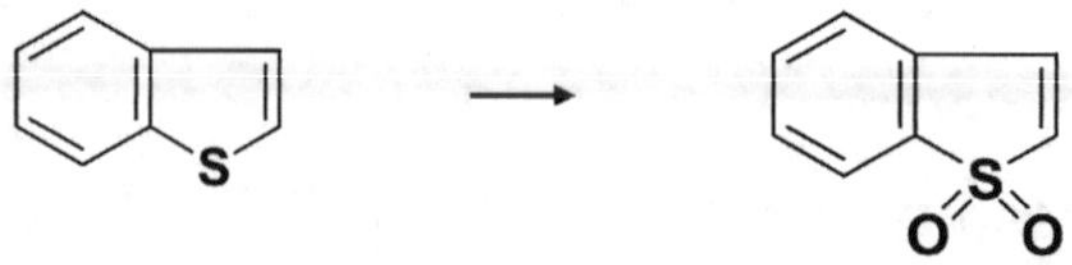

benzo[*b*]thiophene (BT) → benzo[*b*]thiophene-sulfone (BT-S)

dibenzothiophene (DBT) → dibenzothiophene-sulfone (DBT-S)

Due to the change in polarity which accompanies oxidation, the sulfone compounds are more easily extracted/adsorbed than the parent compounds to potentially yield jet fuel with sulfur content less than 10 ppm_w. The critical unit operation of the process is an innovative three-phase slurry reactor - liquid (H_2O_2, formic acid, and water)/liquid (jet fuel)/solid (activated carbon) - which accomplishes the required high rate of reaction using ultrasound. Our ODS process has reached total sulfur levels as low as 17 ppm after a polishing step with aluminum oxide. Fuel recovery after oxidation is greater than 90%, and we have found no evidence of hydrocarbon oxidation using mass spectrometry. Improvements in the aluminum oxide polishing step should allow for sub-10 ppm sulfur concentrations.

Studies were performed to test the relative importance of the ODS components – H_2O_2, formic acid, activated carbon, and power ultrasound. Table 1 presents the ODS performance data in these experiments for 2,3-dimethylbenzothiophene (2,3-DMBT) and 2,3,7-trimethylbenzothiophene (2,3,7-TMBT), as they are two most important fuel sulfur compounds present in JP-8. All other sulfur compounds that we can monitor behave similarly to 2,3-DMBT and 2,3,7-TMBT and the data in Table 1 are therefore representative. First (Run 1), a baseline was established by using the full treatment – ultrasound (60% amplitude), 7 mL of oxidant phase (30 wt% H_2O_2, pH adjusted to 1 using formic acid), and 0.3 g of activated carbon. Run 1 yielded the best results presented in Table 1, with >70% ODS removal of both monitored BT compounds in a single 60 min step. Benzothiophene-sulfones were detected in the fuel treated during Run 1, evidence of ODS. Run 2 was conducted without the addition of formic acid and the low BT removal observed is consistent with the importance of performic acid in the oxidation of BT compounds. Run 3 was conducted in the absence of hydrogen peroxide; it yielded the lowest BT removal observed, showing that the oxidant is crucial to the ODS process and that adsorption to activated carbon alone does not achieve desulfurization. Run 4 was conducted without addition of activated carbon, and the poor removal performance proves the importance of the heterogeneous catalytic surface. Finally, Run 5 was performed in the absence of ultrasound (the reaction mixture was agitated using a PTFE-coated magnetically coupled stir bar). Again, the removal performance of the BT compounds was significantly lower in Run 5 than Run 1,

Table 1. ODS Performance of Select Sulfur Compounds from JP-8 Fuel

Run	pH	H_2O_2	Carbon	Ultrasoun	Fuel	2,3-DMBT	2,3,7-TMBT

#		vol%	Loading (g)	dAmplitude	Recovery %	removal %	removal %
1	1	30	0.3	60%	92±3	71	76
2	3	30	0.3	60%	89±3	6	11
3	1	0	0.3	60%	93±3	<5	<5
4	1	30	None	60%	95±3	20	30
5	1	30	0.3	0%	94±3	28	33

All other conditions: continuous ultrasound at 60% amplitude, 20 kHz; 60 min reaction time; 60±3 °C reaction temperature; water-washed MW99 activated carbon.

emphasizing the importance of micro-mixing and or sonochemical effects on BT ODS. The control experiments prove that: a) desulfurization occurs via an ODS process, and b) all 5 components are needed for efficient ODS performance.

A thorough study revealed that ODS removal performance increases with *decreasing* pH, *increasing* carbon catalyst loading, *increasing* temperature, *increasing* time, *increasing* oxidant concentration, and *increasing* ultrasonic amplitude. Lower pH values catalyze the formation of performic acid,

$$H_2O_2 + HCOOH \rightarrow HCOOOH,$$

believed the be the most active oxidant in our system. Higher temperatures and oxidant concentrations also increase the rate of formation of performic acid, and thus the rate of oxidation of sulfur compounds. Increased carbon loading provides more surface area for the heterogeneous oxidation of sulfur compounds, and higher ultrasound amplitudes result in increased micromixing of the immiscible phases. Interestingly, ODS performance is not reduced by operation at a reduced duty cycle as low as 25%, thereby affording a means to reduce substantially parasitic power loss (<100 W for a 1 L reactor capable of supporting a 2 kW fuel cell). The "saturation" effect can be attributed to the formation of emulsions that persist for a few seconds in the absence of ultrasound. In fact, continuous sonication may have negative effects on micro-mixing as ultrasound may promote de-emulsification coagulation of the water droplets. We also found that ODS removal performance is improved when high-shear macro-mixing (introduced via an impeller connected to an external motor) and power ultrasound are used in combination.

The carbon material used in the ODS process had a pronounced effect on the sulfur removal attained for a given set of conditions. In fact, the reaction rate obtained with top-performing carbon catalysts is more than 8-times greater than that obtained with no catalyst. After screening more than a dozen candidates covering a wide range of carbon feedstocks and preparation techniques, wood-based activated carbons, treated with phosphoric acid, were found to be the best class of activated carbon ODS catalysts. The different carbon materials were characterized using diffuse reflectance Fourier transform infrared spectroscopy (DRIFTS), x-ray photoelectron spectroscopy (XPS), Boehm titration, BET surface analysis, and nitrogen porosimetry. We found that the most active catalysts share in common: surface areas in excess of 1500 $m^2 g^{-1}$, high mesopore volume, high surface oxygen content (~10%), and low surface pH. Surface area, mesopore volume, surface acidity, and surface oxygen content are all greater for the phosphoric acid treated wood based activated carbons than the other types of carbon materials tested (coal, coconut, ammonia treated wood based carbon, carbon black). We anticipate using this knowledge to refine further catalyst selection to achieve ultradeep desulfurization performance at milder conditions.

We tested addition of sodium formate as a novel, inexpensive, and safe solid reagent for effective desulfurization under mild conditions. With formic acid, ultra-deep desulfurization can only be achieved at pH less than 0.5 – too low for a commercial process. By addition of a sodium formate/sulfuric acid buffer, we can achieve similar results at a) milder pH and/or b) lower reagent volumes. The quantities of sulfuric acid required are small and the sodium formate can be stored safely as a solid. Increasing the initial concentration of formate ions in solution is the key to the success of the sodium formate technology. Despite its apparent simplicity, we are unaware of any previous attempts to use a sodium formate buffer to increase ODS performance at mild pH.

Design of Ammonia Oxidation Reactor for Portable Fuel Cell Applications

Vyjayanthi Alagharu, Srinivas Palanki, and Kevin N. West,
Chemical and Biomolecular Engineering, University of South Alabama, Mobile,AL

In this paper, reaction engineering principles are utilized to analyze process conditions for producing sufficient hydrogen in an ammonia decomposition reactor for generating net power of 100 W in a fuel cell. It is shown that operating the reactor adiabatically results in a sharp decrease in temperature due to endothermic reaction, which results in low conversion of ammonia. For this reason, the reactor is heated electrically to provide heat for the endothermic reactions. It is observed that when the reactor is operated non-adiabatically, it is possible to get over 99.5% conversion of ammonia. The weight of absorbent to reduce ammonia to ppb levels is calculated. An energy balance on the reactor exit gas indicates that there is sufficient heat available to vaporize enough water to achieve 100% relative humidity in the fuel cell. A suitable fuel cell stack is designed and it is shown that this stack is able to provide the necessary power to electrically heat the reactor and produce net power of 100 W.

Method to Release Hydrogen From Ammonia Borane for Portable Fuel Cell Applications

Moiz Diwan, School of Chemical Engineering, Purdue University, West Lafayette, IN

David Hanna, School of Chemical Engineering, Purdue University, West Lafayette, IN

Arvind Varma, School of Chemical Engineering, Purdue University, West Lafayette, IN

Portable electronic devices are becoming more widespread and power-demanding. Unfortunately, battery improvements are not advancing rapidly enough to meet the need for increased power. Hydrogen powered fuel cell systems have a potential to significantly improve the performance characteristics of portable electronic devices, including duration of their operation, without recharging/replacing the power source. A more widespread use of hydrogen fuel cells, however, is limited by the lack of availability of a practical, high density fuel source. In this context, chemical hydrides, such as ammonia borane (AB, NH_3BH_3), is considered to be a promising hydrogen storage material as it contains 19.6 wt% hydrogen. It is difficult, however, to release hydrogen from AB. Thermolysis, catalytic hydrolysis and heat generated by additional reactive mixtures are usually employed but these methods have disadvantages that decrease the efficiency of hydrogen storage systems. In this paper, we use a new approach to release hydrogen, which does not require any catalyst and produces relatively high hydrogen yield and environmentally benign byproducts. It involves metal/water combustion reactions, which provide heat for AB dehydrogenation and release additional hydrogen from water [1]. One possible design is to mix AB with metal powder and water. However, it is necessary to create conditions for AB thermolysis and prevent its hydrolysis as it provides a lower hydrogen yield (9.1 wt% based on reacted water) than thermolysis (up to 19.6 wt%). Thus, the best design option is to spatially separate AB and metal/water mixture. This approach currently provides H_2 yield up to 9.7 wt%. Our results show that this method is promising for hydrogen storage and could be used in compact power sources for portable electronic devices.

See more of this Session: Portable Power Systems
See more of this Group/Topical: Fuels and Petrochemicals Division

Sulfur Selectivity of Ag/TiO_2 Sorbents and the Effects of Fuel Chemistry

Sachin Nair and Bruce J. Tatarchuk
Auburn University, Auburn, AL, U.S.A

Introduction

Onsite fuel clean-up capabilities become critical for portable fuel cell applications as the fuel sources and quality vary with location. Use of desulfurization units are inevitable for auxiliary power units running on high sulfur logistic fuels. These systems are also needed due to the variation in sulfur speciation in fuels arising between the source of crude and the various refining operations involved. Variability in fuel chemistry poses risk to reformation units as well as the fuel cell units downstream. Dispersed silver oxides on supports such as TiO_2, γ-Al_2O_3 and SiO_2 were observed to be effective desulfurizing agents for refined fuels at ambient conditions. A unique attribute of these sorbents was that they were thermally regenerable using air as a stripping medium over multiple cycles. Desulfurization characteristics were observed to vary with fuel chemistry. Variation in desulfurization performance between JP5, JP8 and a light fraction JP5 were established and associated with the differences in sulfur composition of these fuels. This variation in sulfur selectivity was further investigated using model fuel compositions.

Two mechanisms have been suggested for adsorptive capacity of sulfur sorbents. One is through π-electron interaction between the aromatic ring and the metal ion on the sorbent [1-9]. The activity is also attributed to interaction between the lone electron pair on the sulfur atom and the active site [10]. The presence of alkyl side chains on sulfur species have been both shown to have little influence on the effectiveness of Ni based adsorbents [1] as well as having a negative influence [11] on sulfur removal for CuCl-Al_2O_3 sorbents. Ni/SiO_2-Al_2O_3 and activated carbon had higher capacity for DBT than 4,6- DMDBT [12].

The goal of this study was to determine the efficacy of the Ag/TiO_2 sorbent as a desulfurization agent for portable power units and the capacity to handle varying fuel chemistry. Studies of the selectivity of these sorbents also bring insight to the active desulfurization center on the sorbent as well as the mechanism of sulfur removal.

Experimental

Incipient wetness impregnation was used to disperse silver on the support using aqueous $AgNO_3$. The support particles were sized to 1.4-0.8mm. The concentration of the impregnating solution was adjusted to obtain the required Ag loading on the support. Typical metal loading on the Ag/TiO_2 sorbent was 4 wt%. The resulting particles were then dried at 110^0C for 6h followed by calcination in air at 400^0C for 2h.

The breakthrough characteristics of the sorbents were determined in a packed column configuration where the challenge fuel flowed from the bottom of the bed to the top. 10.0 g of the sorbent was used in all the breakthrough studies. The bed was contained in quartz tubing supported on both ends by quartz wool. The fuel face velocity was maintained at 0.25 cm/min (LHSV 2.1 h^{-1}) by a peristaltic pump. The bed output was sampled at regular intervals and analyzed for sulfur content using an Antek 9000 total sulfur analyzer. Regeneration was carried out using air as a stripping medium in a two stage process. Initially, the bed was heated to a temperature of 220^0C in 10 min. and held at the temperature for 1h. This step ensured the vaporization of fuel held in the pores. Bed temperature was then raised to 450^0C in 10 minutes and held at temperature for another two hours to complete the regeneration cycle.

JP5 with a total sulfur content of 1172 ppmw and light fraction JP5 with a sulfur content of 582 ppmw were obtained from NAVSEA Philadelphia. JP8 fuel was obtained from TARDEK and had a total sulfur content of 630 ppmw sulfur. Model fuels with aromatic sulfur species as well as other organic components were prepared using n-octane as a solvent. The major sulfur species in the fuels were identified using analytical standards obtained from Chiron AS and a Varian CP3800 gas chromatograph equipped with a pulsed flame photometric detector (PFPD).

Summary of Results

1. Performance comparison: Sulfur sorbents were prepared as described in literature and tested for sulfur removal from JP5. Cu-Y zeolite was prepared through liquid phase ion exchange [3, 13, 14]. $PdCl_2/\gamma\text{-}Al_2O_3$ was prepared using monolayer dispersion technique [15, 16]. Activation steps were also carried out prior to sulfur removal step as described in the respective literature.

The breakthrough characteristics of the sorbent are compared in Figure 1. Ag/TiO_2, Ag/ $\gamma\text{-}Al_2O_3$ and the $PdCl_2$/ $\gamma\text{-}Al_2O_3$ sorbents demonstrated similar capacities for sulfur. The Cu-Y zeolite however was not effective in removing sulfur from JP5. Considering a breakthrough concentration of 10 ppmw, the highest capacity was showed by Ag/TiO_2 at 0.93mg/g followed by $PdCl_2/\gamma\text{-}Al_2O_3$ at 0.69 mg/g. The highest saturation capacity was demonstrated by $PdCl_2\gamma\text{-}Al_2O_3$ sorbent at 7.97mg/g followed by $Ag/\gamma\text{-}Al_2O_3$ at 6.4mg/g. Ag/TiO_2 sorbent showed a saturation sulfur capacity of 6.3mg/g. TiO_2 was chosen for further study over $\gamma\text{-}Al_2O_3$ as a support for Ag because TiO_2 was observed to be a more stable support when the sorbents were taken through multiple cycles of regeneration.

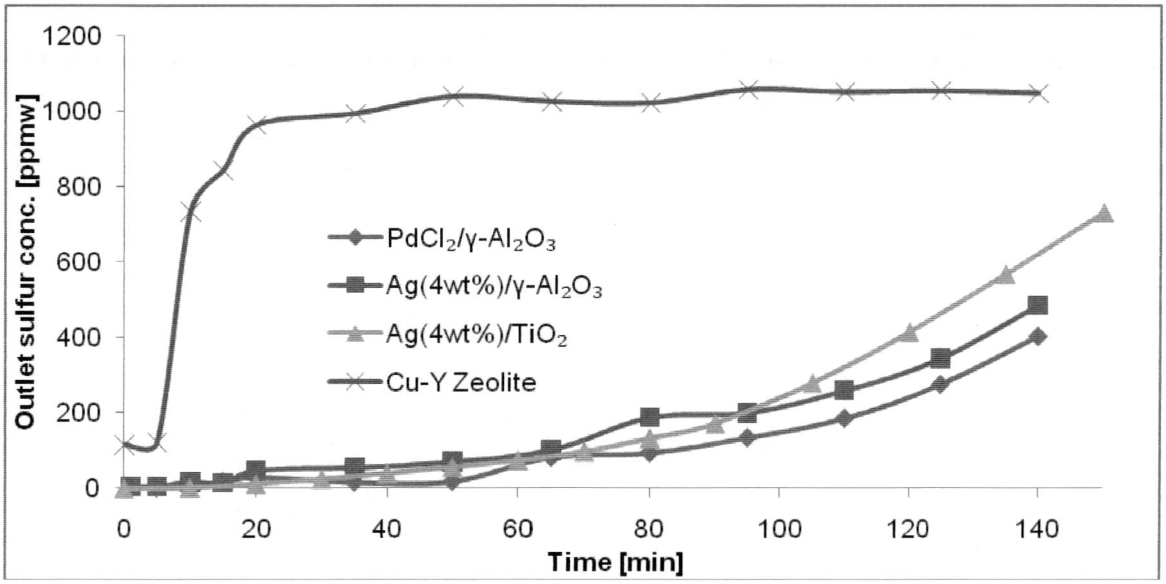

Figure 1. Comparison of breakthrough characteristics between Ag/TiO_2, $Ag/\gamma\text{-}Al_2O_3$, Cu-Y and $PdCl_2/\gamma\text{-}Al_2O_3$ using JP5 with sulfur content of 1172ppmw

2. Multi-cycle Performance: Sulfur sorbents that operate in the liquid phase have relatively low capacity. Thus non-regenerable units become prohibitively large for portable power generation applications. Regeneration maybe carried out on-board or spent sorbent cartridges collected and regenerated offsite. Regenerability is most cost-effectively carried out using air rather than solvents or reducing/inert gases. Ag/TiO_2 was taken through ten cycles of adsorption, regeneration for JP5 fuel. The resulting breakthrough curves are shown in Figure 2. Desulfurization performance was more or less consistent for the ten cycles tested.

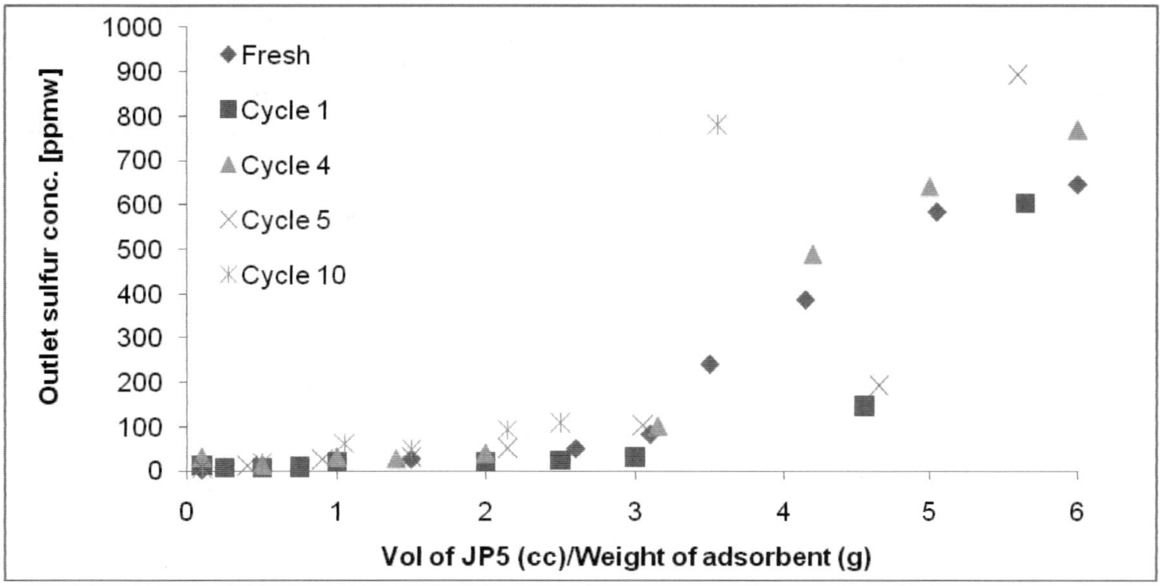

Figure 2. Multi-cycle performance of Ag/TiO_2 sorbent tested with JP5 fuel; regenerated in air for 10 cycles

3. Effect of varying fuel composition: The sulfur content of a fuel varies depending on the origin of the fuel, refining and blending operations. Thus desulfurization performance of any sorbent should vary with fuel chemistry. JP5, JP8 and a light fraction JP5 were studied for desulfurization using the Ag (4.0 wt%)/TiO_2 sorbent. Tri methyl benzothiophenes were found to be in higher concentration in JP8 compared to JP5. Fractionation of JP5 resulted in the separation of a majority of the heavier sulfur aromatics such as 2,3,4 Trimethyl benzothiophenes (TMBT) to the heavier fraction.

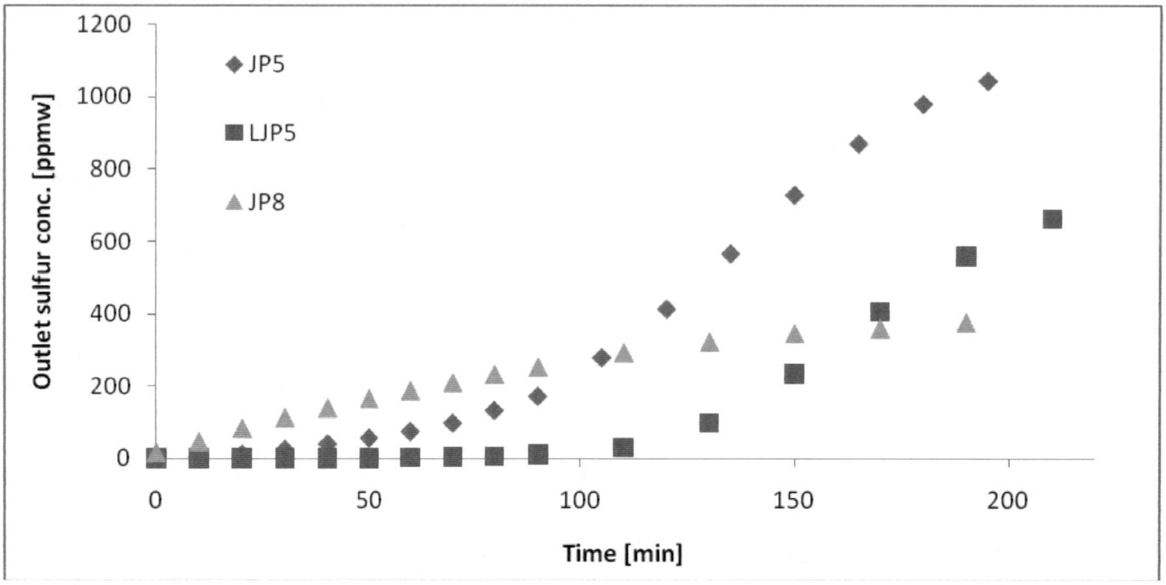

Figure 3. Performance comparison of Ag/TiO_2 sorbent with JP5(1172ppmw), JP8 (693 ppmw) and light fraction JP5 (582ppmw) fuels

A significant difference in breakthrough characteristics was observed between the three fuels as shown in Figure 3. Considering 10 ppmw as the breakthrough concentration, the highest capacity was demonstrated for light fraction JP5 at 4.22 mg/g and the highest saturation capacity was 6.32 mg/g for JP5. The sorbent had no breakthrough capacity for JP8. Analysis of the bed outlet after 30 min. (Figure 4) showed that the tri-methyl benzothiophenes were not removed by the sorbent resulting in an early sulfur breakthrough for JP8.

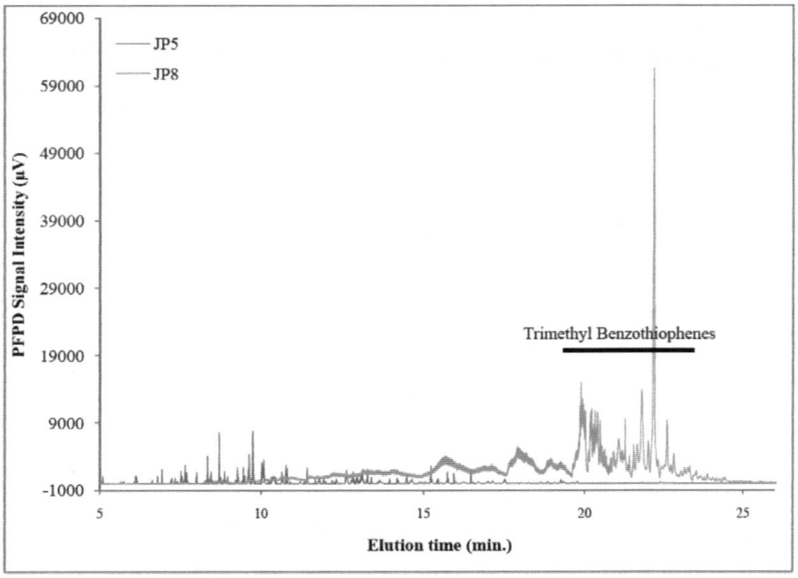

Figure 4. Chromatograms of bed output after 30 min. showing the higher concentration of Trimethyl benzothiophenes in the bed outlet

4. Variation in sulfur affinity with sulfur species: In order to determine the true effects of the structure of a sulfur aromatic and its interaction with the active center on the sorbent, effects of competitive adsorption between the multitudes of components of a natural fuel needed to be removed. Thus model fuels containing methyl thiophene (MT), benzothiophene (BT), dibenzo thiophene (DBT) and 4,6 dimethyl benzothiophene (4,6 DMDBT) were tested for desulfurization performance with Ag/TiO_2 sorbent. The total sulfur content was 0.0774 mmol/ml in all the fuel compositions. Breakthrough characteristics of these sulfur aromatics are shown in Figure 5. Sulfur capacities varied as MT<4,6 DMDBT<BT<DBT. The addition of the aromatic ring was thus observed to increase the interaction between the sulfur species and the sorbent resulting in higher capacities. This suggests a stronger interaction between the delocalized π electrons on the aromatic rings and the active center on the sorbent. A significant drop in capacity was also observed between 4,6 DMDBT and DBT. Stearic hindrance of the alkyl groups on the sulfur atom/π electron cloud may explain this lowering of activity.

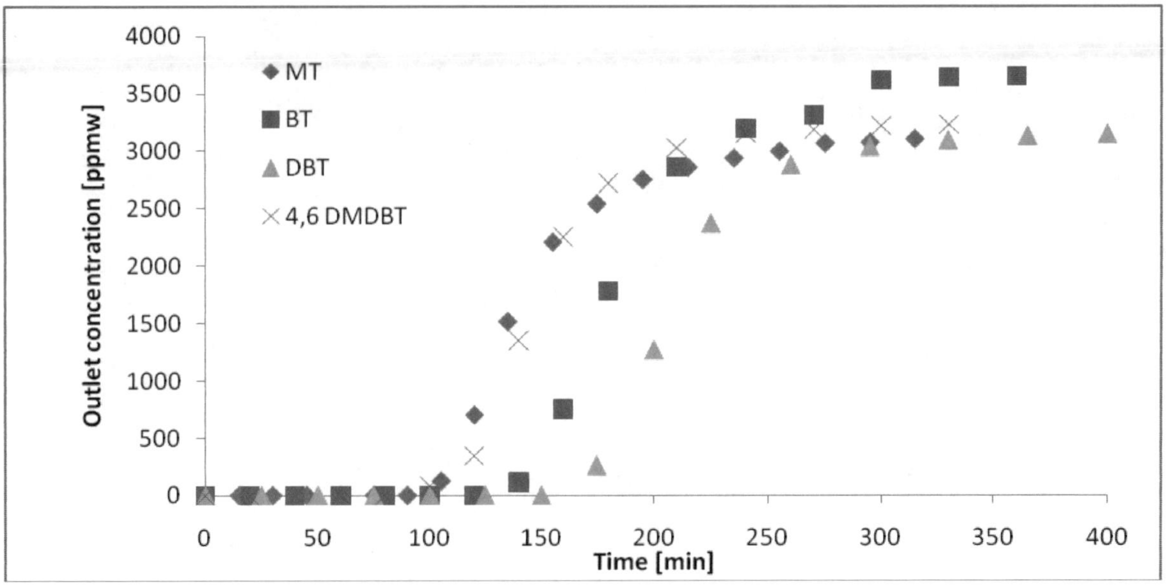

Figure 5. Breakthrough characteristics of methylthiophene (MT), benzothiophene (BT), dibenzothiophene (DBT) and 4,6 dimethyl dibenzothiophene (4,6 DMDBT) containing model fuels using Ag/TiO_2 sorbent

5. Interactions of the sorbent active center: From the above experiment, no decisive information was obtained on the nature of the interaction between the active center and the solute molecules. In order to distinguish between the interaction of the π electron cloud on the aromatic ring and the lone pair of electrons on the sulfur atom with the sorbent, selectivity of the sorbent was studied using model fuels consisting of naphthalene, quinoline and benzothiophene as solutes (approximately 3500 ppmw in n-octane). Despite having similar molecular masses, the presence/absence of sulfur/nitrogen atoms in the heterocyclic molecules would provide some insight on the mechanism of sulfur removal. The breakthrough behavior of naphthalene and benzothiophene was similar as shown in Figure 6. This indicated that the lone pair of electrons on the sulfur atom had minimal interaction with the active center compared to interactions of the π electron cloud. However the sorbent showed approximately twice the capacity for quinoline. This indicated the stronger interaction between the active center on the sorbent and the more basic nitrogen atom in quinoline.

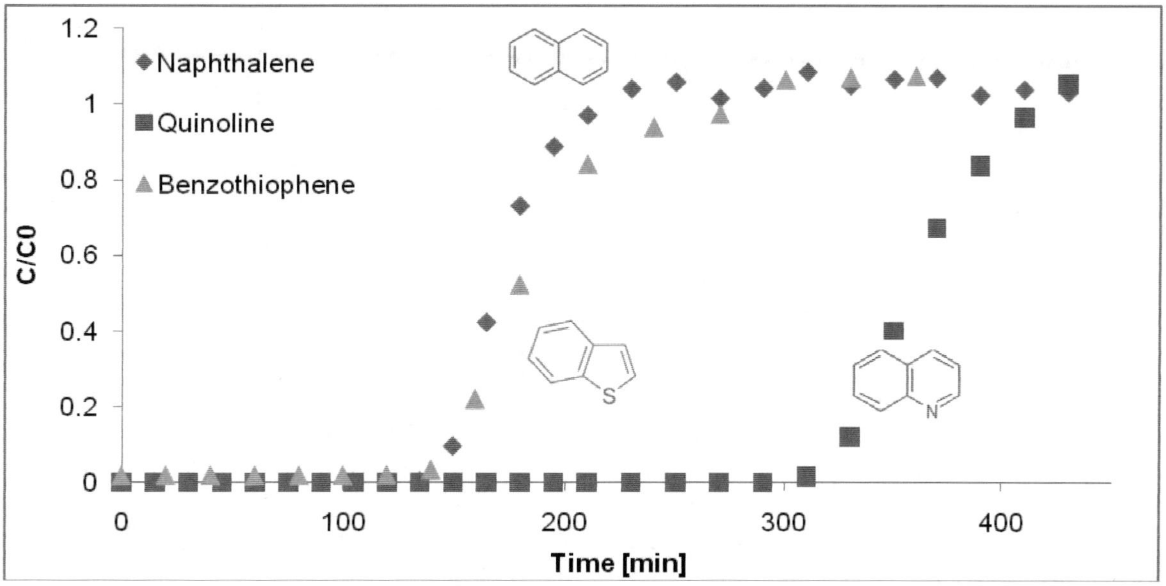

Figure 6. Breakthrough characteristics of benzothiophene, quinoline and benzothiophene using Ag/TiO_2 sorbent

Conclusions

The Ag/TiO_2 sorbent was found to have sulfur capacities comparable to reported sorbents for JP5. The capacity was also demonstrated over multiple cycles of thermal regeneration using air. The significant variation in performance of the sorbent between JP5, JP8 and a light-fraction JP5 was attributed to variation in fuel chemistry. Thus it is generally impossible to formulate a sorbent that is equally effective in desulfurizing a variety of fuels. Interactions of the π electron cloud with the active center on the sorbent were more significant compared to lone pair of electrons on the sulfur atom. The active center was observed to be acidic in nature based on the strength of interaction with basic centers on various solutes.

References

[1] Herna´ndez-Maldonado AJ, Yang RT. Desulfurization of commercial fuels by π-complexation: Monolayer $CuCl/\gamma\text{-}Al_2O_3$. Applied Catalysis B: Environmental 2005; 61: 7.

[2] Herna´ndez-Maldonado AJ, Yang FH, Qi G, Yang RT. Desulfurization of transportation fuels by π-complexation sorbents: Cu(I)-, Ni(II)-, and Zn(II)-zeolites. Applied Catalysis B: Environmental 2005; 56: 17.

[3] Herna´ndez-Maldonado AJ, Stamatis SD, Yang RT. New sorbents for desulfurization of diesel fuels via π complexation: Layered beds and regeneration. Ind Eng Chem Res 2004; 43: 8.

[4] Herna´ndez-Maldonado AJ, Yang RT. Desulfurization of liquid fuels by adsorption via π complexation with Cu(I)-Y and Ag-Y zeolites. Ind Eng Chem Res 2003; 42: 7.

[5] Herna´ndez-Maldonado AJ, Yang RT. Desulfurization of commercial liquid fuels by selective adsorption via π-complexation with Cu(I)-Y zeolite. Ind Eng Chem Res 2003: 8.

[6] Takahashi A, Yang RT. New adsorbents for purification: Selective removal of aromatics. AIChE J 2002; 48.

[7] Jayaraman A, Yang RT. Deactivation of π-complexation adsorbents by hydrogen and rejuvenation by oxidation. Ind Eng Chem Res 2001; 40: 7.
[8] Takahashi A, Yang RT. Cu(I)-Y-zeolite as a superior adsorbent for diene/olefin separation Langmuir 2001; 17: 9.
[9] Takahashi A, Yang FH, Yang RT. New sorbents for desulfurization by π-complexation: Thiophene/ benzene adsorption Ind Eng Chem Res 2001; 41: 10.
[10] Kim HJ, Ma X, Zhou A, Song C. Ultra-deep desulfurization and denitrogenation of diesel fuel by selective adsorption over three different adsorbents: A study on adsorptive selectivity and mechanism. Catal Today 2006; 111: 74-83.
[11] Ma X, Sprague M, Song C. Deep desulfurization of gasoline by selective adsorption over nickel-based adsorbent for fuel cell applications. Ind Eng Chem Res 2005; 44: 5768-75.
[12] Hyung J, Xm K, Zhou A, Song C. Ultra-deep desulfurization and denitrogenation of diesel fuel by selective adsorption over three different adsorbents: A study on adsorptive selectivity and mechanism. Catal Today 2006; 111: 10.
[13] Takahashi A, Yang FH, Yang RT. New sorbents for desulfurization by π-complexation: Thiophene/ benzene adsorption Ind Eng Chem Res 2001; 41: 10.
[14] Yang RT, Herna´ndez-Maldonado AJ, Yang FH. Desulfurization of transportation fuels with zeolites under ambient conditions. Science 2003; 301: 79-81.
[15] Wang Y, Yang FH, Yang RT, Heinzel MJ, Nickens DA. Desulfurization of high-sulfur jet fuel by π-complexation with copper and palladium halide sorbents. Ind Eng Chem Res 2006; 45: 7649-55.
[16] Xie Y, Tang Y. Spontaneous monolayer dispersion of oxides and salts onto surfaces of supports: Applications to heterogeneous catalysis. Adv Catal 1990; 37: 1-43.
[17] Kim JH, Ma X, Song C. Kinetics of two pathways for 4,6-dimethyldibenzothiophene hydrodesulfurization over nimo, como sulfide, and nickel phosphide catalysts. Energy Fuels 2005; 19: 7.
[18] Herna´ndez-Maldonado AJ, Yang RT. Desulfurization of diesel fuels via π-complexation with nickel(ii)-exchanged x- and y-zeolites. Ind Eng Chem Res 2004; 43: 9.

Behavior of Platinum Nanotubes in the Methanol Oxidation Reaction

Shaun Alia, Department of Chemical and Environmental Engineering, University of California, Riverside, Riverside, CA

Yushan Yan, Department of Chemical & Environmental Engineering, University of California, Riverside, Riverside, CA

Platinum supported by amorphous carbon (Pt/C) is often the commercial material used as an electrocatalyst in both the methanol oxidation reaction (MOR) as an anode in direct methanol fuel cells (DMFCs) and the oxygen reduction reaction (ORR) as a cathode in proton exchange membrane fuel cells (PEMFCs). Small platinum particles are used due to the high surface area to volume ratio. The carbon support spaces the particles to prevent agglomeration and increase specific surface area and mass activity. A decrease in particle size, however, causes the specific activity in ORR and organic molecule oxidation to drop due to alterations in the coordination number and crystal planes of the surface platinum atoms. Low durability of commercial catalysts in PEMFCs and DMFCs hinders the commercialization of these devices. Slow kinetics and intermediate adsorption at the anode during MOR, notably of carbon monoxide, further limit the capabilities of DMFCs.

Platinum nanotubes have been synthesized to improve the activity, durability, and carbon monoxide adsorption characteristics of conventional DMFC catalysts. Platinum nanotubes of 5 nm thickness were synthesized by the galvanic replacement of silver nanowires, which were formed by the ethylene glycol reduction of silver nitrate.[1] The catalytic properties of platinum nanotubes were evaluated to demonstrate the improvement in activity and durability of the material to conventional catalysts. MOR, ORR, and durability experiments were conducted to study the activity of platinum nanotubes, platinum nanoparticles, and bulk platinum electrocatalysts. Catalyst activity for MOR was characterized through cyclic voltammetry and chronoamperometry techniques. Bulk platinum electrodes have been previously found to have a higher degree of carbon monoxide adsorption and poisoning when compared to electrodispersed platinum. While platinum nanotubes do not utilize a carbon support, the catalyst exhibits carbon monoxide adsorption characteristics similar to Pt/C. When compared to conventional materials, platinum nanotubes offer significant catalyst improvement in DMFCs.

[1] Z. W. Chen, M. Waje, W. Z. Li, Y. S. Yan, Angewandte Chemie-International Edition 2007, 46, 4060.

See more of this Session: Portable Power Systems
See more of this Group/Topical: Fuels and Petrochemicals Division

Reformates Desulfurization Using Regenerable Sorbent for Logistic SOFC Power Units

Hongyun Yang, IntraMicron Inc., Auburn, AL

Troy J. Barron, IntraMicron Inc., Auburn, AL

Bruce Tatarchuk, Center for Microfibrous Materials Manufacturing, Department of Chemical Engineering, Auburn University, Auburn, AL

Carbon monoxide (CO) and carbon dioxide (CO_2) usually exist in the reformates. They can react with hydrogen sulfide, the major sulfur compound in reformates, and form carbonyl sulfide, which is difficulty to be removed by ZnO based sorbents. As a result, the sorbents usually demonstrate low sulfur capacity or require long packed beds to achieve low sulfur levels. In this study, a specially designed layered bed made of a commercial ZnO sorbent and supported sorbent mixture was investigated in the presence of refomates of high CO and CO_2 concentrations and high face velocities. The layered bed maintained its sulfur capacity and was able to reduce the total sulfur concentration to less than 1ppm (lower than typical sulfur thresholds of most SOFCs) after 30 desulfurization/regeneration cycles. Due to the low weight and small size of the layered bed, it is very suitable to be used as a desulfurizer for logistic power units based on SOFCs.

See more of this Session: Portable Power Systems
See more of this Group/Topical: Fuels and Petrochemicals Division

Algae Products for Biofuels: Isoprenoids

Juergen Polle, Biology, Brooklyn College of the City University of New York, Brooklyn, NY

D. Brogun, Brooklyn College of the City University of New York, Brooklyn, NY

D. Tran, Brooklyn College of the City University of New York, Brooklyn, NY

Microalgae represent a very diverse group of photosynthetic organisms that have recently received renewed attention regarding their potential as feedstock for renewable fuels production. Algae metabolites that could be used for fuel production are for example lipids, isoprenoids, polysaccharides, and hydrogen. Isoprenoids are a large and diverse family of natural products including many hydrocarbon molecules of different chain lengths that are of interest for oil production. Molecules of the class of isoprenoids have in common that they are made up of the five-carbon containing molecule isoprene as a basic building block. In algae either the HMG-CoA reductase pathway and/or the 1-deoxy-D-xylulose 5-phosphate pathway exist for synthesis of the precursor to isoprenoids, isopentenyl pyrophosphate and its isomer dimethylallyl pyrophosphate. This presentation will review the biosynthesis of isoprenoids in algae.

See more of this Session: Solar Biofuels II
See more of this Group/Topical: Topical H: Solar Topical

Effects of Two-Stage Semi-Continuous Culture and CO2 Aeration On Cell Growth and Lipid Accumulation of Nannochloropsis Oculata

Fang Zhang, Department of Chemical and Biochemical Engineering, Zhejiang University, Hangzhou, China

Jianming Lu, Department of Chemical and Biochemical Engineering, Zhejiang University, Hangzhou, China

Lihua Cheng, Department of Chemical and Biochemical Engineering, Zhejiang University, Hangzhou, China

Lin Zhang, Department of Chemical and Biochemical Engineering, Zhejiang University, Hangzhou, China

Huanlin Chen, Department of Chemical and Biochemical Engineering, Zhejiang University, Hangzhou, China

As a result of current high oil prices, depleting fossil oil reserves and growing concerns about increased levels of atmospheric carbon dioxide, algae have emerged as one of the most promising sources for biofuel production. In this paper, Nannochloropsis oculata culture was studied aiming at the impact evaluation of operational conditions over cellular growth and lipid production associated with the CO2 sequestration on focus. Concerning about nitrogen source effect, semi-continuous two-stage strategy was applied to obtain high biomass and lipid productivity. The first N-sufficient biomass production stage was followed by a lipid accumulation stage under N-deficient condition, and harvest culture volume (40%) in three-day replacement with N-deficient medium. The microalgal special growth rate and lipid productivity was 0.87d-1 and 0.33g/L, respectively. N. oculata were cultured with 1%, 5% and 10% CO2. With 1% CO2 aeration, the highest biomass concentration was 0.84g/L. Different CO2 concentrations have an apparent effect on lipid and other organic compounds content in microalgae. To optimize the condition for long-term biomass and lipid yield, N. oculata was suggested to grow in the semi-continuous two-phase system aerated with 1% CO2 and operated by three-day replacement.

See more of this Session: Solar Biofuels II
See more of this Group/Topical: Topical H: Solar Topical

Direct Monitoring of Oil Production in Individual Microalgae *in Situ* by Raman Spectroscopy

Huawen Wu, Biomass Sci & Conversion Technology, Sandia National Laboratories, Livermore, CA

Joanne Volponi, Biomass Sci & Conversion Technology, Sandia National Laboratories, Livermore, CA

Seema Singh, Biomass Sci & Conversion Technology, Sandia National Laboratories, Livermore, CA

Microalgae, producing oils that can be converted into biodiesel, are one of the promising candidates for the next generation of viable fuel to help solve the global energy crisis. However, the recent booming of research on microalgae, so far appears not benefiting from the booming of modern high-throughput, high-resolution, real-time, and *in vivo* biophysical techniques. Traditionally, these microalgae were treated in bulk, lyophilized or in extracted forms, making it impossible to assess the information on fundamental biological processes in the single-cell or sub-cellular level. In this work, we focus on *in situ*, *in vivo* and label-free Raman characterizations of modeled algal lipids, extracted algal oil, and most importantly, single living algae (Figure 1). Our study has demonstrated a label-free and direct method to obtain quantitative information of chain length and degree of unsaturation of the oil produced inside algae. It also connects with the important issues of the cloud point and the quality of algal biodiesel. Our single-cell, real-time, and *in vivo* study of algae, with various lipid-triggering mechanisms, enables the possibility of researching and engineering of the best conditions and the best species for algal growth and oil production.

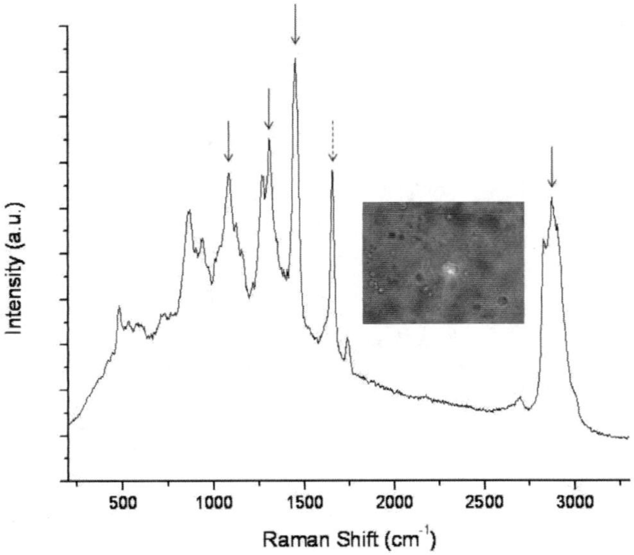

Figure 1. A single *Neochloris oleoabundans* algal cell is held by an optical trap and its Raman spectrum is obtained. Black arrows are pointing at the Raman bands of oil identification. A dashed arrow is pointing at the Raman band of unsaturation (C=C bonds) of the oil inside the alga.

See more of this Session: Solar Biofuels II
See more of this Group/Topical: Topical H: Solar Topical

Diurnal Cycling of Triacylglycerides in a Chlorophyte sp.: Implications for the Production of Algal Biofuels.

Keith E. Cooksey and Robert M. Thomas, Algal Biotechnology Group, Montana State University-Bozeman, MT.

There is no doubt that crude oil will be depleted eventually. The concept of global warming is now well accepted. Given these twin constraints, it behooves us to find alternative fuels, especially liquid ones for transportation. Microalgae have the potential to both reduce increases in greenhouse gases and provide an energy dense fuel. The world-wide nascent industry in algal biofuels lacks basic physiological information concerning organisms that might be used in an oil production effort. Neutral lipids such as triacylglycerides are formed in algal batch cultures as growth progresses. Biology in general utilizes such molecules as alternative stored energy sources when the primary energy source, e.g., photosynthesis, is not available such as in darkness. We postulate that what we see in algae is the result of TAG *accumulation* which is the remainder after utilization has taken place. It is unlikely that this phenomenon will be recognized unless cell division synchronized cells are investigated. Our data using Chlorella indicate that TAG accumulation, as evidenced by Nile Red fluorescence, increases in the light and decreases in the dark and is tied to the time when the cells divide. Most aerobic organisms use the Krebs tricarboxylic cycle pathway to metabolize C-2 fragments derived from the fatty acids found in TAGs. Inhibition of this pathway by monofluoroacetate in Chlorella and some diatoms inhibits TAG utilization leading to an increase in TAG accumulation. This phenomenon is important in deciding when an oil accumulating algal culture should be harvested and provides at least two points where metabolic regulation could be altered to increase oil accumulation. A potential mechanism for the effect of nitrogen starvation on TAG accumulation in algae will be presented as well as some thoughts on the type of standards the nascent algal biofuels industry needs.

Physical and Nutrient Stresses for Stimulating Algal Biofuel Production: Quantitative Control of Fermentation

Damian Carrieri, Princeton University, Princeton, NJ

Gennady Ananyev, Princeton University, Princeton, NJ

Dariya Momot, Princeton University, Princeton, NJ

Ian Brasg, Princeton University, Princeton, NJ

G.Charles Dismukes, Princeton University, Princeton, NJ

Oliver Lenz, Princeton University, Princeton, NJ

Oliver Lenz, Institute for Biology, Humboldt University, Berlin, Germany

Microalgae and cyanobacteria are phototrophs that have important advantages over land-based crops as potential feedstocks for energy production. They do not compete with food-crops for arable land, can often be cultured in saline or brackish waters, and produce non-recalcitrant biopolymers. One strategy for conversion of these feedstocks to fuel molecules is through anaerobic fermentation. We have studied the filamentous cyanobacterium Arthrospira (Spirulina) maxima for its ability to produce hydrogen and ethanol via fermentation of its own internal energy storage molecules, mainly carbohydrates. Growth of filaments in highly saline medium increases A. maxima's production of carbohydrates from 20% to 50% of dry weight. Subsequent fermentation of these filaments in hypotonic buffers stimulates by 50-fold the fermentative carbon products excreted relative to cells cultured in less saline growth medium. This stress stimulates production of ethanol at a rate of 1.56 mmol / gDW / day, or 121-fold higher than cultures that are not hypotonically stressed during fermentation. Unexpectedly, dark hydrogen production does not correlate with the increased rate of catabolism of soluble sugars. A second strategy for increasing fermentative yields from cells was investigated. Replacement of sodium with potassium in the fermentation buffer of A. maxima stimulates by 2-fold ethanol and hydrogen production, apparently due to increased energy demand arising from loss of energy-coupled sodium ion gradient. Supported by the AFOSR-MURI program.

See more of this Session: Solar Biofuels II
See more of this Group/Topical: Topical H: Solar Topical

Algal Feedstocks for Biogas Production: What Are We Waiting For?

Mark B. Smith[1], Jun Cheng[1,2], Minghui Zhang[2], Damian Carrieri[1], G. Charles Dismukes[1, 3]

[1]*Princeton University, Department of Chemistry and Princeton Environmental Institute, Princeton, NJ 08544 USA*

[2]*Zhejiang University, State Key Lab of Clean Energy Utilization, Hangzhou, 310027 China*

[3] *Rutgers University, Waksman Institute for Microbiology, Piscataway, NJ 08854 USA*

Biogas energy (mainly hydrogen and methane) from renewable sources can replace fossil fuels and address issues of environmental pollution and energy security, however, this resource has been scarcely developed in the USA. The choice of biomass feedstock is being debated nationally as it is a major determinant of energy conversion efficiency (ECE = [heating value of gas]/[heating value of biomass]). Herein we illustrate that microalgal biomass feedstocks offer a significant ECE advantage over traditional terrestrial plant feedstocks as they are devoid of recalcitrant biopolymers (cellulose, hemicellulose & lignin) and can be grown at high specific yields.

A novel two-stage dark fermentation process for cogenerating H_2 and CH_4 from microalgae biomass was used in this paper, ultilizing a mixed consortia of hydrogenogenic and methanogenic bacteria described in Xie et al. 2007[1]. The filamentous cyanobacterium *Arthrospira maxima* was selected as a model substrate for biogas generation due to its robust growth and established commercial significance.

The two-stage biogas production process involved primary fermentation by a hydrogenogenic consortia and secondary fermentation by methanogens. Standard cultivation conditions were shown to yield *A. maxima* cells with 14.5 wt% carbohydrate, and an ECE of 15.1% with our two-stage fermentation system. Compositional engineering through nitrate depletion and high sodium induced physiological stresses led to cells with much higher carbohydrate compositions (41 wt%) and ECEs (24.3%) as illustrated in Figures 1-3 below.

The ECEs achieved in this early work are still only a fraction of the theoretical ECE (62%) calculated for the two-stage conversion of *A. maxima* biomass described, however these results still compare favorably to the yields achieved from terrestrial crops due to both the high conversion efficiency and productivity of organisms like *A. maxima*. For comparison, annual biogas production from whole corn plants yield approximately 5,800 cubic meters of methane per hectare[2] while biogas produced from *A. maxima* is reported here as 9,200 cubic meters of methane *and* 2,000 cubic meters of hydrogen gas per hectare, about 70% more on an energy equivalence basis than biogas produced from corn[3].

The concerted effects of our compositional engineering approach and two-stage energy recovery system demonstrate the promise of generating valuable biogas fuels from aquatic phototroph substrates.

Supported by grants from the DOE-GTL program and AFOSR-MURI program.

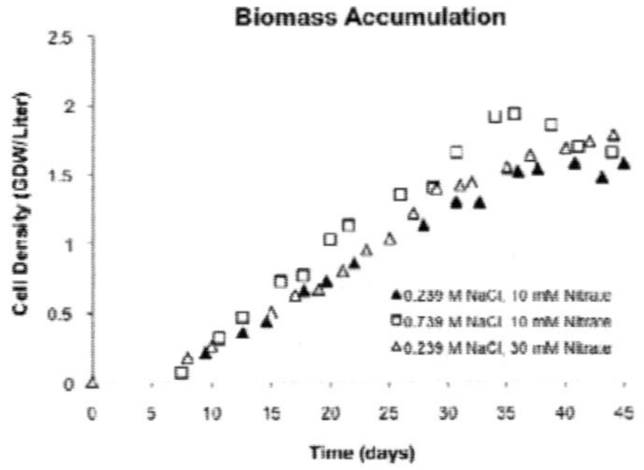

Figure 1: *Biomass accumulation measured gravimetrically from 20 L bioreactors. Similar biomass yields and temporal dynamics are measured across conditions suggesting that stress conditions do not significantly inhibit growth.*

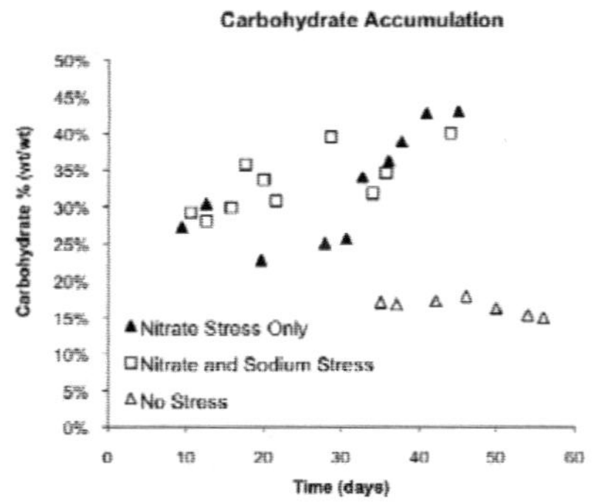

Figure 2: *Carbohydrate composition in A. maxima growing at 10 mM nitrate and either 239 mM sodium or 739 mM sodium. % Carbohydrate corresponds to the fraction of total cell dry mass. Terminal yields are similar although the temporal dynamics are somewhat different between the two conditions.*

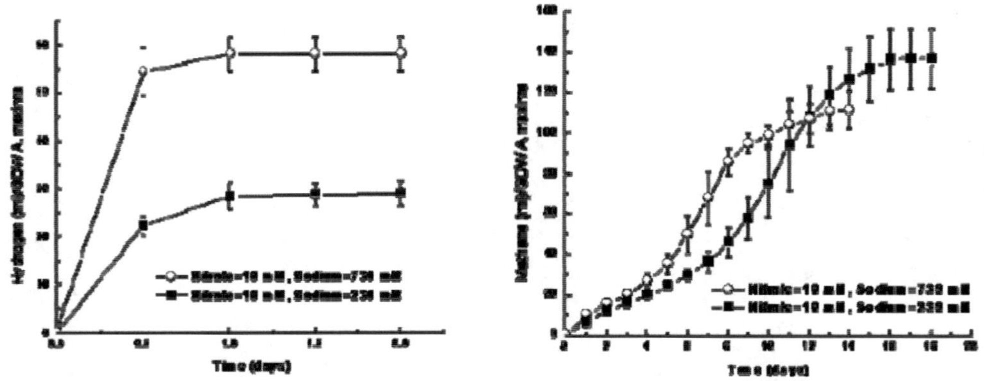

Figure 3: *Dynamic H_2 and CH_4 cogeneration from A. maxima cells grown under different sodium stress*

Table 1 H_2 and CH_4 cogeneration and ECE from A. maxima cells

Growth conditions		Carbohydrates in dried cells (wt%)	Fermentative H_2 and CH_4 cogeneration			
Nitrate (mM)	Sodium (mM)		H_2 yield (mL/g)	CH_4 yield (mL/g)	ECE (only H_2)	ECE (H_2+CH_4)
30	239	14.5	21	81	1.1 %	15.1 %
10	239	41.0	29	132	1.5 %	24.3 %
10	739	40.0	58	111	3.0 %	22.2 %

1. Xie, B. et al. Production of hydrogen and methane from potatoes by two-phase anaerobic fermentation. *Bioresource Technology* **99**, 5942-5946 (2008).

2. Weiland, P. Production and energetic use of biogas from energy crops and wastes in Germany. *Applied Biochemistry and Biotechnology* **109**, 263-274 (2003).
3. Dismukes, G.C., Carrieri, D., Bennette, N., Ananyev, G.M. & Posewitz, M.C. Aquatic phototrophs: efficient alternatives to land-based crops for biofuels. *Curr.Opin.Biotechnol.* **19**, 235-240 (2008).

The Role of Molecular Level Modeling in Gas Hydrate Studies

Brian J. Anderson, National Energy Technology Laboratory, 3610 Collins Ferry Road, West Virginia University, Department of Chemical Engineering, Morgantown, WV

For decades, researchers have been able to model and simulate systems of different scales independent of each other; however, the ability to bridge multiple time and length scales between such models has only recently begun to be intensively investigated. Natural gas hydrates (GH) provides an ideal model system on which to develop multiscale modeling techniques. The formation properties of hydrates are governed by molecular-level interactions between the gas and water molecules as well as the host media, yet hydrate reservoirs in nature can span tens of meters vertically and kilometers horizontally. Over the past few decades, many studies using Molecular-Level Modeling (MLM) have been conducted on gas hydrates and molecular dynamics (MD) simulations have been shown to correlate with experimental data in many areas such as phase equilibria, inhibition, cage occupancy, diffusion, and vibrational spectra to name a few examples.

Many successful research programs have been based on the constant integration of accurate theoretical modeling and laboratory studies. Concepts developed through simulations can enable a fundamental understanding of the processes in the laboratory while the lab can expose gaps in the fundamental knowledge eager for computational enlightenment. A few examples of how researchers at NETL, The University of Pittsburgh, and West Virginia University are using MLM for hydrate systems and their connection to hydrates, and their recoverability, in nature are presented. This work was performed as part of NETL's Institute for Advanced Energy Solutions (NETL-IAES).

Models of gas hydrate responses to production stimulation do not include expansion due to either thermal effects or guest effects for gas hydrates. Currently, the compressibility and thermal expansivity of ice are used in the absence of any data on the subject. To address this problem, NETL, Pitt, and WVU has examined the temperature, pressure, and compositional dependence of the crystal lattice distortion of gas hydrates using molecular dynamics[1]. MD simulations were used to study the lattice dimensions (or lattice constant) of various sI and sII hydrates at different temperatures and pressures. Temperature and guest size have a large influence on the lattice dimensions while pressure and guest-guest interaction energy have a relatively smaller but significant effect. The lattice constant, ***l***, has been shown to be a function of hydrate guest composition, the fractional occupancy of the cavities and the repulsive nature of the guest molecule. This work highlights the importance of precise lattice dimensions in calculating hydrate equilibria and demonstrates the role that molecular dynamic simulations can play in determining the effect of temperature, pressure and guest size on ***l***. The primary function of such calculations is to create better thermodynamic models for hydrate equilibria, but the calculations can also provide a way to determine the lattice expansion or contraction that occurs during production methods that involve, for example, temperature change or gas exchange (CO_2 for CH_4).

The thermal behavior of hydrates has attracted considerable attention since the discovery that the thermal conductivities of hydrates are very different from that of ice. In addition, gas hydrate thermal conductivity is anomalously low, and general thermal properties are relatively insensitive to temperature, in contrast to the behavior expected for crystalline materials.[2] The Jordan Group at NETL and the University of Pittsburgh has recently carried out non-equilibrium MD (NEMD) simulations of the thermal conductivity of methane hydrate and for the empty hydrate cage. They have found that, except at low ($\boldsymbol{T} \leq 50$ K) temperatures, the calculated thermal conductivities of methane hydrate and the empty cage system are nearly the same, ruling out host-guest coupling as a dominant factor for the anomalous thermal conductivity of methane hydrate.[2] This finding is remarkable in that although the thermal conductivity of methane hydrate can differ from ice Ih by as much as 15 times, the difference in thermal conductivity between an empty and a methane-filled hydrate is not significant.

Recent work at NETL and WVU on the equilibrium of mixed CO_2-CH_4 hydrates has illustrated that molecular-level processes result in macroscopic properties. The phenomenon of mixed hydrate lattice strain is a perfect example of a phenomenon in which macroscopic properties are governed by molecular-level events, and can be studied carefully using molecular-level modeling. In the case of CO_2-CH_4 hydrate equilibrium, these equilibrium pressure errors could lead to large differences in predictions of CO_2-CH_4 exchange processes and should be incorporated into reservoir simulation algorithms. As recently reported by Anderson and Velaga[6], most thermodynamic models for hydrate dissociation pressure drastically overpredict the occupancy of CO_2 in the small cage of the hydrate lattice compared to NMR data obtained near the hydrate equilibrium pressure[7]. Through the use of ***ab initio*** calculations we have been able to develop a CO_2-H_2O potential that results in improved accuracy of CO_2 and CO_2-CH_4 thermodynamic predictions[6].

Currently, MD simulations are being performed at NETL and WVU to determine the dissolution rate and mechanism for CH_4 and CO_2 hydrates. In 2006, Nihous and Masutani[8] suggest that the concentration of the hydrate guest species at the interface between the desorption film and diffusive boundary layer may be much lower than ambient solubility. Preliminary MD simulations support this conclusion. Additionally, in the area near the hydrate-water interface, an increase in the solubility of CO_2 may aid in increasing the rate of CO_2 dissolution compared to CH_4. An understanding of the dissolution of gas hydrates in contact with undersaturated water is important to determining the long-term stability of gas hydrate reservoirs.

At the nano-, or molecular-, scale, molecular modeling can be utilized to estimate key parameters such as thermal expansion, isothermal compressibility, thermal conductivity, and dissolution rates necessary for larger-scale models. These models can be extended to include relevant real-world conditions such as sediment composition and porosity and integrated into the reservoir simulation models at various degrees of detail, allowing the reservoir models to compute methane hydrate production scenarios over decade time scales.

References

(1) Zhang, M.; Anderson, B. J.; Warzinski, R. P.; Holder, G. D. "Molecular dynamics simulation of hydrate lattice distortion"; Prepr. Pap.-Am. Chem. Soc., Div. Fuel Chem., 2009, Salt Lake City, UT.

(2) Jiang, H.; Myshakin, E. M.; Jordan, K. D.; Warzinski, R. P. ***The Journal of Physical Chemistry B*** **2008**, ***112***, 10207.

(3) Klauda, J. B.; Sandler, S. I. ***Chemical Engineering Science*** **2003**, ***58***, 27.

(4) Sloan, E. D.; Koh, C. A. ***Clathrate hydrates of natural gases***, 3rd ed.; CRC Press: Boca Raton, FL, 2008.

(5) Anderson, B. J. ***Fluid Phase Equilibria*** **2007**, ***254***, 144.

(6) Velaga, S.; Anderson, B. J. "Phase equilibrium and cage occupancy calculations of CO_2 hydrates using an ab initio intermolecular potential"; Prepr. Pap.-Am. Chem. Soc., Div. Fuel Chem., 2009, Salt Lake City, UT.

(7) Ripmeester, J. A.; Ratcliffe, C. I. ***Energy & Fuels*** **1998**, ***12***, 197.

(8) Nihous, G. C.; Masutani, S. M. ***Chemical Engineering Science*** **2006**, ***61***, 7827.

See more of this Session: Methane Hydrate: Modeling, Experiment and Exploration
See more of this Group/Topical: Fuels and Petrochemicals Division

Molecular Dynamics Simulations of WATER Cavity Distortion for Determining Clathrate Hydrate Equilibria

Krishnadeo Jatkar, Chemical and Natural Gas, Texas A&M University-Kingsville, Kingsville, TX

Sangyong Lee, Texas A&M University-Kingsville, Kingsville, TX

Jae W. Lee, Department of Chemical Engineering, The City College of New York, New York, NY

Gas hydrate is a non-stoichimetric crystalline compound with hydrogen bonded water molecules and gas molecules connected with van der Waals force. The research on gas hydrate became emphasized because of its potential to be used as a future energy resource (Natural gas hydrate contains over 90% of methane and can be used as the source of natural gas) and as an energy storage medium. An accurate thermodynamic model is needed to predict the dissociation condition of natural gas hydrate to produce the natural gas from hydrate deposit or to predict the storage condition of hydrogen or natural gas in hydrate. In the present work, equilibrium conditions of structure I and structure II gas hydrates are predicted using Molecular Dynamic (MD) Simulation. The simulations are worked on a highly portable C-program package called 'Moldy®' and conducted at the High Performance Cluster Computer at Texas A&M University-Kingsville. The equilibrium conditions were calculated based on the distortion theory. Instead of using a rigid lattice assumption, lattice expansion or stretching assumption is employed for the prediction of hydrate equilibria and the unit cell size of each gas hydrate was adjusted to fit the experimental equilibrium conditions at various temperature conditions. A constant temperature and volume ensemble (canonical ensemble) is used to obtain equilibrium pressures of hydrates at any given temperature through MD simulation. As the size of gas molecules in water cavities increase, the unit cell size of gas hydrate is increased. The unit cell volume expansion on temperature was also observed. Based on the optimized unit cell of gas hydrate at 273K, chemical potentials for theoretical empty cavity for each single gas hydrate was calculated. The calculation results were compared with empirical values. The error between calculated value and experimental values was around 20 to 30%.

See more of this Session: Methane Hydrate: Modeling, Experiment and Exploration
See more of this Group/Topical: Fuels and Petrochemicals Division

Comparison of Numerical Simulators for Methane Production From Hydrate Reservoirs

Yong Liu, National Energy Technology Laboratory, U.S. Department of Energy, Pittsburgh, PA

Isaac K. Gamwo, National Energy Technology Laboratory, US Department of Energy, Pittsburgh, PA

Several numerical simulators for methane hydrate dissociation in porous media such as ToughFx, Tough+Hydrate, HydrateResSim, FLUENT, COMSOL, CMG STARS, MH-21, and STOMP-HYD are available. These simulators are useful in developing a fundamental understanding of complex hydrate dissociation systems in which heat, mass, and momentum transports interact with the hydrate kinetics and or thermodynamics. Particularly, HydrateResSim is a free open-source numerical reservoir simulator software available from the National Energy Technology Laboratory.

We have investigated the dissociation and production of methane induced by depressurization of hydrate sediment in a laboratory-scale reactor using four numerical simulators namely HydrateResSim, ToughFx, Tough+Hydrate, and FLUENT. Our predictions show that the free HydrateResSim software interfaced with PetraSim can produce, with some modifications, comparable results obtained using the commercial software ToughFx, Tough+Hydrate, and FLUENT. Furthermore, experimental data of methane production in the open literature compare reasonably well with HydrateResSim predictions as well as with the commercial codes tested.

The effects of radiative heat transfer and coordinate systems on hydrate dissociation are investigated. Our simulations show that neglecting the thermal radiation (Stefan-Boltzmann law) lead to negligible variation on the hydrate production. Predicted two-dimensional results using the Cartesian coordinates differ from simulated methane production using the Cylindrical coordinates.

See more of this Session: Methane Hydrate: Modeling, Experiment and Exploration
See more of this Group/Topical: Fuels and Petrochemicals Division

Mechanisms for Thermal Conduction in Various Polymorphs of Methane Hydrate

Niall J. English, Chemical and Bioprocess Engineering, University College Dublin, Dublin, MT, Ireland

John S. Tse, Physics and Engineering Physics, University of Saskatchewan, Saskatoon, SK, Canada

Extensive equilibrium molecular dynamics (MD) simulations have been performed to investigate thermal conduction mechanisms via the Green-Kubo approach for fully-occupied type I, II and H methane hydrates. We have shown that the crystal structure of the clathrate framework and damping of guest-host interactions in type I methane hydrate contribute to a lower thermal conductivity vis-à-vis ice Ih and its glass-like temperature dependence. We have found that substantially less damping in guest-host energy transfer is present in type II and H methane hydrate polymorphs at higher temperatures, giving rise to somewhat larger thermal conductivities relative to type I methane hydrate with a more crystal-like temperature dependence.

See more of this Session: Methane Hydrate: Modeling, Experiment and Exploration
See more of this Group/Topical: Fuels and Petrochemicals Division

Investigation of Methane Hydrate Growth Using Molecular Dynamics Simulation

Matt Walsh, Chemical Engineering, Colorado School of Mines, Golden, CO

David T. Wu, Chemical Engineering and Chemistry Departments, Colorado School Of Mines, Golden, CO

Amadeu K. Sum, Chemical Engineering, Colorado School of Mines, Golden, CO

Carolyn A. Koh, Chemical Engineering, Colorado School of Mines, Golden, CO

E. Dendy Sloan, Chemical Engineering, Colorado School of Mines, Golden, CO

For the first time, molecular dynamics simulation is used to investigate steady-state methane hydrate growth in a system larger than 50 sI hydrate unit cells for durations greater than 100ns without employing artificial mass or heat transport constraints. Molecular dynamics simulations are initiated from partially ordered configurations, and as solid hydrate grows it replaces areas of the simulation box previously occupied by a methane and water vapor/liquid system. At the interface of growth, partially formed cages are observed to persist for up to 10 nanoseconds while halo-like hydrogen-bonded networks of water molecules eventually form the remainder of the cages. The estimated growth rates for these simulations are between 1.3 cm/s and 3.4 cm/s.

See more of this Session: Methane Hydrate: Modeling, Experiment and Exploration
See more of this Group/Topical: Fuels and Petrochemicals Division

Biodiesel Production From Algae by Supercritical Methanol and Liquefaction Method

Prafulla Patil, New Mexico State University, Las Cruces, NM

Shuguang Deng, Chemical Engineering Department, New Mexico State University, Las Cruces, NM

Biodiesel derived from oil crops is a potential renewable and carbon neutral alternative to petroleum fuels. Unfortunately, biodiesel from oil crops, waste cooking oil and animal fat cannot realistically satisfy even a small fraction of the existing demand for transport fuels. As demonstrated here, microalgae appear to be the only source of renewable biodiesel that is capable of meeting the global demand for transport fuels. Biodiesel from microalgae using supercritical methanol (one-step) and liquefaction method followed by transesterification (two-step) were studied in detail. A non-catalytic biodiesel production route with supercritical methanol has been developed for dry algae that allow a simple process and high yield because of simultaneous transesterification of triglycerides and methyl esterification of fatty acids. The liquefaction method followed by the transesterification was implemented for wet algae for the biodiesel production.

See more of this Session: Biodiesel From Microorganisms
See more of this Group/Topical: Fuels and Petrochemicals Division

The Impacts of Light, CO2 Concentration, Temperature, Medium On Growth of Microalgae for Biofuels Production

Haiying Tang, Department of Chemical Engineering and Materials Science, Wayne State University, Detroit, MI

Nadia J. Abunasser, Chemical Engineering and Materials Science, Wayne State University, Detroit, MI

M. E. D. M. E. D. Garcia, Chemical Engineering, Wayne State University, Detroit, MI

Meng Chen, Chemical Engineering, Wayne State University, Detroit, MI

Ansheng Li, Chemical Engineering, Wayne State University, Detroit, MI

Hongzhi Ma, Chemical Engineering, Wayne State University, Detroit, MI

Steven O. Salley, Department of Chemical Engineering and Materials Science, Wayne State University, Detroit, MI

K. Y. Simon Ng, Department of Chemical Engineering and Materials Science, Wayne State University, Detroit, MI

Developing biomass feedstocks for biofuels production is vital for our energy security and national economy, as well as for our environment. Algae are among the most promising non-food-crop-based biomass feedstocks. However, there are a very limited number of large-scale commercial operations that grow and harvest algae and these mainly produce relatively high value products such as food supplements and none produce algae with potential for biofuels use economically. The algal production was limited by oil content, growths rates, and cultivation. Microalgal cell growth rates are affected by environmental parameters such as temperature, light intensity, period, trace medium, and gas composition in the culture system, as well as the combination of all these factors. The accumulation of algae oil is strongly dependent on the cell morphology and status of the green cells produced. However, the relationship of operating parameters, algal growth rate, and lipid content for biofuels has not been fully evaluated. Therefore, it is necessary to perform a systematic study to gain a fundamental relationship of the appropriate combination of environmental parameters and algae production system with a large number of cells, rapid turnover rate, and higher lipid content.

In this study, three types of microalgae, Dunaliella tertiolecta, Chlorella Minutissima, and Botryococcus braunii were investigated to better understand and define the operating parameters for the mass production under controlled condition, and to be used towards identification the feasibility of microalgae for producing algal oil based biodiesel. Specifically, the objectives of this research are: (1) Elucidate the effect of light intensity and light wavelength on microalgal growth rate; (2) Identify the effect of CO2 and O2 concentration on microalgal growth rate; (3) Investigate the effect of temperature and PH on microalgal growth rate; and (4) Evaluate the lipid content in the status of microalgal growth.

The results will provide the fundamental engineering knowledge to contribute to develop the large scale of algal production for biofuels production.

See more of this Session: Biodiesel From Microorganisms
See more of this Group/Topical: Fuels and Petrochemicals Division

Coordination of Growth Rate and Lipids Biosynthesis by Nutrients in Microalgae

Meng Chen, Chemical Engineering, Wayne State University, Detroit, MI

Haiying Tang, Chemical Engineering and Materials Science, Wayne State University, Detroit, MI

Hongzhi Ma, Chemical Engineering, Wayne State University, Detroit, MI

Danton Garcia, Chemical Engineering, Wayne State University, Detroit, MI

Ansheng Li, Chemical Engineering, Wayne State University, Detroit, MI

Nadia J. Abunasser, Chemical Engineering and Materials Science, Wayne State University, Detroit, MI

Kapila Wadumesthrige, National Biofuel Energy Laboratory, NextEnergy, USA, Detroit, MI

Thomas Holland, Immunology and Microbiology, Wayne State University, Detroit, MI

Steven O. Salley, Department of Chemical Engineering and Materials Science, Wayne State University, Detroit, MI

Simon K. Y. Ng, Chemical Engineering and Materials Science, Wayne State University, Detroit, MI

Algae have been recognized as a potential source of biomass for production of biofuels because some algal species can produce and accumulate high levels of oil and other valuable energy storage molecules. The efficiency of biofuels production mainly depends on the rates of algae growth and lipids biosynthesis. Nutrients uptake plays an important role in algae growth and lipids biosynthesis. However, the effect of trace nutrients on algae growth and lipids biosynthesis has not been fully understood. In this study, inductive couple plasma (ICP) was use to examine the nutrients content: Phosphorus (P), Zinc (Zn), Calcium (Ca), Cobalt (Co), Molybdenum (Mo), Manganese (Mn), Iron (Fe), Sodium (Na), Sulfur(S), Silicon (Si) in the algae media that are required for the algae growth as a function of time. Spectrophotometer (HACH) was used to measure the nitrogen content. Moreover, algae growth rate and lipids content are monitored by using optical density and Nile Red methods, respectively. The effects of phosphorus, nitrogen, silicone content on algal growth and lipid content were investigated. This research will identify the mechanism how nutrient uptake will affect the algae growth and lipids biosynthesis and is critical for the bioreactor design and biofuel production.

See more of this Session: Biodiesel From Microorganisms
See more of this Group/Topical: Fuels and Petrochemicals Division

Extraction of Microalgae Using An Ionic Liquid Co-Solvent System.

Gregory Young[1], Franz Nippen[2], Sebastian Titterbrandt[2], and Michael J. Cooney[2].

[1]Current address: Department of Chemical and Materials Engineering, San Jose State University, San Jose, CA 95192

[2]Hawaii Natural Energy Institute, University of Hawaii, Honolulu, HI 96822

INTRODUCTION

Ionic liquids (IL) are non-aqueous solutions of relatively large asymmetric organic cations coupled with a small inorganic or organic anion salts that remain liquids at moderate to room temperatures (0 °C to 140 °C)[11]'[12] [13]. As they are charged species that strongly self associate they also possess extremely low vapor pressures[14]. The cations are generally composed of a nitrogen containing ring structure (e.g., immidazolium or pyrinnidine) onto which a broad range of functional side groups can be attached[14]. Depending upon the side group added, the polarity of the resulting IL solution can be varied[15]. The anions can vary from single ions such as chloride, to larger molecules such as $[N(SO_2CF_3)_2]^-$ [16]. As demonstrated by Freire and co-workers the anion and cation can also be modified to alter the relative hydrophobicity/hydrophilicity of the resulting IL[16]. Combined, these features contribute to their growing reputation as "green solvents" that possess relatively low toxicity and that can tailored for a specific solubility, polarity, electrical conductivity, and relative hydrophobicity[16]. IL's have also been recently shown to extract protein from aqueous two phase systems[17], extract hormones from urine samples[18], esterify lipids[19], and used as a "green" solvent with biodegradable copolymers[20].

In this paper we study the ability of a co-solvent system comprised of a hydrophilic IL and polar covalent molecule to extract and auto partition lipids from various biomass sources. In our approach the volatile organic solvent is replaced with an ionic liquid (IL) to create a co-solvent system that both dissolves and extracts the biomass, but also auto-partitions the extracted lipids into a separate immiscible phase. This is in contrast to traditional co-solvent systems, comprised of a volatile organic solvent and a polar covalent molecule (PCM), which both dissolves and extracts the bio-oil into a single phase. The IL-PCM co-solvent system is also evaluated for its ability to facilitate the direct transesterification of intercellular lipids to fatty acid methyl esters which are likewise auto-partitioned to a separate immiscible phase. In both cases we propose that the action of the PCM is largely to disrupt the cell wall and to improve the efficiency at which the lipid is contacted with the solvent, and that the action of the strongly self-associating ionic liquid is to rapidly transfer the lipids to the surface interface where they auto partition into a self associating and separate phase. An added benefit is that the IL, which possesses negligible vapor pressure due to the strong self association between the cation and anion molecules, should help to reduce the vapor pressure of the PCM (e.g. methanol) co-solvent. Finally, the capacity of this system to co-extract protein is also discussed.

EXPERIMENTAL

Biomass material

The yeast and *Chlorella* microalgae were heterotrophically grown on sugar in the absence of sunlight. The *Duniella* microalgae was obtained from an outdoor microalgae production company in the Imperial valley of California. The oil seeds were obtained from local growers as a generous gift from Dr. Goro Uehara at the University of Hawaii. All samples were either oven dried or freeze dried and then ground with mortar and pestle prior to all extraction and transesterification experiments. The exceptions were Jatropha and Kamani oil seeds which were removed from their shells and husk's prior to being ground and the Canola seeds which were ground whole with the shell material.

Extraction and Direct Transesterification.

The co-solvent 1-ethyl-3-methyl immidazolium methyl sulfate (EMIM) and methanol was used in all experiments at a mass ratio of 1.2:1 (methanol to EMIM) and total mass of 6.5 g. In all cases the biomass was added to the co-solvent mixture and the mixture agitated under magnetic stirring for 18 hours at 65 °C in a sealed test tube. Biomass samples for the direct transesterification experiments were added to the co-solvent mixture, as described above, except that the test tube was chilled on ice to minimize the volatilization of the methanol and the hydrochloric acid product as the acetyl chloride (5% v/v, acetyl chloride to methanol) was added drop-wise to the reaction mixture. Complete conversion was achieved by running the reaction for 18 hours at 65°C.

Product Isolation

The extraction or direct transesterification products were centrifuged, the resulting two phase liquid solution (i.e., the auto-partitioned lipid layer and the co-solvent solution) decanted, washed three times with 5 ml aliquots of hexane, pooled together, and then water washed to remove any last traces of polar material. The lipid or FAME product was then concentrated using a Büchi™ Rotoevaporator and the residue weighed to determine the gravimetric yield.

H-NMR analysis

10 μl of sample (~8 mg) and 2 μl of a p-xylene internal standard were added to $CDCl_3$ a total volume of 1 ml. The triglyceride doublet peak was set to 4.2 ppm and the methyl ester peak of the FAMEs product to 3.7 ppm.

Results and Discussion

Co-solvent extraction with Ionic liquid-Methanol

The results from biomass extracted using the EMIM-methanol co-solvent system are presented in Table 1. The experimental results, which agree well with reported literature values and independent laboratory testing [21], suggest that the IL-methanol co-solvent is successful in complete extraction of the lipids from these biomass sources. A typical case is presented in

Figure 1. Figure 1(a) shows the extraction mixture at the start of the extraction process. Figure 1(b) shows the typical auto-partitioned lipid layer at the end of the extraction period. Figure 1(c) presents concentrated lipid product. The yield of the lipid product recovered was 44%. The lipid product recovered was free flowing, slightly yellow, and containing only trace amounts of FAME product (NMR data not shown).

Direct transesterification in Ionic liquid-Methanol co-solvent

Direct transesterification of internal lipids was executed on Canola oil seeds in EMIM-methanol co-solvent and a typical reaction sequence is shown in Figure 1d-f. NMR spectra of the recovered product revealed that the transesterification reaction was carried out to 100% conversion of triglycerides to FAMEs (data not shown). Direct transesterification was likewise applied to several other biomass sources in the same EMIM-methanol co-solvent system and the results are presented in Table 2. With the exception of the *Chlorella* microalgae, the conversion from FAMEs to triglycerides was complete (100%). As the gravimetric yields of extraction (Table 1) and direct transesterification (Table 2) are similar, the co-solvent is equally efficient at both reactions for all biomass sources considered (*Duniella*, *Chlorella*, Canola and Jatropha).

Effect of Methanol on Transesterification

The molar ratio of alcohol to triglycerides (assuming an average triglyceride molecular weight of 900 g/mol) used in the direct transesterification experiments equates to 3000:1. In this process, the methanol served not only as a reactant but also as a component of the co-solvent extraction mixture. To determine the importance of the methanol volume to the conversion efficiency of the transesterification reaction, the amount of methanol used in the direct transesterification of *Duniella* biomass was varied. The yield of the extracted product as a function of methanol:triglyceride molar ratio is plotted in Figure 2. Also plotted is the percent of the reaction product that is actually FAMEs (as determined by NMR analysis). The extraction yield remained constant at 7.5% as the molar ratio was reduced from 3000:1 to 1000:1. As the molar ratio was reduced below 1000:1, however, the yield continuously decreased to a lower value of 1% at a molar ratio of 200:1. By contrast, the NMR spectra for all the recovered products showed that the conversion of extracted lipids to FAMEs remained constant at 100% over the same range of molar ratios (NMR data not shown). Consequently, the only effect of a reduction in the methanol:triglyceride molar ratio appears to decrease the amount of lipids extracted.

Effect of PCM

Several other PCMs (Table 3) were paired with EMIM to evaluate their potential extraction value on *Duniella* biomass. Methanol (in this work considered our baseline PCM) and isopropyl alcohol produced similar results. DMSO and acetic acid resulted in lower lipid yields. Acetone and chloroform resulted in a larger gravimetric yield than the alcohols. However, the recovered product were also found to contain a solid precipitate that was not present in the other samples. Consequently, acetone and chloroform were considered to "over extract" more polar material, giving unrealistically high yields.

Extraction of Wet Biomass

Solvent extraction of lipids has generally been applied to dried biomass, a process that requires the dewatering of the algae prior to its extraction[22]. To determine the impact of water on the efficiency of the extraction co-solvent, the IL co-solvent solution was used to extract lipids from a *Chlorella* microalgal paste that was 70% water by weight (30% biomass). Based upon the results presented in Table 1, the lipid content in freeze dried *Chlorella* was 38% (wt% of dry biomass). Therefore, the lipid content in a wet sample that is only 30% biomass would be approximately 11% (wt% of wet biomass sample). As such the ionic liquid –methanol co-solvent mixture was able to extract 8.2 % from the wet sample, giving an extraction efficiency of 75%.

CONCLUSIONS

The proposed IL-methanol co-solvent system differs from traditional organic co-solvent systems which both dissolve and solubilize the extracted lipid and thus suffer extraction efficiencies limited by the solvent's carrying capacity. Due to their molecule structure, which possess regions that are both polar and non polar, IL based co-solvents yield a heterogenous solvent space of polar and non-polar regions on a microscopic level despite the fact that overall solution is polar on a macroscopic scale. As the IL-PCM co-solvents systems used in this study were highly polar and thus immiscible with the target lipid, we propose that the action of the solvent system was more to push the extracted lipid through and out of phase, a feature that avoids the limitations related to low solvent carrying capacity. The IL-PCM co-solvent mixture will also support extraction of lipids in the presence of moderate water concentrations and the direct transesterification of biomass if an appropriate catalyst is added. In terms of methanol content, the limiting step in the direct transesterification process appears to be the extraction of the lipids from the biomass. Once the lipids are free of the biomass the amount of methanol present is more than sufficient to carry the transesterification reaction to 100% conversion.

The IL-PCM solvent system is particularly robust in terms of the potential for additional applications. Due to the solvent's ability to penetrate cellular membranes as well as extract and auto-partition hydrophobic material, it is possible that this solvent system can be extended to other valuable components within biomass. Proteins, which are polar in nature, could be extracted into the co-solvent and subsequently recovered. In our work protein (Coomassie blue) assays of IL-methanol co-solvent solutions applied to Jatropha oil seeds regularly reported reasonable levels of extracted protein (~ 8 to 10 % w/w, data not shown). Although the selection of an optimal PCM for protein extraction has yet to be determined, it may become apparent from future studies that other combinations of ILs and/or PCMs are superior at extracting more polar materials like protein. IL's also possess virtually no vapor pressure which classifies them as a green solvent with no emissions.

ACKNOWLEDGEMENTS

This work was supported by funding from LiveFuels and BioEcoTek Hawaii. The authors would also like to thank Goro Uehara for the donation of oil seed samples and Jim DeMattia for the Duniella microalgal samples. The authors would also like to thank Wes Yoshida for his help with the NMR analysis.

Table 1: Lipid extraction results with Ionic liquid co-solvent

Biomass Type	Sample preparation	Reported yield	Experimental yield
Duniella-microalgae	Freeze dried	11.1%[21]	8.6%
Chlorella-microalgae	Freeze dried	11% - 23% [23]*	38.0%
Canola-oil seed	Oven dried	42.9%[21]	44.0%
Jatropha-oil seed	Removal of shell and husk	55%-58% [24]	49.9%
Kamani-oilseed	Removal of shell and husk	49% [25]	38.0%
Pongamia-oil seed	Removal of shell and husk	30-40% [26]	11.0%

*Literature report range of *Chlorella* is highly dependent upon growth conditions.

Table 2. Direct transesterification yield from various biomass sources

Biomass Type	Sample preparation	Conversion	Gravimetric yield
R. toruloides-yeast	Freeze dried	100%	28.8%
Duniella-microalgae	Freeze dried	100%	8.0%
Chlorella-microalgae	Freeze dried	85%	36.0%
Canola-oil seed	Oven dried	100%	34.5%
Jatropha-oil seed	Removal of shell and husk	100%	49.1%

Table 3 The biomass used in this study, *Duniella* microalgae, had a lipid content of approximately 8-11% (wt %).

PCM	Yield (wt %)	Comments
Dimethyl sulfoxide	6.0	Low yield, highly viscous product
Acetic Acid	5.6	Low yield, highly viscous product
Methanol	7.9	Baseline PCM
Acetone	9.2	High yield, with a solid precipitate
Chloroform	8.4	High yield, with a solid precipitate
Isopropyl Alcohol	8.5	Product very similar to methanol

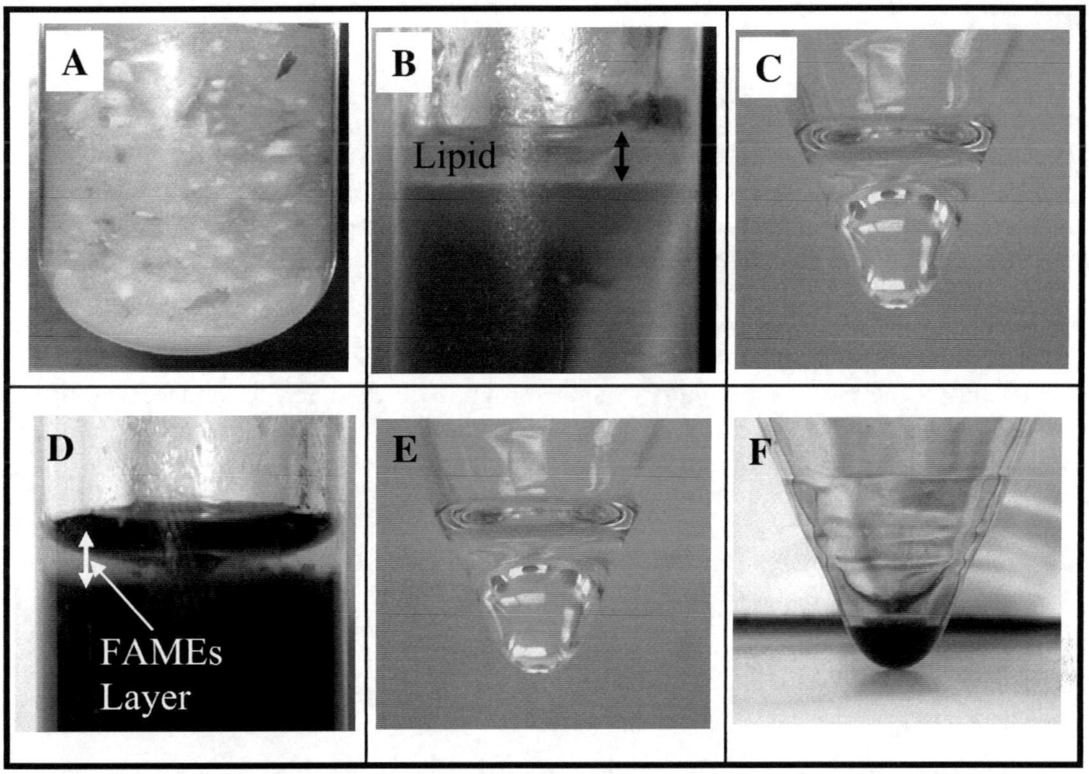

Figure 1. (a) Canola oil seeds suspended in co-solvent system. (b) Lipid layer on the top of co-solvent extraction of canola oil seeds. (c) Recovered lipid from canola oil seed extraction. (d) FAMEs product layer on top of co-solvent. (e) Canola oil direct transesterification FAMEs Product. (f) *Duniella* direct transesterification FAMEs product

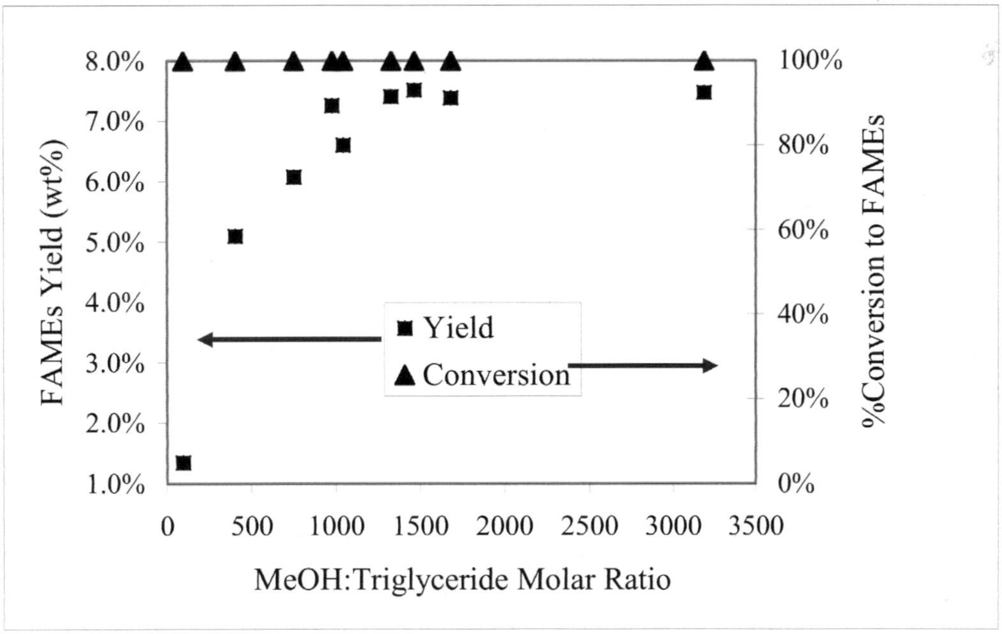

Figure 2 Yield of direct transesterification as a function of methanol to triglyceride molar ratio.

REFERENCES

1. Folch J, Lees M, Sloane Stanley GH: A simple method for the isolation and purification of total lipids from animal tissues. Journal of Biological Chemistry 1956; 226: 497-504.
2. Bligh EG, Dyer WJ: A rapid method for total lipid extraction and purification. Canadian Journal of Biochemistry and Physiology 1959; 37: 911-917.
3. USDA Da: Biomass as Feedstock for a Bioenergy and Bioproducts Industry: The Technical Feasibility of a Billion-Ton Annual Supply. 2005.
4. Fajardo AR, Cerdain LE, Medina AR, Fernandex FGA, Molina Grima E: Lipid extraction from the microalgae Phaeodactylum tricornutum. European Journal of Lipid Science and Technology 2007; 109: 120-126.
5. Canoira L, Alcantara R, Garcia-Martinez M, Carrasco J: Biodiesel from Jojoba oil-wax: Transesterification with methanol and properties as a fuel. Biomass and Bioenergy 2006; 30: 76-81.
6. Pernet F, Tremblay R: *Effect of ultrasonication and grinding on the determination of lipid class content of microalgae harvested on filters*. Lipids 2003; 38(11): 1191-1196.
7. Eikani MH, Golmohammad F, Rowshanzamir S: Subcritical water extraction of essentail oils from coriander seeds (Coriandrum sativium L). Journal of Food Engineering 2007; 80: 735-740.
8. Luque de Castro MD, Valcarcel M, Tena MT: Supercritical Fluid Extraction. Heidelberg: Springer Verlag, 1994.
9. Herrero M, Cifuentes A, Ibanez E: Sub - and supercritical fluid extraction of functional ingredients from different natural sources: Plants, food-by-products, algae and microalgae. A review. Food Chemistry 2006; 98: 136 - 148.
10. Chueng PCK: Temperature and pressure effects on supercritical carbon dioxide extraction of n-3 fatty acids from red seaweed. Food Chemistry 1999; 65: 399 - 403.
11. Martyn E, Seddon K: Ionic liquids. Green solvents for the future. Pure Appl. Chem 2000; 72(7): 1391-1398.
12. Johnson K: What's an Ionic Liquid. The Electrochemical Society Interface 2007(Spring): 38-41.
13. Marsh K, Boxall J, Lichtenthaler R: Room temperature ionic liquids and their mixtures-a review. Fluid Phase Equilibria 2004; 219: 93-98.
14. Huddleston J, Visser A, Reichert W, Willauer H, Broker G, Rogers R: Characterization and comparison of hydrophilic and hydrophobic room temperature ionic liquids incorporating the imidazolium cation. Green Chemistry 2001; 3: 156-164.
15. Dzyuba SV, Bartsch RA: Expanding the polarity range of ionic liquids. Tetrahedron Letters 2002; 43: 4657-4659.
16. Freire MG, Santos LMNBF, Fernandes AM, Coutinho JAP, Marrucho IM: An overview of the mutual solubilities of water-imidazolium-based ionic liquids systems. Fluid Phase Equilibria 2007; 261: 449-454.
17. Pei Y, Wang J, Wu K, Xuan X, Lu X: Ionic liquid-based aqueous two-phase extraction of selected proteins. Separation and Purification Technology 2009; 64: 288-295.
18. He C, Li S, Liu H, Li K, Liu F: Extraction of testosterone and epitestosterone in human urin using aqueous two-phase systems of ionic liquids and salt. Journal of Chromotography 2006; 1082: 143-149.
19. Lue M, Guo Z, Xu X: Lipid Processing In Ionic Liquids. Lipid Technology 2007; 19(9): 204-207.
20. Zhua J, Wang W, Wang X, Li B, Wang Y: Green synthesis of a novel biodegradable copolymer base on cellulose and poly(p-dioxanone) in ionic liquid. Carbohydrate Polymers 2008; 76(1): 138-144.

21. Lab TNF: Analytical report CN1144. 2008.
22. Grima EM, Belarbi EH, Fernández FGA, Medina AR, Chisti Y: Recovery of microalgal biomass and metabolites: process options and economics. Biotechnology Advances 2003; 20: 491-515.
23. Illman AM, Scragg AH, Shales SW: Increase in Chlorella strains calorific values when grown in low nitrogen medium. Enzyme and Microbial Technology 2000; 27: 631-635.
24. Martý´nez-Herrera J, Siddhuraju P, Francis G, Da´vila-Ortiz G, Becker K: Chemical composition, toxic/antimetabolic constituents, and effects of different treatments on their levels, in four provenances of Jatropha curcas L. from Mexico. Food Chemistry 2006; 96: 80-89.
25. Ajayi IA, Oderinde RA, Taiwo VO, Agbedana EO: Short-term toxicological evaluation of Terminalia catappa, Pentaclethra macrophylla and Calophyllum inophyllum seed oils in rats. Food Chemistry 2008; 106: 458-465.
26. Karmee SK, Chadha A: Preparation of biodiesel from crude oil of Pongamia pinnata. Bioresource Technology 2005; 96: 1425-1429.

BIODIESEL PRODUCTION FROM MICROALGAE

Nkongolo Mulumba, Ihab H. Farag
Chemical Engineering Department, University of New Hampshire (UNH) W315 Kingsbury Hall, 33 Academic Way, Durham, NH 03824-3591, phone 603-862-2313, ihab.farag@unh.edu, nkongolo.mulumba@unh.edu

ABSTRACT: Biodiesel production from algae is a promising technique. Microalgae have the potential to produce 5,000-15,000 gallons of biodiesel/(acre-year). However, there are challenges; these include high yield of algae biomass with high lipid content and the effective technique to harvest the grown algae, extract the algal oil and transesterify the oil to biodiesel.

In this project Tubular PhotoBioReactor (TPBR) was designed and achieved a ten times increase in algae concentration. It produced 1g of dry algal biomass per liter of medium within 12 days, with a lipid content of 12% approximately. Healthy algal culture grew well in the TPBR reaching 56×10^6 cells/mL of culture medium. The 10 fold increase is higher than those reported for open ponds and helical photobioreactor.

1 INTRODUCTION

All humankind has been relying on non-renewable fuel such as petroleum-based fuel since its discovery more than four thousand years ago. Increasing fuel demand worldwide, limited reserve and its effect on environment push scientists to look for a clean and renewable fuel in replacement of petroleum based fuel. Biodiesel has been loomed as a renewable fuel and potential replacement for petro-fuel.

1.1 Background: Biodiesel is a group of fatty methyl- esters produced by a transesterification reaction between fatty acids (triglyceride) and alcohol in presence of a catalyst, e.g.,

$$RCOO\text{-}CH_2CH(\text{-}OOCR')CH_2\text{-}OOCR + 3R'COH \rightarrow 3\ R\text{-}COR' + CH_2OH\text{-}CHOH\text{-}CH_2OH$$

where R represents fatty acids chains[2]. Fatty acids (FA) used in biodiesel production come from animal fat or vegetal oil. One of the biggest challenges in biodiesel production is finding sufficient feedstock capable of meeting the demand. Both oil and alcohol are extracted from food crops such as corn, canola, soybean, palm, etc. In effect this process is diverting a food crop to be an energy crop.

There is already an increase rejection of food for bio-fuel. The United Nations (UN), for instance, is expressing a strong disapproval of using food crops for bio-fuel, which they believe to be the key in foods price increase and food shortage, although none of these crops could produce enough oil to be converted in biodiesel able to meet worldwide demand

The United States alone consume about 70 billion gallons of diesel per year. This cannot be met with oil from corn, soybean nor canola, which yield nearly 50, 60 and 90 gallons of biodiesel per acre per year respectively. Nonetheless, microalgae could produce 5000 to 15000 gallons of biodiesel per acre per year in open pond[1]. Obviously, algal biodiesel could potentially replace petro-diesel.

Microalgae are unicellular microorganism, like simple plants with no root and leaves that grow through photosynthesis process. They capture carbon dioxide during photosynthesis and convert it into biomass which can be used as food, fertilizer, a source of medicine and biodiesel[6].

Growing algal cells in open ponds raise several concerns such as impossibility to control growth conditions and contamination risks. Algal cells in open ponds are exposed to the environment, subject to risk of contamination, light deficiency, and heterogeneous medium depending upon the mixing mechanism, the shape of the ponds and the depth of the culture. On the other hand, closed ponds (photobioreactor) mitigate fluid culture contamination, and enhance a full control of algal growth parameters such as light penetration, homogeneous culture, pH and carbon dioxide input. They would use less space with high algal biomass yield. However, they are expensive to build and maintain.

1.2 **Biodiesel production from microalgae:** Production of biodiesel from algae has emerged as a possible unique way to compete petro-diesel. However, biodiesel yield depends upon lipid content of the algae strain used as a source of oil. Some algae strains could contain lipid up to 60% of dry biomass[1,7]. Medium composition and growth conditions are likely to be the most affecting parameters on oil content in algae.

Following the growth and harvesting of algae, the biodiesel production process consists of two steps: extracting oil using either solvent or mechanical extraction followed by biodiesel production through transesterification reaction using alcohol and catalyst[2]. Certainly, two steps process increases production cost, and impact the environment negatively with the use of excessive solvent. In this project, the feasibility of producing biodiesel in one step process (in situ transesterification) will be explored.

2 OBJECTIVES

The objectives of this research project are to: 1-Select a microalgae-strain that, not only grows fast, but also contains high lipid contain. 2-Design a photobioreactor, which facilitates the control of algal growth conditions for high productivity. 3-Explore an alternative biodiesel production by combining oil extraction and transesterification process into one step (in situ algal biomass transesterification).

3 METHODS

3.1 Algal strain selection: Microalgae are unicellular photoautotrophic or photoheterotrophic microorganism[3], which grow like plant through photosynthesis process during which they capture carbon dioxide and photonic energy, and convert them into lipid. One of the biggest challenges in algae culture for biodiesel is to find a suitable strain with high lipid content and growth rate. Seven strains were selected based on their lipid content and growth rate information. First, they were cultured in 2L glass flask using fresh medium at room temperature with light supplied 24hr per day during the entire culture period. The medium contained soluble nutrients subdivided into two categories: 1-macronutrients rich in N, P, K and NaCl; 2- micronutrients containing metals traces especially Fe, Cu, Co, Mo and Zn. In addition, vitamins B1 and B7 were supplied to some strains as needed for a health growth.

The growth rate and lipid content for all seven batches were monitored over time. Lipid content was measured by Nile Red fluorescence[5] using a spectrofluorometer .The turbidity (absorbance) of the culture was measured using a Bausch and Lomb spectronic spectrophotometer.

3.2 Photobioreactors: The initial study of algae as feedstock for bio-fuel was done using open ponds. This exposes algal cultures to high risk of contamination and impact of weather. In addition, mixing mechanism has been very difficult for large ponds resulting in heterogeneous culture medium and limited exposure to light[7]. As a result, algal cultures in open ponds have less productivity and enormous unhealthy batches. In recent years, researchers began studying closed system (photobioreactor) as a way to mitigate all treats encountered in open ponds. On the other hand, the construction as well as maintenance of a photobioreactor have shown high cost compared to an open pond.

Figure 1 shows the tubular photobioreactor (TPBR) designed in this project for algal cells growth. It has a main tank connected to two spiral tubes set in series. Both spiral parts were clear PVC tubes of 1" external diameter and 3/4" internal diameter. The volume of both spiral parts was 3.4 gal. The main tank served as a feeding point of medium to the PVC tubes with a maximum capacity of 5 gallons.

Culture medium was pumped into the tubings at set at fixed flow rate. These tubes provided an area of 20 ft^2 exposed to the fluorescence light. Air compressor supplied air to the system for aeration and to serve as a source of carbon dioxide. The air flow rate was set in the ranges of 193 – 210 gal/hr.

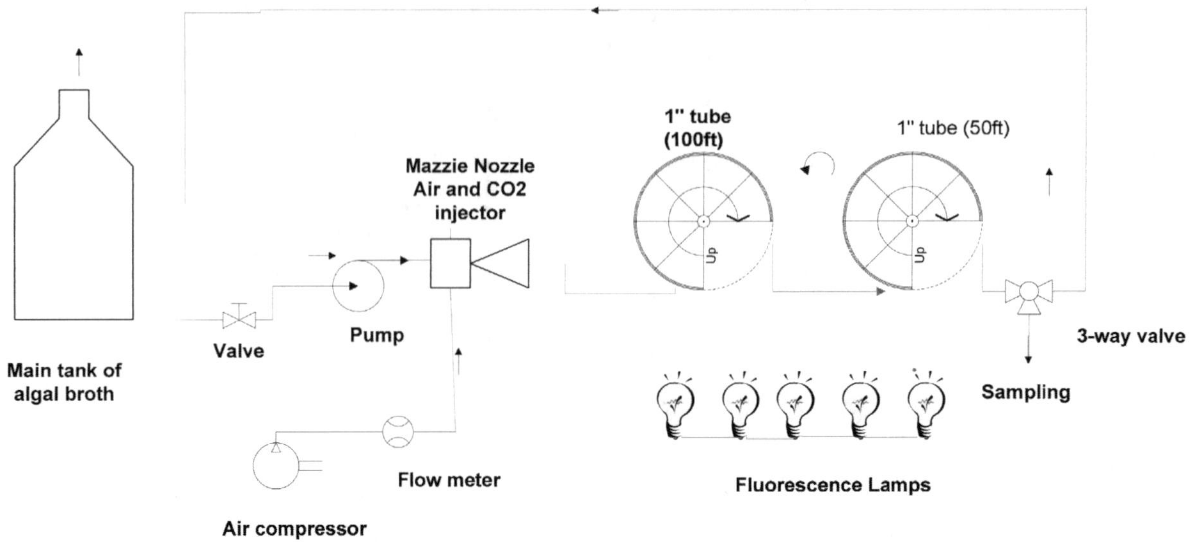

Figure 1: Tubular Photobioreactor

Culture flow rate was in turbulent regime with a Re >2000 to assure a good mixing and agitation. This TPBR could be used indoor with artificial source of photonic energy such as fluorescence lamps with low light intensity compared to sunlight. If weather permitted, the system could be set outdoor using sunlight as a source of photonic energy to minimize production cost.

3.3 Algal culture and monitoring: The selected algal strain was cultured in the TPBR to assess its functionality using fresh medium with no modification. Culture concentration and pH were measured over period of time varying between 12 to 14 days. A sample was taken every two day to measure the culture turbidity using a spectrophotometer (Bausch and Lomb spectronic) at 682 nm and cells count using a microscope. The pH of culture was measured as well using pH test strip. It is believed that strain released OH^- during its growth, which followed photoautotrophic and mixotrophic mechanism[6] according to the following reactions:

Reaction 1: Photoautotrophic mechanism:

$$H_2O + HCO_3^- \rightarrow C\ (\text{biomass}) + \frac{1}{2} O_2 + 3\ OH^-$$

Reaction 2: Mixotrophic mechanism

$$a\ HCO_3^- + b\ CH_2O \rightarrow c\ C\ (\text{biomass}) + 3\ OH^- + d\ CO_2$$

The selected strain exhibited the typical growth curve of other micro-organisms, which included the following phases: lag, exponential, stationary and lysis. The length of each phase depended mixing mechanism, nutrients concentration, light penetration and the solubility of oxygen in the medium. Nutrients concentration and light penetration were the only parameters taken in consideration in this project. Nutrient composition was changed by varying the concentrations of KNO_3 and NaCl to assess their impact on algal biomass yield. After reaching a stationary or lysis phase, algal culture was harvested by centrifugation followed by lyophilization to produce dry algal biomass.

Crude lipid was extracted from dried algal biomass using either modified Folsch method[5] or soxhlet extractor[9]. In both methods, polar and non-polar solvent such as methanol and chloroform/hexane were used. The combination of polar and non-polar solvents enhances the extraction of both polar and non-polar lipid.

3.3 In situ algal biomass transesterification: Traditionally biodiesel is produced by transesterification of oil extracted from food crops such as soybean, canola and corn. This process included oil extraction then conversion into biodiesel. This two steps process is not only time and space consuming, but also increases the price of resulting product downstream.

Transesterifying algal lipid without oil extraction process could potentially reduce production cost. In 2006, Haas and Scott successfully conducted in situ transesterification[10]. A year later, Ferrentino and Farag initiated a study of in situ transesferification technique on dried algal biomass[4, 12, 13]. The current project explores an integrated transesterification of algal biomass into biodiesel.

Lyophilized algal biomass was pulverized using a mortar and pistol to release as much lipid as possible. Then a sample was mixed with methanol in the presence of an alkaline catalyst such as KOH or NaOH to produce biodiesel. A sonicator was used not only to break algal cell wall for lipid release, but also to mix released lipid with alcohol/catalyst.

4 RESULTS AND DISCUSSION

4.1 Algal strain selection: Each of the seven algal strains was cultured in 2L-glass flask operating in batch mode. Algae concentration was monitored by measuring the turbidity, and the algae oil content was monitored by the Red Nile method. Figure 2 is a plot of algal concentration and lipid content in the broth. It shows that that x_7 is the best. Therefore this strain was retained for the rest of the project experiments.

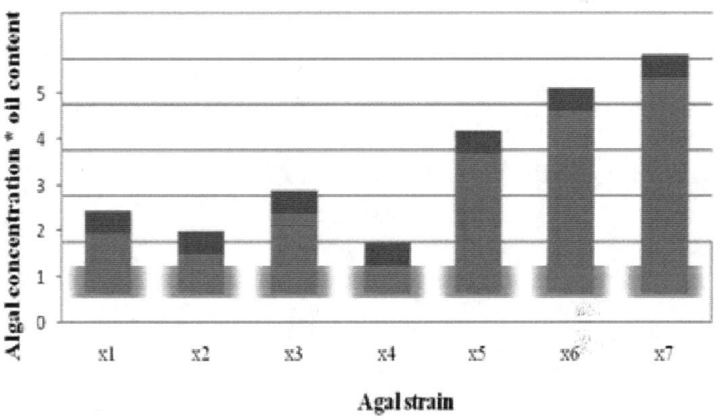

Figure 2 Lipid Content of different Algae broth solution

4.2 Algae growth and culture monitoring: Figure 3 is the growth curve of the selected algae strains. It plots the number of cells per ml of broth solution versus time. It shows that the cell count increases by factor of 10, since it started at $4x10^6$ Cells/mL (inoculums concentration of 0.1g/L) and reached about $56x10^6$ Cell/mL after 12 days. Based on these measurement it was possible to determine the parameters of the Monod Equation:

$$\mu = \mu_{\max} * \frac{S}{K+S}$$

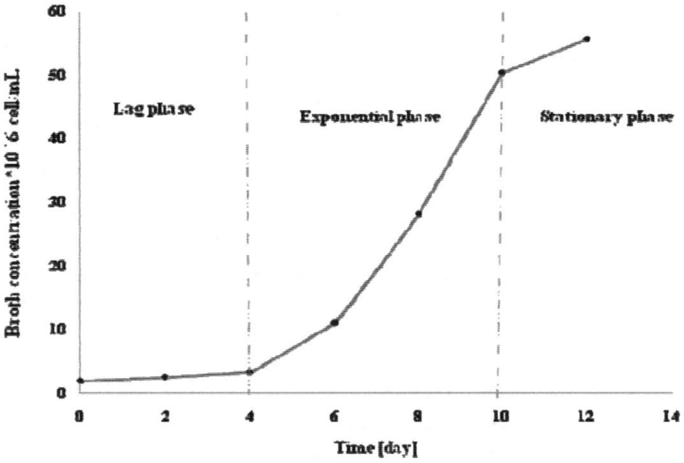

Figure 3 Growth Curve of Selected Algae Strain

where μ is the specific growth rate of the cells, μ_{max} is the maximum specific growth rate for the algae species, S is the concentration of the limiting nutrient, and K is the half saturation coefficient. The maximum growth rate of algae was estimated to be 0.6 per day.
The doubling time of the algae strain in TPBR was calculated as follows and showed that the broth concentration was doubling every 26 hr with a maximum growth rate of 0.6 per day.

$$T_d = \frac{Ln2}{\mu_{max}} = 1.16 \text{ day or } 26\text{hr}$$

Figure 4 is the plot of pH versus time. It shows that the pH of the selected algae strain culture started around 6.5 at the beginning of the culture and attained a plateau of pH 9 when the culture reached a stationary phase. The two batches (batch I and II) cultured at different time using same medium composition and same inoculums concentration, showed that the pH varied in the same way and reached a plateau at pH 9. The increase of pH was due to the release of OH^- as shown in reactions 1 and 2.

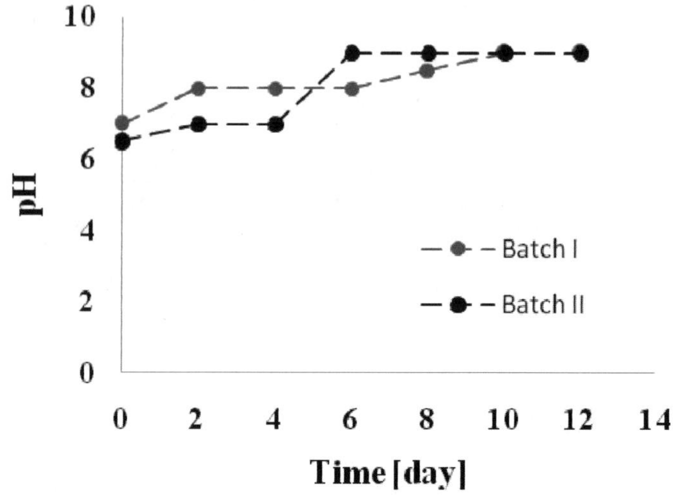

Figure 4 pH of Selected Algae Strain vs time

4.3 Algae growth rate: Four batches of same algae were grown in the TPBR and harvested after 12 days. Algal Broth for each batch was centrifuged and lyophilized after discarding excess water. The weight of each dry algal biomass was measured. The concentration of harvested broth plotted versus batch in Figure 5 shows that an average yield of algal biomass of 1.1g of algal biomass per L of broth with a dilution rate of 0.083 day^{-1}.
It is important to note that the starting concentration for the broth was 0.1g of algal biomass per mL of broth. It is also important to note that algal biomass concentration reached in TPBR is twice the yield in raceway ponds[8].

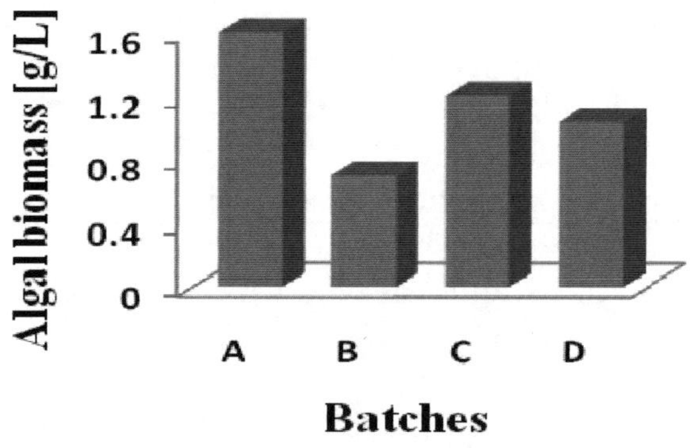

Figure 5 Algae concentration for harvested broth versus batch

Thus the ratio of initial concentration to the final concentration was 11, which was higher than the ratio 4 reported for open ponds[2] or helical phobioreactors[11].

The volumetric productivity[11] for this TPBR was determined as P=DX, where D was dilution rate in $[hr^{-1}]$ and X was the biomass concentration in [g/L].

$$P=0.004 \text{ g.L}^{-1}\text{hr}^{-1}$$

A low algal biomass concentration shows a high impact on volumetric productivity in TPBR being 10 times lower than the volumetric productivity in helical photobioreactor[11]. However, if a culture starts with an inoculum at higher concentration, for instance 1.0 g/L, the volumetric productivity in TPBR is expected to be much higher compared to those of other photobioreactors

4.4 Algae Oil Content: Algal oil content (g oil per g of dry algal biomass) of harvested algae was determined by measuring weights of extracted oil using soxhlet extractor and hexane. Figure 6 presents algal oil content (in %) for each batch. The oil content varies from 7 to 13 % with an average of 11% or 0.11 g of oil per g of dry algal biomass.

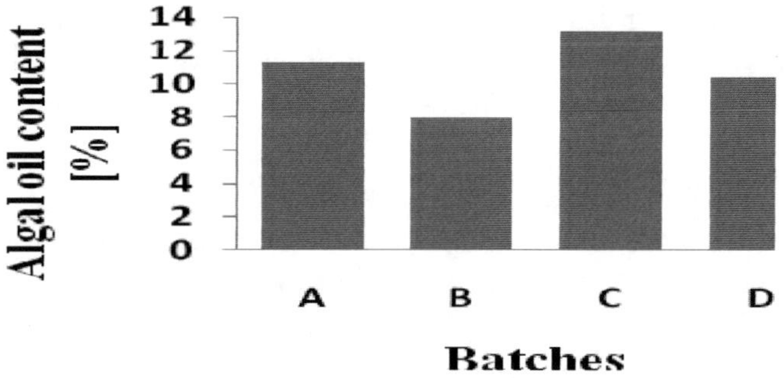

Figure 6 Algae oil content broth versus batch

4.5 Effect of KNO_3 and NaCl on algal growth: Five difference batches (flask A, B, C, D and E) of the same algae strain were cultured using different composition of KNO_3 and NaCl in the medium as follow: flask A as a control, flask B with no NaCl, Flask C with double concentration of NaCl, flask D with no KNO3 and flask E with double concentration of KNO3. Figure 7 shows the variation of turbidities (Optical Density) of each of five algal cultures with respect to time. It shows that KNO_3 and NaCl concentration are directly proportional to the algal concentration; doubling the concentration of KNO_3 or NaCl increases the concentration of algal culture. In fact, no growth was observed in flask D, which did not contain any KNO_3

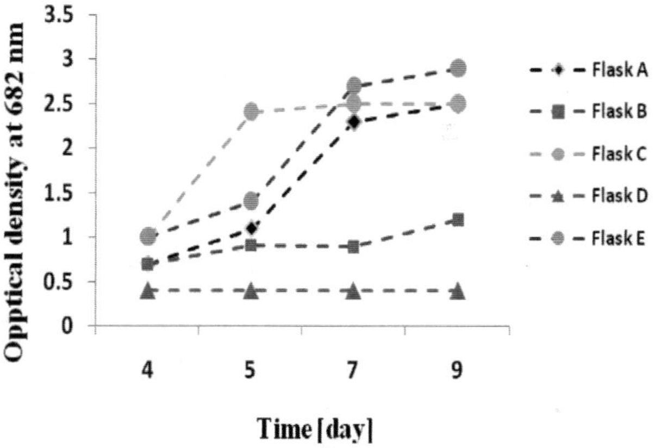

Figure 7 Effect of KNO3 and NaCl on algal absorbance versus time

4.6 Light effect on algal growth in TPBR: The light penetration distance was evaluated using the following empirical formula[14].

$$d_p = \frac{6000}{C}$$

Where d_p and C are respectively light penetration distance (in cm) and algal broth concentration (in mg/L). The light penetration for the TPBR with an algal biomass concentration of 1.1g/L or 1100 mg/L is 5.5 cm. The outside diameter of the PVC tubes used in TPBR was 2.5 cm or 1"; therefore, light was not a limited factor.

The maximum algal concentration that could be reached in current TPBR and lead to light limitation would be 3.6 g/L or 3600 mg/L according to the following formula[14].

$$C_{max} = \frac{9000}{\text{external diameter}} = \frac{9000}{2.5} = 3600 \text{ mg/L}$$

5 CONCLUSION

Seven strains were tested and the best was selected for its highest oil content and per liter of broth. The selected strain was cultured in the TPBR, which was designed to achieve high productivity. Four different batches cultured in the TPBR showed that a high yield of algal biomass could be attained by comparing the ratio of initial broth concentration to the final concentration being high (11) to those of existing close or open system.

In addition to the yield in algal biomass being 1.1g/L with a volumetric productivity of 0.04 $gL^{-1}hr^{-1}$ or 0.096 $kg.m^{-3}.day^{-1}$, the dry algal biomass contained an average lipid concentration of 11% in mass. Also, the modification of medium composition showed that doubling the concentration of KNO_3 and NaCl significantly increases algal biomass yield. It is important to note that light was not a limited factor in algal grown in TPBR.

AKNOWLEDGEMENT

The authors would like to thank the UNH Chemical Engineering Department for the support of the project, Dr Leland Jahnke and Dr Samuel Smith for their advice and help with different instruments. Also authors would like to thank Jonathan Grohs and Louis Jakob.

REFERENCES

1. Shehan, J. et. Al (1998) "A *Look Back at the U.S Department of Energy's Aquatic Species Program-Biodiesel from Algae".* Golden, CO, National Renewable Energy Lab, NREL/TP-580-24190,255
2. Knothe, Gerhard,:"*The Biodiesel Handbook*"AOCS press, Champaign, Illinois, 2005
3. Ferrentino, J. and Farag I.H (2007),"*Microalgae Oil Extraction and In Situ Transesterification*" MS. Thesis, Chemical Engineering Department, University of New Hampshire, Durham, New Hampshire
4. Ferrentino, J, Farag, I.H, Jahnke, L.S. (2006), "Microalgae Oil Extraction and In Situ Transesterification," AIChE Annual meeting, Nov 11-13, 2006, San Francisco, California
5. Cooksey, K.E., Guckert, J.B., Williams, S.A., and Callis, P.R. (1987) "*Fluorometric determination of the neutral lipid content of microalgal cells using Nile Red.*" *Journal of Microbiological Methods, 6*, 333-345
6. Chojnaka, K, et. Al: "*Kinetic and Stoichiometric Relationships of the Energy and Carbon metabolism in the Culture of Microalgae.*" Journal of Biotechnology, Vol 3-2004. Pp21-34
7. Chisti, Yusuf (2007),"Biodiesel from Microalgae." ScienceDirect, Biotechnology Advances 25: 294-306
8. Chisti, Yusuf (2008),"Biodiesel from Microalgae beat Bioethanol," Trends in Biotechnology 3, 126-131
9. Schafer, K. (1998)"*Accelerated solvent extraction of Lipid for Determining Fatty Acid Composition of Biological Material*", anal Chem. Acta 358:68-77
10. Haas, Michael and Scott Karen (2007),"*Moisture Removal Substantially improves the Efficiency of In Situ Biodiesel Production from Soybean.*" J. Maer Oil Chem. Society 84:197-204
11. Nerantzis, E. et. al. ' '*Continuous production of Arthrospira (=spirulina) platansis in a helical photobioreactor.*
12. Farag, I.H. (2007) "*Biodiesel, The Renewable Alternative to Diesel,*" 7th International Conference and Exhibition on Chemistry in Industry CHMINDIX 2007, March 23-25,2007, Manama, Bahrain
13. Farag, I.H (2008), "*Biodiesel: Challenges and New Fronts,*" Keynote Paper, International Biodiesel Workshop, National Research Center, March 23-25, 2008, Cairo, Egypt
14. Andersen, A. Robert, "Algal Culturing Techniques," Elsevier Academic Press, Burlington, Massachusetts, 2005

Identification and Reduction of Expression of the Delta-12 Desaturase in Microalgae Dunaliella Tertiolecta

Meng Chen, Chemical Engineering, Wayne State University, Detroit, MI

Steven O. Salley, Department of Chemical Engineering and Materials Science, Wayne State University, Detroit, MI

K. Y. Simon Ng, Chemical Engineering and Materials Science, Wayne State University, Detroit, MI

Thomas Holland, Immunology and Microbiology, Wayne State University, Detroit, MI

Algae have been recognized as a potential source of biomass for production of biofuels because some species can produce and accumulate high levels of oil and other valuable energy storage molecules. Dunaliella tertiolecta is a marine green algae species known to accumulate oil when stressed. However, the majority of fatty acids in Dunaliella tertiolecta contain 2 or more double bonds, which are not preferred for biodiesel production due to their oxidizability. Unsaturated fatty acids are produced by the action of fatty acid desaturases (FADs), which convert specific C-C bonds in fatty acid chains to double bonds. The delta-12 fatty acid desaturase has a key role in the production of fatty acids with multiple double bonds. Using degenerate PCR primers, we will clone the D. tertiolecta delta-12 FAD. The PCR products will be fully sequenced and the function of complete cDNA will be verified in yeast. RNA interference (RNAi) will be used to reduce the expression of delta-12 FAD in D. tertiolecta. It is expected that this research will result in the production of algal oil with improved properties for biodiesel production.

See more of this Session: Biodiesel From Microorganisms
See more of this Group/Topical: Fuels and Petrochemicals Division

AUTHOR INDEX

AUTHOR INDEX

AUTHOR INDEX

AUTHOR INDEX

American Institute of Chemical Engineers
3 Park Avenue
New York, New York 10016-5991

CURRAN ASSOCIATES INC
proceedings.com